Lymphangiogenesis in Cancer Metastasis

Cancer Metastasis – Biology and Treatment

VOLUME 13

Recent Volumes in this Series

Volume 6: Bone Metastasis and Molecular Mechanisms
Editors: Gurmit Singh and William Orr
ISBN 1-4020-1984-X

Volume 7: DNA Methylation, Epigenetics and Metastasis
Editor: Manel Esteller
ISBN 1-4020-3641-8

Volume 8: Cell Motility in Cancer Invasion and Metastasis
Editor: Alan Wells
ISBN 1-4020-4008-3

Volume 9: Cell Adhesion and Cytoskeletal Molecules in Metastasis
Editors: Anne E. Cress and Raymond B. Nagle
ISBN 1-4020-5128-X

Volume 10: Metastasis of Prostate Cancer
Editors: Richard J. Ablin and Malcolm D. Mason
ISBN 1-4020-5846-2

Volume 11: Metastasis of Breast Cancer
Editors: Robert E. Mansel, Oystein Fodstad and Wen G. Jiang
ISBN 1-4020-5866-7

Volume 12: Bone Metastases: A Translational and Clinical Approach
Editors: Dimitrios Kardamakis, Vassilios Vassiliou and Edward Chow
ISBN 1-4020-9818-5

Volume 13: Lymphangiogenesis in Cancer Metastasis
Editors: Steven A. Stacker and Marc G. Achen
ISBN 90-481-2246-2

Lymphangiogenesis in Cancer Metastasis

Edited by

Steven A. Stacker
Ludwig Institute for Cancer Research, Royal Melbourne Hospital, Parkville, Victoria, Australia

and

Marc G. Achen
Ludwig Institute for Cancer Research, Royal Melbourne Hospital, Parkville, Victoria, Australia

Springer

Editors
Dr. Steven A. Stacker
Ludwig Institute for Cancer Research
Royal Melbourne Hospital
Parkville, Victoria 3050
Australia
Steven.Stacker@ludwig.edu.au

Dr. Marc G. Achen
Ludwig Institute for Cancer Research
Royal Melbourne Hospital
Parkville, Victoria 3050
Australia
Marc.Achen@ludwig.edu.au

ISBN: 978-90-481-2246-2 e-ISBN: 978-90-481-2247-9

DOI 10.1007/978-90-481-2247-9

Library of Congress Control Number: 2009921270

Printed on acid-free paper

9 8 7 6 5 4 3 2 1

springer.com

Contents

Contributors

Marc G. Achen
Ludwig Institute for Cancer Research, Royal Melbourne Hospital, Parkville, Victoria 3050, Australia, Marc.Achen@ludwig.edu.au

Kari Alitalo
Molecular/Cancer Biology Laboratory, Biomedicum Helsinki, FI-00014 Helsinki, Finland, kari.alitalo@helsinki.fi

Chris Boshoff
CR-UK Viral Oncology Group, UCL Cancer Institute, UCL, WC1E 6BT, London, UK, c.boshoff@ucl.ac.uk

Peter L. Choyke
Molecular Imaging Program, National Cancer Institute (NCI), Bethesda, MD 20892-1088, USA, pchoyke@mail.nih.gov

Michael Detmar
Institute of Pharmaceutical Sciences, Swiss Federal Institute of Technology, ETH Zurich, HCI H303, CH-8093 Zurich, Switzerland, michael.detmar@pharma.ethz.ch

Luc Y. Dirix
Translational Cancer Research Group, Laboratory of Pathology, University of Antwerp/University Hospital Antwerp; Oncology Centre, General Hospital Sint-Augustinus, 2610 Wilrijk, Belgium, luc.dirix@gra.be

Victoria Emuss
CR-UK Viral Oncology Group, UCL Cancer Institute, UCL, WC1E 6BT, London, UK, v.emuss@ucl.ac.uk

Mark B. Faries
Division of Surgical Oncology, The Roy E. Coats Research Laboratories, John Wayne Cancer Institute, Saint John's Health Center, 2200 Santa Monica Blvd., Santa Monica, CA 90404, USA, fariesm@jwci.org

Paula Haiko
Molecular/Cancer Biology Laboratory, Biomedicum Helsinki, FI-00014 Helsinki, Finland, phaiko@mappi.helsinki.fi

Benjamin M. Hogan
Hubrecht Institute (KNAW), University Medical Center Utrecht, Uppsalalaan 8, Utrecht 3584 CT, The Netherlands, b.hogan@niob.knaw.nl

Nicole C. Johnson
Department of Genetics and Tumor Cell Biology, St Jude Children's Research Hospital, 332 N. Lauderdale, Memphis, TN 38105-2794, USA, nicole.johnson@stjude.org

Hisataka Kobayashi
Molecular Imaging Program, National Cancer Institute (NCI), Bethesda, MD 20892-1088, USA, kobayash@mail.nih.gov

Toshio Kumasaka
Human Pathology, Juntendo University School of Medicine, 2-1-1 Hongo, Bunkyo-Ku, Tokyo 113-8421, Japan, kumasaka_toshio@med.jrc.or.jp

Keiko Mitani
Human Pathology, Juntendo University School of Medicine, 2-1-1 Hongo, Bunkyo-Ku, Tokyo 113-8421, Japan, m-kei@juntendo.ac.jp

Donald L. Morton
Division of Surgical Oncology, The Roy E. Coats Research Laboratories, John Wayne Cancer Institute, Saint John's Health Center, 2200 Santa Monica Blvd., Santa Monica, CA 90404, USA, mortonD@jwci.org

Guillermo Oliver
Dept. of Genetics and Tumor Cell Biology, St. Jude Children's Research Hospital, 262 Danny Thomas Place, Memphis, TN 38105-3678, guillermo.oliver@stjude.org

Gregory Ravizzini
Molecular Imaging Program, National Cancer Institute (NCI), Bethesda, MD 20892-1088, USA, ravizzini@msn.com

Matthias Rinderknecht
Institute of Pharmaceutical Sciences, Swiss Federal Institute of Technology, ETH Zurich, HCI H303, CH-8093 Zurich, Switzerland, matthias.rinderknecht@pharma.ethz.ch

Teruhiko Sato
Departments of Respiratory Medicine, Juntendo University School of Medicine, 2-1-1 Hongo, Bunkyo-Ku, Tokyo 113-8421, Japan, satoteru1112@mac.com

Stefan Schulte-Merker
Hubrecht Institute (KNAW), University Medical Center Utrecht, Uppsalalaan 8, Utrecht 3584 CT, The Netherlands, s.schulte@niob.knaw.nl

Kuniaki Seyama
Department of Respiratory Medicine, Juntendo University, School of Medicine, 2-1-1 Hongo, Bunkyo-Ku, Tokyo 113-8421, Japan, kseyama@med.juntendo.ac.jp

Jacqueline D. Shields
Institute of Bioengineering, École Polytechnique Fédérale de Lausanne (EPFL), CH-1015 Lausanne, Switzerland, mailto:jacqueline.shields@epfl.ch

Steven A. Stacker
Ludwig Institute for Cancer Research, Royal Melbourne Hospital, Parkville, Victoria 3050, Australia, Steven.Stacker@ludwig.edu.au

Melody A. Swartz
Institute of Bioengineering, École Polytechnique Fédérale de Lausanne (EPFL), CH-1015 Lausanne, Switzerland, melody.swartz@epfl.ch

Baris Turkbey
Molecular Imaging Program, National Cancer Institute (NCI), Bethesda, MD 20892-1088, USA, turkbeyi@mail.nih.gov

Ilse Van der Auwera
Translational Cancer Research Group, Laboratory of Pathology, University of Antwerp/University Hospital Antwerp; Oncology Centre, General Hospital Sint-Augustinus, Oosterveldlaan 24, 2610 Wilrijk, Belgium, ilse.van.der.auwera@telenet.be

Peter B. Vermeulen
Translational Cancer Research Group, Laboratory of Pathology, University of Antwerp/University Hospital Antwerp; Oncology Centre, General Hospital Sint-Augustinus, 2610 Wilrijk, Belgium, peter_vermeulen@gvagroup.be

Chapter 1

LYMPHANGIOGENESIS IN HEALTH AND DISEASE – AN OVERVIEW

Steven A. Stacker[1]**, Marc G. Achen**[1]**, Paula Haiko**[2] **and Kari Alitalo**[2]
[1]*Ludwig Institute for Cancer Research, Royal Melbourne Hospital, Parkville, Victoria 3050, Australia, e-mail: Steven.Stacker@ludwig.edu.au*
[2]*Molecular/Cancer Biology Laboratory, Biomedicum Helsinki, FI-00014 Helsinki, Finland, e-mail: phaiko@mappi.helsinki.fi*

Abstract: The blood and lymphatic vascular networks combine to facilitate immune function and maintain tissue fluid homeostasis in the body. Although these two systems share many common structural and molecular features, recent advances in our understanding of the molecular control of the lymphatics have identified distinct molecular pathways responsible for the formation and function of the lymphatic network. These advances have led to the characterisation of lymphatic-specific markers and growth factors which control lymphatic development and function. Insights gained from in vitro and in vivo studies over the past decade have highlighted the importance of the lymphatic system in human diseases such as lymphedema, inflammatory disorders and cancer. The lymphatic vasculature is an important route for the metastatic spread of tumor cells, and recent studies based on animal models of cancer indicated that lymphangiogenic growth factors, secreted by tumor cells or components of the tumor stroma, can induce formation of new lymphatic vessels in the vicinity of a primary tumor. These studies, as well as clinicopathological data, suggest that this process of tumor lymphangiogenesis can be associated with enhanced metastastic spread – hence tumor lymphangiogenesis is being explored as a therapeutic target for restricting the metastatic spread of cancer.

Key words: Lymphangiogenesis · Growth factors · Growth factor receptors · Metastasis · Cancer

The characterization of the anatomy and physiology of the lymphatic system has been ongoing over centuries, however, advances during the past decade in identifying molecular markers of the lymphatics have accelerated this process (see chapter by Shields and Swartz). The lymphatic vasculature begins as blind-ended, thin-walled

S.A. Stacker, M.G. Achen (eds.), *Lymphangiogenesis in Cancer Metastasis*,
Cancer Metastasis – Biology and Treatment 13, DOI 10.1007/978-90-481-2247-9_1,

capillaries that collect extravasated fluid and cells from tissues. The lymph fluid then drains into pre-collecting lymphatics, located in the deep dermis, which in turn drain into the collecting lymphatics located in the subcutaneous tissue. The collecting lymphatics, which are invested with smooth muscle cells and pericytes, are capable of propelling lymph fluid, are studded with lymph nodes and coalesce into lymphatic trunks which drain lymph fluid back to the blood circulation via intra-thoracic ducts [1,2]. The lymphatic vasculature plays crucial roles in immune function, tissue fluid homeostasis and the absorption of dietary fat.

The development of the lymphatic vascular system during embryogenesis begins with sprouting of lymphatic endothelial precursor cells from the cardinal vein, giving rise to the lymph sacs – lymphatic endothelial cells then sprout from these sacs to form the primary lymphatic plexus, and further sprouting, proliferation and migration generates the lymphatic networks of tissues and organs [3] (see chapter by Johnson and Oliver). Differentiation of lymphatic endothelial cells to generate the distinct types of lymphatic vessels is an important aspect of lymphatic development. Elegant developmental studies, utilizing traditional and emerging animal models of embryonic development (see chapter by Hogan and Schulte-Merker), have mapped the initial events in the formation of the lymphatics, as well as of blood vessels, showing their origins in the embryo. Some of the early markers of the lymphatic system such as Prox-1 [4], podoplanin [5], and vascular endothelial growth factor receptor-3 (VEGFR-3) [6, 7] are important for the development or function of the lymphatics [8]. VEGF-C and VEGF-D are significant as they are ligands for VEGFR-3 [9–11], which are capable of inducing lymphangiogenesis when delivered to adult tissues [12, 13]. VEGF-C is indispensible for development of the lymphatic vasculature during embryogenesis [14].

The metastatic spread of tumor cells from the primary tumor to establish metastases at distant sites in the body is the most lethal aspect of cancer. The importance of the lymphatic vasculature in the metastatic spread of cancer has been appreciated for centuries, and the extent of lymph node metastasis is a major determinant for prognostic assessment and planning of treatment (see chapter by Faries and Morton). Recently, the involvement of tumor lymphangiogenesis in lymph node metastasis has become an important focus of study, and the molecular mechanisms underlying lymph node metastasis are being revealed (see chapter by Rinderknecht and Detmar). The VEGF-C/VEGF-D/VEGFR-3 signalling axis is closely linked to the formation and function of the lymphatics in cancer [15]. Initial studies showed that over-expression of VEGF-C or VEGF-D in mouse tumor models led to formation of lymphatic vessels in and/or around the primary tumor, and to increased metastatic spread to regional lymph nodes, as well as increased tumor growth in some cases [16–18]. Some of the underlying mechanisms are schematically shown in Fig. 1.1. Furthermore some of these effects could be inhibited by antibodies [18–20], soluble receptors [21, 22] or small molecule protein tyrosine kinase inhibitors [23] which targeted signalling via VEGF receptors. These experiments were further substantiated by clinicopathological data showing a correlation of VEGF-C or VEGF-D expression levels in human primary tumors with clinical parameters and patient outcomes [24]. Significantly, these correlations were seen over a range of different tumor types including tumors originating from the colon, lung, breast,

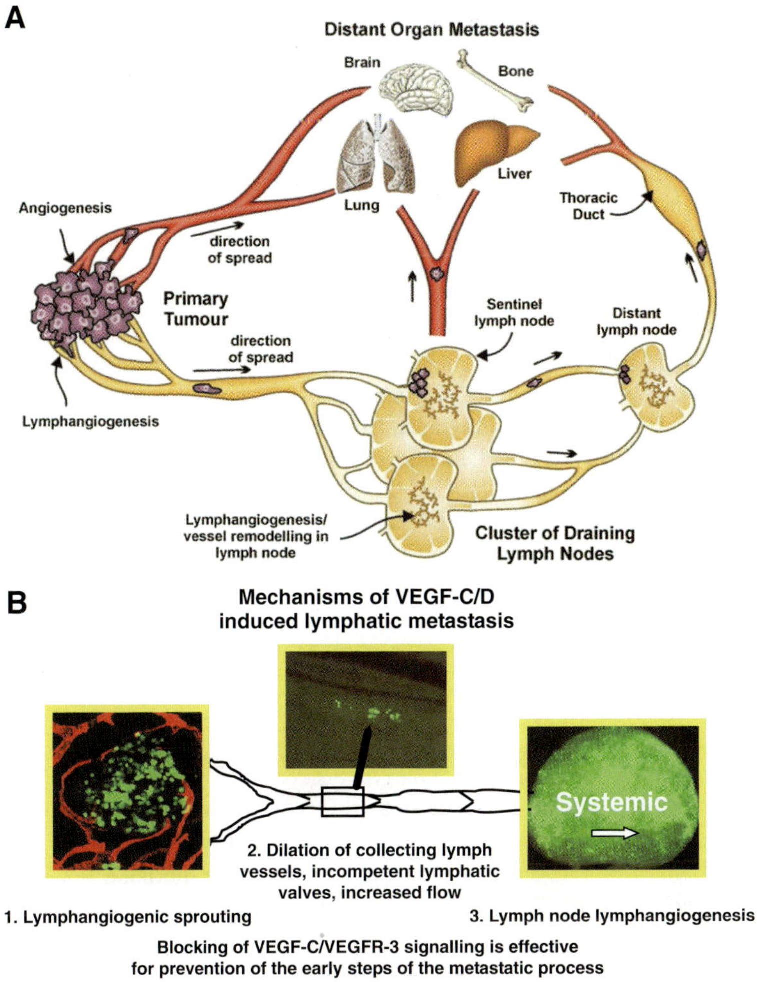

Fig. 1.1 Schematic illustration of the various steps in lymphatic metastasis and effects of lymphangiogenic growth factors. (A) Schematic diagram showing the potential routes of spread of primary tumor cells via blood vessels, and via lymphatic vessels to regional lymph nodes. Tumor cells drain to the sentinel lymph node and may then pass to other lymph nodes or potentially to distant organs via the thoracic duct or perhaps through abnormal connection made between lymphatic vessels and arteries or veins within the regional lymph node. Figure reproduced from *The Lymphatic Continuum Revisited*, Edited by Stanely Rockson, Ann NY Acad Sci, Vol 1131, M.G. Achen & S.A. Stacker, Molecular Control of Lymphatic Metastasis, pp. 225–234, 2008. **(B)** Tumors that secrete VEGF-C or VEGF-D (VEGF-C/D) stimulate lymphatic vessels to grow and form new sprouts. In this process, the button-type junctions between lymphatic endothelial cells are lost and the tumor cells gain better access to the vessel lumen. Intraluminal VEGF-C/D also stimulates growth of the lymphatic endothelial cells in the vessel wall, leading to increased lumen size that facilitates the transit of metastatic cells via increased net flow to the lymph node, where tumor cells first arrive in the marginal sinus. High VEGF-C/D, and apparently also VEGF levels in lymph can further induce the process of lymph node lymphangiogenesis, whereby the sinusoidal endothelial cells within the lymph node undergo capillary phenotypic changes and proliferate. To what extent this last process also contributes to distant metastasis is not yet clear

gut, pancreas, gall bladder and kidney. This data suggested that at least some of the molecules involved in lymphatic vessel development could be clinically and biologically important markers for disease progression in cancer. Nevertheless, some results from different clinical centres regarding the role of lymphangiogenesis-related parameters and their influence on cancer prognosis have been contradictory, so this area requires further study (see chapter by Van der Auwera et al.).

The advancement of knowledge about molecular aspects of the lymphatic network and the process of lymphangiogenesis, which occurred over the past decade, has impacted on other diseases [25]. For example, lymphedema, which is linked to the dysfunction of the lymphatic vasculature, is now being more readily analysed in light of genes which predispose people to this condition [1, 26, 27]. Further, research on the causes and treatment of lymphedema has benefited from animal models that involve manipulation of lymphangiogenic signalling pathways [28, 29]. Some of the mechanisms behind lymphedema distichiasis and Milroy's disease, the two commonest forms of lymphedema are illustrated in Fig. 1.2. Conditions such as lymphangioleiomyomatosis (LAM) [30–32] (see chapter by Seyama et al.) and Kaposi's sarcoma [33] (see chapter by Emuss and Boshoff) have also been shown to have links with lymphatic biology, and this has provided novel avenues for exploring the aetiology of these diseases.

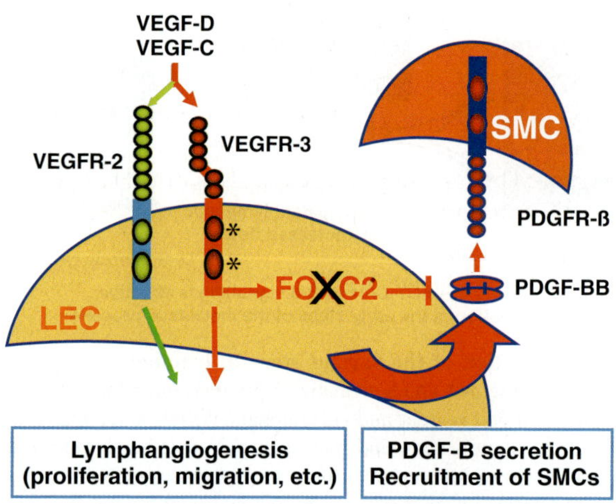

Fig. 1.2 Signaling mechanisms affected by genetic damage in lymphedema. VEGFR-3 missense point mutations, indicated as (*) in the kinase domain of the receptor, lead to defective receptor signaling in Milroy's disease, resulting in hypoplastic and poorly functional lymphatic vessels [48]. Deficiency in forkhead transcription factor FOXC2 (Indicated schematically by a "X" through FOXC2) in lymphoedema-distichiasis (LD) leads to abnormal recruitment of mural cells, deposition of basement membrane and lack of valves in lymphatic vessels, resulting in disturbed lymph flow and lymphedema. Unlike the blood capillaries, the lymphatic capillaries in wild-type mice do not normally have a covering of smooth muscle cells (SMCs), whereas FoxC2 deficiency in LD leads to PDGF-B secretion from lymphatic capillary endothelial cells and recruitment of PDGF receptor-ß expressing SMCs. Genetic experiments indicate that the VEGFR-3 signal transduction pathway regulates FoxC2 mediated transcription [49]

The detection of diseased tissue or cells within the lymphatic system can be of great clinical relevance, e.g. sentinel lymph node (SLN) biopsy which assesses the presence of tumor cells in regional lymph nodes and has become standard surgical practice for some tumors. Therefore, advances in our ability to image the lymphatic network, and to detect diseased tissue/cells located within or passaging through the lymphatics by imaging technology, is a high priority. A range of new lymphatic imaging technologies are currently being assessed that allow systemic or targeted lymphatic imaging – these may not only improve diagnosis but also facilitate intra-lymphatic treatment for cancer (see chapter by Turkbey et al.). Such technologies may be relevant for other diseases involving lymphatic dysfunction, such as lymphedema.

The lymphangiogenic signalling pathway involving the VEGF-C/VEGF-D/VEGFR-3 axis provides potential molecular therapeutics for stimulating lymphangiogenesis, and possibly the repair of damaged lymphatic vessels, which could be beneficial for treatment of lymphedema [34]. VEGF-C and VEGF-D have been successfully delivered in animal models via adenoviruses to promote lymphangiogenesis [35–37], VEGF-C protein has also been used for this purpose [38], and the optimal delivery approach for therapeutic settings is being established. Characterization of alternative lymphangiogenic signaling pathways may provide alternative opportunities for lymphedema therapies. Both VEGF-C and VEGF-D, when proteolytically activated, are also capable of promoting angiogenesis [36,39], presumably due to signalling via the angiogenic receptor VEGFR-2, although VEGFR-3 has recently been shown to also play a role in sprouting angiogenesis [40]. The capacity of these growth factors to promote angiogenesis, repair damaged blood vessels and to restrict stenosis or restenosis of large blood vessels [41,42] could lead to a range of clinical applications.

The lymphangiogenic signalling pathways also offer the opportunity for therapeutic approaches designed to restrict tumor lymphangiogenesis and the metastatic spread of cancer. Inhibitors targeting the VEGF-C/VEGF-D/VEGFR-3 signalling axis, that have been examined in animal models of cancer, include a monoclonal antibody (mab) to VEGFR-3, which blocks activation of this receptor by VEGF-C and VEGF-D [43] – this mab has been reported to restrict lymphatic metastasis in a range of tumor models, and to restrict the rate of primary tumor growth in some models [19,20,40]. Likewise, a VEGF-D mab, that blocks the binding of this growth factor to both VEGFR-2 and VEGFR-3 [44], restricted tumor growth and spread in a mouse model of cancer [18]. A soluble form of the VEGFR-3 extracellular domain, a so-called receptor-trap, has been used in a range of animal models of cancer to prevent VEGF-C and VEGF-D from activating endogenous VEGFR-3 on lymphatic endothelial cells – this resulted in a reduction of lymphatic metasatasis [21,22,45]. In addition, a range of small molecules which inhibit the tyrosine kinase activity of both VEGFR-2 and VEGFR-3 have been developed [15,23], although their effects on tumor lymphangiogenesis and lymph node metastasis in animal models of cancer have not been well characterised. Some of these small molecule inhibitors are already in clinical use and others are being evaluated as anti-cancer therapeutics in clinical trials [15].

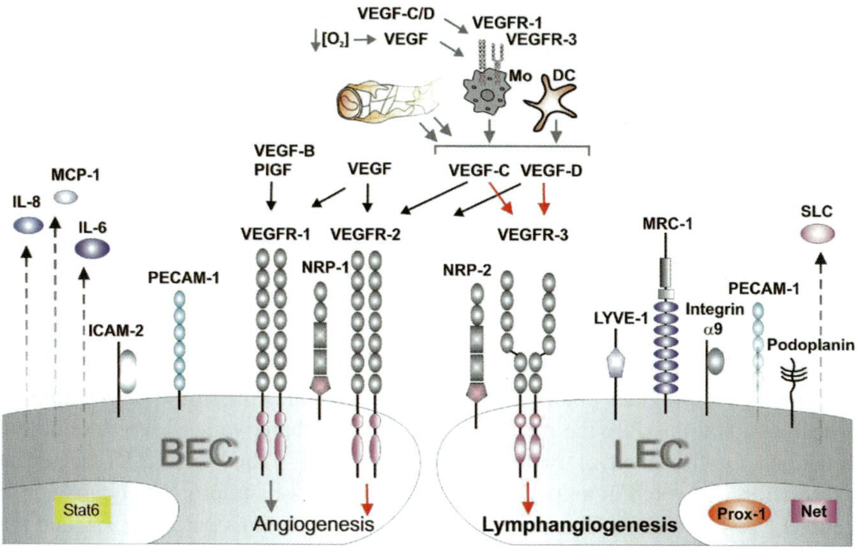

Saharinen et al., Trends Immunol.

Fig. 1.3 Molecular characteristics of blood vascular endothelial cells (BECs) and lymphatic endothelial cells (LECs). VEGF-C is secreted by smooth muscle cells and pericytes but also by BECs, which can recruit LECs to the vicinity of newly formed blood vessels. Hypoxic conditions in tumors induce VEGF expression, and many tumors also produce VEGF-C and VEGF-D, resulting in recruitment of VEGFR-1 and VEGFR-3 positive macrophages (Mø). Macrophages and inflammatory dendritic cells (DCs) secrete VEGFs and other factors that can boost the angiogenic and lymphangiogenic responses, especially because the expression of VEGF and VEGF-C is induced by pro-inflammatory cytokines [50, 51]. VEGFR-2, which binds VEGF, VEGF-C and VEGF-D, is expressed mainly in BECs. VEGFR-1 is expressed in BECs but also in macrophages and monocytes and binds VEGF-B, PlGF and VEGF. VEGFR-3 is expressed mainly in LECs, and VEGF-C- and VEGF-D-induced VEGFR-3 signaling is one of the proximal regulators of lymphangiogenesis. Among members of the VEGF family, VEGF is the main inducer of angiogenesis, which is mediated by VEGFR-1 and VEGFR-2. Neuropilin-2 (NRP-2) binds VEGF-C, whereas NRP-1 functions as a co-receptor for VEGF. BECs express the intercellular adhesion molecule-2 (ICAM-2) and platelet endothelial cell adhesion molecule-1 (PECAM-1) that is also weakly expressed by LECs. BECs also express Stat6 (signal transducer and activator of transcription 6), whereas Prox-1, a homeodomain transcription factor (TF), and Net, an Ets family TF, are specific for LECs. BECs secrete interleukin-6 (IL-6), IL-8 and monocyte chemoattractant protein-1 (MCP-1), whereas LECs secrete secondary lymphoid tissue chemokine (SLC). LECs also express the lymphatic vessel endothelial hyaluronan receptor-1 (LYVE-1), integrin α9 and podoplanin, a mucin-type transmembrane protein originally characterized in kidney podocytes. The mannose receptor C-1 (MRC-1) is highly expressed in macrophages, where it mediates the endocytosis of glycoproteins. MRC-1 is also expressed in LECs, and in tumor associated lymphatic vessels [52]. Abbreviation: PlGF, placental growth factor. Reproduced from Saharinen P, Tammela T, Karkkainen MJ, Alitalo K. Lymphatic vasculature: development, molecular regulation and role in tumor metastasis and inflammation. Trends Immunol. 2004 Jul;25(7): 387–95 with permssion of the publisher and the authors

This past decade has seen exciting developments in lymphatic biology as novel lymphatic markers and regulators of lymphangiogenesis and lymphatic development have been identified. Some of the characteristic proteins that differ between blood vascular and lymphatic endothelial cells are indicated in Fig. 1.3. Nevertheless, it is almost certain that there are other markers and key regulatory signalling pathways yet to be discovered. New biological models will be required to facilitate the rapid identification and validation of novel pathways important for lymphatic development, growth and differentiation (see chapter by Hogan and Schulte-Merker). Model vertebrate organisms such as the Zebrafish will be important, allowing large-scale genetic screening for phenotypes relevant to lymphatic biology, in a high throughput manner [46, 47]. Another promising approach is based on high-throughput image acquisition systems allowing rapid, large-scale screening of purified lymphatic endothelial cells for various biological responses to siRNA or drug libraries. Based on these and other technologies plus systems biology analysis, this exiting period of discovery for lymphatic biology is set to continue for years to come.

References

1. Baldwin ME, Stacker SA, Achen MG. Molecular control of lymphangiogenesis. Bioessays 2002, 24: 1030–40.
2. Karpanen T, Alitalo K. Molecular biology and pathology of lymphangiogenesis. Annu Rev Pathol Mech Dis 2007, 3: 367–97.
3. Oliver G. Lymphatic vasculature development. Nat Rev Immunol 2004, 4: 35–45.
4. Wigle JT, Oliver G. Prox1 function is required for the development of the murine lymphatic system. Cell 1999, 98: 769–78.
5. Breiteneder-Geleff S, et al. Podoplanin, novel 43-kd membrane protein of glomerular epithelial cells, is down-regulated in puromycin nephrosis. Am J Pathol 1997, 151: 1141–52.
6. Karpanen T, Alitalo K. Molecular biology and pathology of lymphangiogenesis. Annu Rev Pathol 2008, 3: 367–97.
7. Aprelikova O, et al. FLT4, a novel class III receptor tyrosine kinase in chromosome 5q33-qter. Cancer Res 1992, 52: 746–8.
8. Alitalo K, Carmeliet P. Molecular mechanisms of lymphangiogenesis in health and disease. Cancer Cell 2002, 1: 219–27.
9. Joukov V, et al. A novel vascular endothelial growth factor, VEGF-C, is a ligand for the Flt-4 (VEGFR-3) and KDR (VEGFR-2) receptor tyrosine kinases. EMBO J 1996, 15: 290–8.
10. Achen MG, et al. Vascular endothelial growth factor D (VEGF-D) is a ligand for the tyrosine kinases VEGF receptor 2 (Flk-1) and VEGF receptor 3 (Flt-4). Proc Natl Acad Sci USA 1998, 95: 548–53.
11. Veikkola T, et al. Signalling via vascular endothelial growth factor receptor-3 is sufficient for lymphangiogenesis in transgenic mice. EMBO J 2001, 20: 1223–31.
12. Rissanen TT, et al. VEGF-D is the strongest angiogenic and lymphangiogenic effector among VEGFs delivered into skeletal muscle via adenoviruses. Circ Res 2003, 92: 1098–106.
13. Jeltsch M, et al. Hyperplasia of lymphatic vessels in VEGF-C transgenic mice. Science 1997, 276: 1423–5.
14. Karkkainen MJ, et al. Vascular endothelial growth factor C is required for sprouting of the first lymphatic vessels from embryonic veins. Nat Immunol 2004, 5: 74–80.
15. Achen MG, McColl BK, Stacker SA. Focus on lymphangiogenesis in tumor metastasis. Cancer Cell 2005, 7: 121–7.
16. Mandriota SJ, et al. Vascular endothelial growth factor-C-mediated lymphangiogenesis promotes tumour metastasis. EMBO J 2001, 20: 672–82.

17. Skobe M, et al. Induction of tumor lymphangiogenesis by VEGF-C promotes breast cancer metastasis. Nat Med 2001, 7: 192–8.
18. Stacker SA, et al. VEGF-D promotes the metastatic spread of tumor cells via the lymphatics. Nat Med 2001, 7: 186–91.
19. Laakkonen P, et al. Vascular endothelial growth factor receptor 3 is involved in tumor angiogenesis and growth. Cancer Res 2007, 67: 593–9.
20. Roberts N, et al. Inhibition of VEGFR-3 activation with the antagonistic antibody more potently suppresses lymph node and distant metastases than inactivation of VEGFR-2. Cancer Res 2006, 66: 2650–7.
21. Lin J, et al. Inhibition of lymphogenous metastasis using adeno-associated virus-mediated gene transfer of a soluble VEGFR-3 decoy receptor. Cancer Res 2005, 65: 6901–9.
22. He Y, et al. Vascular endothelial cell growth factor receptor 3-mediated activation of lymphatic endothelium is crucial for tumor cell entry and spread via lymphatic vessels. Cancer Res 2005, 65: 4739–46.
23. Heckman CA, et al. The tyrosine kinase inhibitor cediranib blocks ligand-induced vascular endothelial growth factor receptor-3 activity and lymphangiogenesis. Cancer Res 2008, 68: 4754–62.
24. Stacker SA, Williams RA, Achen MG. Lymphangiogenic growth factors as markers of tumor metastasis. APMIS 2004, 112: 539–49.
25. Alitalo K, Tammela T, Petrova TV. Lymphangiogenesis in development and human disease. Nature (London) 2005, 438: 946–53.
26. Brice G, et al. Milroy disease and the VEGFR-3 mutation phenotype. J Med Genet 2005, 42: 98–102.
27. Karkkainen MJ, et al. Missense mutations interfere with VEGFR-3 signalling in primary lymphoedema. Nat Genet 2000, 25: 153–9.
28. Mäkinen T, et al. Inhibition of lymphangiogenesis with resulting lymphedema in transgenic mice expressing soluble VEGF receptor-3. Nat Med 2001, 7: 199–205.
29. Karkkainen MJ, et al. A model for gene therapy of human hereditary lymphedema. Proc Natl Acad Sci USA 2001, 98: 12677–82.
30. Kumasaka T, et al. Lymphangiogenesis in lymphangioleiomyomatosis: its implication in the progression of lymphangioleiomyomatosis. Am J Surg Pathol 2004, 28: 1007–16.
31. Seyama K, et al. Vascular endothelial growth factor-D is increased in serum of patients with lymphangioleiomyomatosis. Lymphat Res Biol 2006, 4: 143–52.
32. Hayashida M, Seyama K, Inoue Y, Fujimoto K, Kubo K. The epidemiology of lymphangioleiomyomatosis in Japan: a nationwide cross-sectional study of presenting features and prognostic factors. Respirology 2007, 12: 523–30.
33. Wang HW, et al. Kaposi sarcoma herpesvirus-induced cellular reprogramming contributes to the lymphatic endothelial gene expression in Kaposi sarcoma. Nat Genet 2004, 36: 687–93.
34. Tammela T, et al. Therapeutic differentiation and maturation of lymphatic vessels after lymph node dissection and transplantation. Nat Med 2007, 13: 1458–66.
35. Enholm B, et al. Adenoviral expression of vascular endothelial growth factor-c induces lymphangiogenesis in the skin. Circ Res 2001, 88: 623–9.
36. Rissanen TT, et al. VEGF-D is the strongest angiogenic and lymphangiogenic effector among VEGFs delivered into skeletal muscle via adenoviruses. Circ Res 2003, 92: 1098–106.
37. Saaristo A, et al. Lymphangiogenic gene therapy with minimal blood vascular side effects. J Exp Med 2002, 196: 719–30.
38. Szuba A, et al. Therapeutic lymphangiogenesis with human recombinant VEGF-C. FASEB J 2002, 16: 1985–7.
39. Cao Y, et al. Vascular endothelial growth factor C induces angiogenesis in vivo. Proc Natl Acad Sci USA 1998, 95: 14389–94.
40. Tammela T, et al. Blocking VEGFR-3 suppresses angiogenic sprouting and vascular network formation. Nature 2008, 454: 656–60.
41. Hiltunen MO, et al. Intravascular adenovirus-mediated VEGF-C gene transfer reduces neointima formation in balloon-denuded rabbit aorta. Circulation 2000, 102: 2262–8.

42. Rutanen J, et al. Gene transfer using the mature form of VEGF-D reduces neointimal thickening through nitric oxide-dependent mechanism. Gene Ther 2005, 12: 980–7.
43. Pytowski B, et al. Complete and specific inhibition of adult lymphatic regeneration by a novel VEGFR-3 neutralizing antibody. J Natl Cancer Inst 2005, 97: 14–21.
44. Achen MG, et al. Monoclonal antibodies to vascular endothelial growth factor-D block its interactions with both VEGF receptor-2 and VEGF receptor-3. Eur J Biochem 2000, 267: 2505–15.
45. Karpanen T, et al. Vascular endothelial growth factor C promotes tumor lymphangiogenesis and intralymphatic tumor growth. Cancer Res 2001, 61: 1786–90.
46. Yaniv K, et al. Live imaging of lymphatic development in the zebrafish. Nat Med 2006, 12: 711–16.
47. Kuchler AM, et al. Development of the zebrafish lymphatic system requires vegfc signaling. Curr Biol 2006, 16: 1244–8.
48. Karkkainen MJ, Jussila L, Ferrell RE, Finegold DN, Alitalo K. Molecular regulation of lymphangiogenesis and targets for tissue oedema. Trends Mol Med 2001, 7: 18–22.
49. Petrova TV, et al. Defective valves and abnormal mural cell recruitment underlie lymphatic vascular failure in lymphedema distichiasis. Nat Med 2004, 10: 974–81.
50. Enholm B, et al. Comparison of VEGF, VEGF-B, VEGF-C and Ang-1 mRNA regulation by serum, growth factors, oncoproteins and hypoxia. Oncogene 1997, 14: 2475–83.
51. Ristimaki A, Narko K, Enholm B, Joukov V, Alitalo K. Proinflammatory cytokines regulate expression of the lymphatic endothelial mitogen vascular endothelial growth factor-C. J Biol Chem 1998, 273: 8413–18.
52. Irjala H, et al. Mannose receptor is a novel ligand for L-selectin and mediates lymphocyte binding to lymphatic endothelium. J Exp Med 2001, 194: 1033–42.
53. Enholm B, Paavonen K, Ristimäki A, Kumar V, Gunji Y, Klefstrom J, Kivinen L, Laiho M, Olofsson B, Joukov V, Eriksson U, Alitalo K. Comparison of VEGF, VEGF-B, VEGF-C and Ang-1 mRNA regulation by serum, growth factors, oncoproteins and hypoxia. Oncogene 1997 May 22, 14(20): 2475–83.
54. Ristimäki A, Narko K, Enholm B, Joukov V, Alitalo K. Proinflammatory cytokines regulate expression of the lymphatic endothelial mitogen vascular endothelial growth factor-C. J Biol Chem 1998 Apr 3, 273(14): 8413–18.
55. Irjala H, Johansson EL, Grenman R, Alanen K, Salmi M, Jalkanen S. Mannose receptor is a novel ligand for L-selectin and mediates lymphocyte binding to lymphatic endothelium. J Exp Med 2001 Oct 15, 194(8): 1033–42.

Chapter 2

INSIGHT INTO LYMPHATIC VASCULATURE DEVELOPMENT

Nicole C. Johnson and Guillermo Oliver
Department of Genetics and Tumor Cell Biology, St Jude Children's Research Hospital, Memphis, TN 38105-2794, USA, e-mail: guillermo.oliver@stjude.org

Abstract: Our understanding of the genes and mechanisms controlling the formation of the lymphatic vasculature during embryonic development has improved a great deal during the last decade. The availability of molecular markers that allow us to distinguish the lymphatic vasculature from the blood vasculature and the generation of mouse models with various degrees of lymphatic defects have been instrumental to the progress in this field. In this chapter, we highlight some of the key molecular players that regulate the development of the lymphatic vasculature and some available mouse models of lymphatic disorders.

Key words: Lymphatic · Development · Mouse · Prox1 · Embryo

2.1 Introduction

The lymphatic vasculature is composed of a vascular network of thin-walled capillaries and larger collecting vessels that drain protein-rich interstitial fluids from the extracellular spaces within organs. The lymphatic vasculature is composed of a continuous single-cell layer of overlapping endothelial cells (*ECs*) that form loose intercellular junctions. In addition to the lymphatic vasculature, the lymphatic system includes lymphoid organs such as the lymph nodes, tonsils, Peyer's patches, spleen, and thymus, all of which play an important role in immune surveillance and response. With the exceptions of the epidermis, cornea, and central nervous system, the lymphatic vasculature covers all regions of the body where blood vasculature is also present. Therefore, this extensive network is the primary conduit for tumor metastasis to the regional lymph nodes, and malignant tumors can stimulate the growth of lymphatic vessels (*lymphangiogenesis*) [56].

Congenital or acquired malfunction of the lymphatic vasculature results in *lymphedema*, a disfiguring and disabling disorder [64]. Recently, we showed that a

S.A. Stacker, M.G. Achen (eds.), *Lymphangiogenesis in Cancer Metastasis*,
Cancer Metastasis – Biology and Treatment 13, DOI 10.1007/978-90-481-2247-9_2,

defective lymphatic vasculature can also promote late-onset obesity in mouse models of lymphatic dysfunction [18].

In this chapter, we will present a brief historic overview of the field, discuss early and current models of lymphatic vasculature development in mammals, and briefly discuss some of the clinical consequences of lymphatic vasculature malfunction. Finally, we will discuss future perspectives.

2.2 Historic Overview

The earliest references to the lymphatic vessels can be traced back to the Greek anatomist Herophilus in the third century B.C. (Fig. 2.1) [2]. Then, in the second century A.D., the Greek physician Galen described the *lacteals* (lymphatic vessels of the intestine) and mesenteric lymph nodes [2]; however, no comprehensive morphological or functional studies were undertaken until centuries later.

During the Renaissance, the lymphatic vasculature was rediscovered. In 1563, Bartolomeo Eustachi described for first time the *thoracic duct*, the main collecting lymphatic vessel in mammals [2]. Then in 1622, Gaspar Aselli identified in dogs a structure that he named the *venae alba et lacteae*, the lymphatic vessels of the intestines, which as mentioned above are currently known as the lacteals [3]. Almost a century later, William Hewson published a detailed morphological comparison of the lymphatics of birds, fishes, reptiles, and mammals [20].

The first two models describing the development of the lymphatic vasculature were proposed at the beginning of the 20th century and were conflicting. First, Florence Sabin [49] proposed that the lymphatics have a venous origin. In an alternative model, Huntington and McClure [23] proposed that the lymphatics originate in the mesenchyme independent of the veins and secondarily establish venous connections. The lack of lymphatic-specific markers made it difficult to reliably document the development of the lymphatic vasculature and, therefore, to settle this

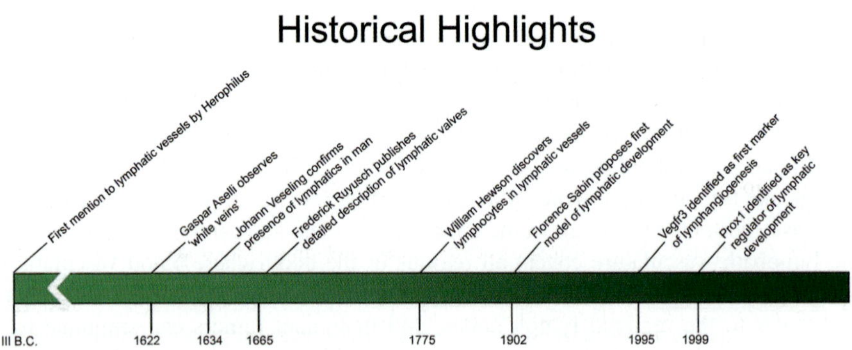

Fig. 2.1 Historic highlights. Although the existence of the lymphatics was first recognized in the III century B.C., only recently have scientists acquired the tools necessary to elucidate the mechanisms and processes that lead to this system's development

controversy for more than 100 years. Only recently have the identification of molecular markers specific for the lymphatic vasculature and the generation of animal models carrying mutations in those genes enabled us to advance our understanding of this poorly characterized process.

2.3 Lymphatic Vessels: Types and Function

The lymphatic vasculature is responsible for the removal of excessive extravasated tissue fluid and consists of two main types of vessels, blind-ended lymphatic capillaries with little or no basement membrane and highly structured, larger collecting lymphatic vessels surrounded by a basement membrane and smooth muscle. ECs of lymphatic capillaries are linked together by loose intercellular discontinuous button-like junctions [6], and are connected to the extracellular matrix by anchoring filaments. As the surrounding interstitial pressure changes, the anchoring filaments tighten and relax, causing the lymphatics to expand and fill or contract and push fluid (*lymph*). Under high interstitial pressure, EC junctions open, anchoring filaments extend, and fluid moves into the vessel (Fig. 2.2) [53]. The smaller lymphatic capillaries drain into larger, secondary collecting lymphatic vessels whose ECs exhibit continuous zipper-like junctions [6].

The large collecting lymphatic vessels are invested with smooth muscle cells (SMCs) that fortify the vessel and provide contractile activity to assist the flow of lymph. These larger vessels also contain loose, intraluminal, one-way valve-like junctions that prevent fluid back flow (Fig. 2.2). Tissue fluid contained in the larger collecting lymphatics will drain into the thoracic duct and be returned to the blood circulation through lymphatic-vasculature connections established at the junction of the jugular and subclavian veins.

In the digestive tract, lacteals inside the intestinal villi absorb and transport dietary fat and fat-soluble vitamins secreted by enterocytes in the form of chylomicrons and transport them to the venous circulation. Chylomicrons are small particles composed of a single lipid membrane impregnated with the proteins apolipoprotein A and B-48 [9]. After the chylomicrons enter the bloodstream, they are disassembled, and the lipids are liberated by a lipoprotein lipase present only in the blood vessel endothelium, cardiac or skeletal muscle, and adipose tissue [19, 39]. The heparin-binding protein GPIHBP1 (glycosylphosphatidylinositol-anchored high-density lipoprotein-binding protein 1) is crucial for the binding of chylomicrons and lipoprotein lipase to the luminal side of blood vessels [7].

The lymph (or chyle) transported by these mesenteric lacteals is composed of 60–70% lipids, mostly long-chain fatty acids. It also contains approximately 25% of the protein concentration contained in plasma [35].

In addition to the lymphatic vessels, a few other specialized lymphatic structures have been identified in different species. For example, in mammals the *cisterna chyli*, a structure that sits at the base of the vertebral column, just below the thoracic duct was originally discovered almost simultaneously in 1653 by Jean Pequet and

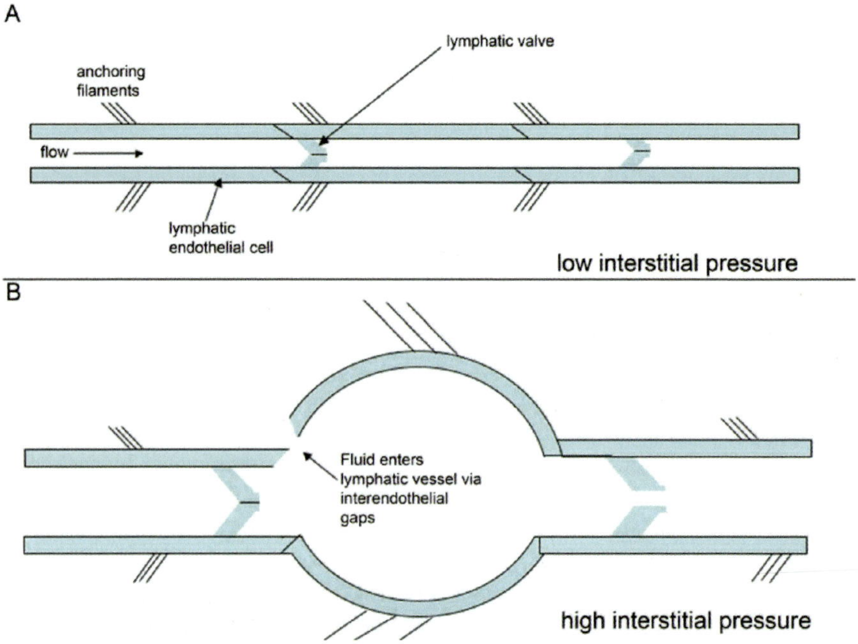

Fig. 2.2 Diagram of lymph flow. A In resting state, the interstitial pressure surrounding lymphatic capillaries is low, allowing the anchoring filaments to remain lax and the endothelial cells to remain in close proximity. **B** Once the interstitial pressure increases due to the pumping action of nearby blood vessels or the extravasation of fluid, the anchoring filaments pull tight, allowing fluid to enter the lymphatic vessel. One-way secondary valves within the vessel close as fluid is pushed towards the trunk to prevent backflow

Olof Rudbeck [45, 48] and serves as a receptacle for the lymph collected by the lacteals. In reptiles, amphibians, and some birds, interstitial fluid pressure and gravity cannot drive the uptake and movement of lymph fluid; therefore, these animals retain the more primitive *lymph hearts* [26]. These small sacs normally develop in pairs along the body trunk and contain contractile tissue that enables the sacs to pulse and generate the required contractile motion to move the fluid through the lymphatic vasculature [21].

2.4 Development of the Lymphatic Vasculature

2.4.1 Early Models of Lymphatic Development

The origin of lymphatic endothelial cells (*LECs*) has been one of the most highly contested issues related to the development of the lymphatic vasculature. One of the first models addressing this question was put forward by Florence Sabin. On the basis of a series of dye-injection studies in fetal pigs, she proposed the *centrifugal model* of lymphatic development [49]. According to her model, lymphatic vessels

Fig. 2.3 The venous origin of lymphatic vessels. In 1904, Sabin's traceable dye studies of fetal pig veins led her to propose the centrifugal model of lymphatic development. She postulated that lymphatic cells originate in the veins, from where they bud off into the surrounding tissue to form the primitive lymph sacs. This scheme is a representation of one of her original drawings of a pig embryo in which lymphatic cells are leaving the anterior cardinal vein (scheme was generated by Joshua R. Stokes)

develop secondary to blood vessels and are venous derived (Fig. 2.3). Sabin proposed that isolated primitive *lymph sacs* originate from ECs that bud from the veins during early embryonic development. Then from the primary lymph sacs, LECs sprout into the surrounding tissues and organs to give rise to the entire lymphatic network.

Sabin's model was challenged by Huntington and McClure [23], who proposed a *centripetal model* of lymphatic development. In this alternative model, the lymphatic network is derived from specialized mesenchymal tissue called "lymph clefts." Primary lymph sacs arise in the mesenchyme independent of the veins and secondarily establish venous connections [23]. This model of lymphatic development has been supported by recent work in chicken and frog embryos, suggesting that in these organisms LECs arise not only from venous-derived ECs but also from mesenchymal lymphangioblasts [41, 62]. Work in mammals has suggested that venous-derived LECs, hematopoietic cell-derived circulating endothelial progenitors, and transdifferentiating leukocytes and macrophages are putative sources of LECs during embryonic and adult lymphangiogenesis [30, 34, 47, 51]. Although these different sources could have a role in the formation of new LECs during adult lymphangiogenesis and pathologic conditions, recent, detailed genetic lineage-tracing analyses performed in mouse embryos has conclusively demonstrated that, as suggested by Sabin, the mammalian lymphatic vasculature is exclusively venous derived [55]. Furthermore, work in zebrafish embryos revealed that LECs of the main thoracic duct–like vessel also arise from primitive veins [65].

2.4.2 Current Model of Lymphatic Development in Mammals

In the last few years, a number of review articles have described recently identified genes whose expression profile and functional activities are important during the different stages of lymphatic vasculature development [32, 42–44, 54]. Therefore, here we will summarize only a few key genes and processes that are ultimately responsible for the development and maturation of the lymphatic vasculature.

The main (if not sole) source of LECs in the mammalian embryo is the veins [55], and the initial step in the process leading to the formation of the lymphatic vasculature is that venous blood endothelial cells (*BECs*) acquire an *LEC identity (LEC specification)* [61]. In mouse embryos, this initial step is identified by the polarized expression of the homeobox gene *Prox1* in a subpopulation of BECs located on one side of the anterior cardinal veins at around E9.5 to E10.0 (Fig. 2.4A) [60, 61]. As development of the lymphatic vasculature progresses, *Prox1* expression is detected in the whole lymphatic network during embryonic and postnatal stages [18]. *Prox1*-null embryos die at around E14.5 and are devoid of lymphatic vasculature, because LEC specification does not take place in these mutant embryos [60]. This work demonstrated that Prox1 activity is necessary for the specification of the LEC fate in default venous BECs [61].

In mice, *LEC progenitors* start emerging from the embryonic veins after the initial LEC-specification step (at around E10.5) (Fig. 2.4C, D) to form the eight primitive lymph sacs (three paired and two unpaired) [50, 61], which are scattered along the anteroposterior axis of the mammalian embryo (Fig. 2.4E, F). This step requires the activity of the vascular endothelial growth factor-c (Vegf-c), a protein expressed

Fig. 2.4 Lymphatic vasculature development is a stepwise process. A As early as E9.5, Prox1 (*green*) expression starts to be detected in venous endothelial cells (stained with Pecam, *red*) located on the dorsolateral side of the anterior cardinal vein. This expression is necessary and sufficient for the specification of the LEC identity. **B** A schematic representation of **A**, with Prox1-positive cells indicated in green. Nt-neural tube, da-dorsal aorta, cv-cardinal vein. **C** Around E10.5, Prox1-expressing (*blue*) LEC progenitors start to exit the veins by following rather precise paths into the surrounding mesenchyme; these emerging LECs form the primary lymph sacs. *Vefg-c* signaling is required for LECs to migrate from the veins. **D** Schematic representation of **C**. **E** As early as E11.5, Prox1-expressing lymph sacs start to form and can be detected along the anteroposterior axis of the developing embryo. As the sacs form, LECs sprout from the sacs, proliferate, and migrate along the developing embryo. During this stage, the blood and forming lymphatic vasculatures separate. **F** Schematic representation of **E**. jls-jugular lymph sacs. **G** At around E14.5, most of the lymph sacs have formed, and an extensive network of lymphatic vessels and capillaries (*blue*) intermingled with the blood vasculature (*red*) can be detected. This image shows dermal lymphatics. **H** An E15.5 *Prox1*$^{+/-}$ embryo stained with X-gal to show the entire primary dermal lymphatic network. **I** During late embryogenesis and early postnatal stages, the primary lymphatic network expands (Lyve1, *green*), and the forming lymphatics mature and remodel and remain separated from the blood vasculature (SMA, *red*). **J** As lymphatic vessels in the mesentery mature, the lymphatic valves in collecting lymphatics form (*arrow*). Scale bars in **A**, **C**, **E**, **G**, **I**, and **J** represent 100 μM; that in H represents 1 mM. Some images in this panel were contributed by Miriam Dillard and R. Sathish Srinivasan

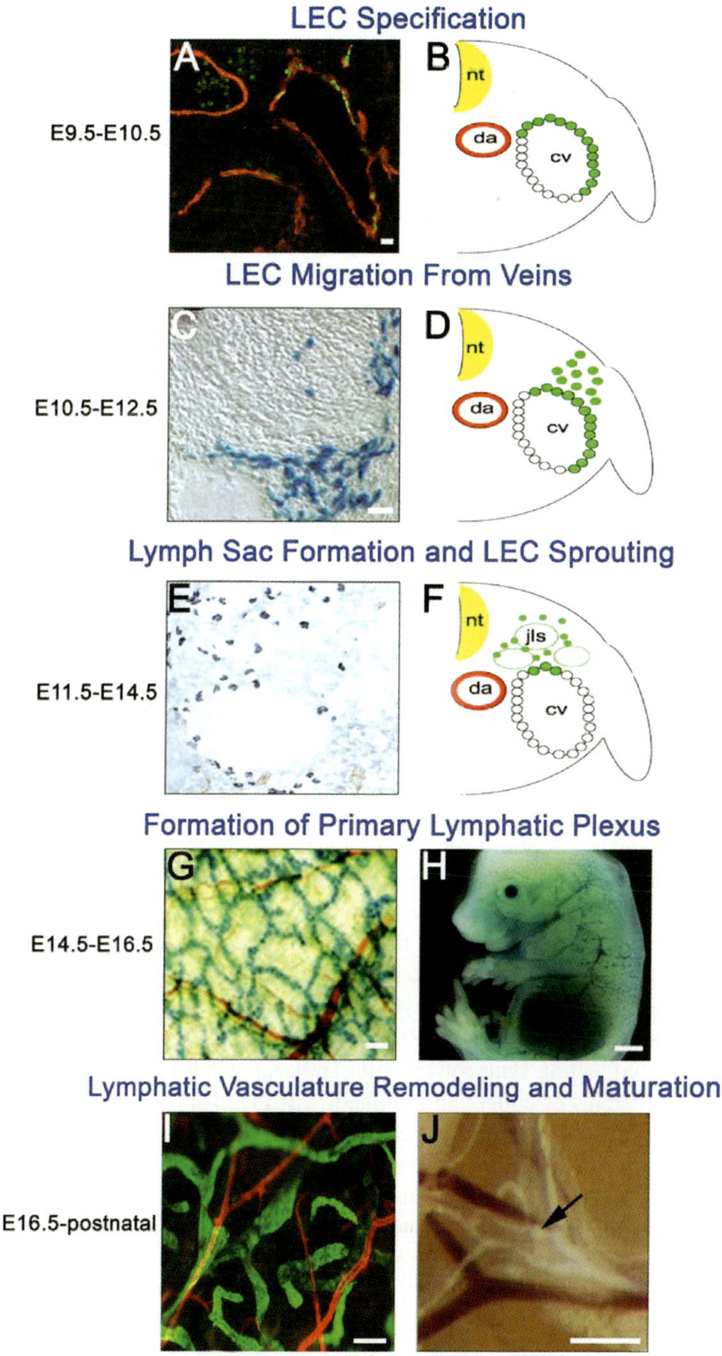

Fig. 2.4 (continued)

in the mesenchyme surrounding the cardinal veins and required to promote and guide the specified LECs to move from the veins toward the ligand source [29]. In *Vegfc*-null embryos, this process is arrested, and Prox1-expressing LEC progenitor cells remain abnormally confined around the wall of the cardinal veins [29].

As the lymph sacs form, LECs sprout from those structures to produce the primary lymphatic plexus (Fig. 2.4G, H) [61]. What promotes and guides this sprouting is not yet known. Recent detailed lineage-tracing analyses determined that local lymph sacs are the source of LECs that by sprouting, proliferation, and migration give rise to the lymphatic vasculature of nearby tissues and organs [55]. Furthermore, those analyses confirmed earlier work by van der Putte [58], who showed that the rostral lymphatics form first. The more posterior lymphatics form later from LEC progenitors located in the posterior veins [55].

As embryonic development progresses, individual sprouting LECs proliferate, migrate, and aggregate to form the entire network of lymphatic capillaries and vessels (Fig. 2.4I). Although not much is yet known about the cellular and molecular mechanisms regulating the sprouting of LECs from the primary lymph sacs, a few players have been identified whose expression is initiated in LECs at around these embryonic stages (E11.0–E13.0) and whose function appears to be necessary for proper lymphatic network formation.

T1α/Podoplanin, a gene encoding a mucin-type transmembrane glycoprotein, is predominantly expressed by the lymphatic endothelium. In the mouse, this protein is detected in specified LECs emerging from the veins as early as E12.0 [10, 42, 52]. *T1α/Podoplanin*$^{-/-}$ pups die soon after birth and exhibit severe lymphedema resulting from defects in lymphatic vascular patterning and function [52].

Neuropilin 2 (*Nrp2*) is a receptor for class III semaphorins; *Nrp2* can also interact with VEGFR2 and VEGFR3 [40, 66]. Although at around E10.0 *Nrp2* is mainly expressed in the veins, starting at E13.0, its expression becomes restricted to LECs [66]. Functional inactivation of *Nrp2* in mice reduces LEC proliferation and results in the transient absence or severe reduction of small lymphatic vessels and capillaries [66]. This result suggests that *Nrp2* differentially controls the formation of large- and small-caliber lymphatic networks [66].

The forkhead transcription factor *Foxc2* is expressed between E12.5 and E14.5 in the jugular lymph sacs and sprouting LECs. Later, *Foxc2* is expressed in the thoracic duct, collecting lymphatics, and capillaries [11]. *Foxc2* is also expressed in the luminal valves of collecting lymphatic vessels [46]. *Foxc2*-null mice exhibit abnormal specification of lymphatic capillaries versus collecting lymphatics, valve agenesis in collecting lymphatics, abnormal accumulation of SMCs and pericytes by lymphatic capillaries, and abnormal expression of BEC markers such as *Pdgfb*, *Endoglin*, and *Collagen IV* [46]. A functionally relevant cooperation between *Foxc2* and *Vegfr-3* in the establishment of a pericyte-free lymphatic capillary network has been proposed [46].

As lymphatic vessels and capillaries spread along the developing embryo, the forming lymphatics separate from the blood vasculature. These two vascular networks remain separated, except at the junction between the thoracic duct and the left subclavian vein. The *blood/lymphatic separation* (Fig. 2.4C, D) step is controlled by at least two molecules expressed mainly by circulating hematopoietic cells, the adaptor protein Slp76 and the tyrosine kinase Syk [1]. *Slp76* expression was recently also detected in a subpopulation of circulating EC progenitors [57]. Functional inactivation of these genes in mice results in a blood-filled lymphatic phenotype [1]. The mechanisms through which these signaling pathways mediate the blood/lymphatic separation remain unknown. Different degrees of defective lymphovenous separation have also been reported in other mouse mutant models in which genes such as *Fasting-induced adipose factor* (*Fiaf*), *ephrinB2*, or *Foxc2* have been inactivated [5, 16, 33, 46].

During late embryogenesis and early postnatal stages, lymphatic vasculature *maturation* and *remodeling* (Fig. 2.4I, J) take place. Expression of additional gene products whose function appears to be required to control different aspects of these processes starts to be detected in the forming lymphatics. The transmembrane ligand ephrinB2 and its receptor, EphB4, are important regulators of embryonic blood vascular morphogenesis. EphrinB2 is expressed in collecting lymphatics, and EphB4 is expressed in collecting lymphatics and lymphatic capillaries [33]. Deletion of the cytoplasmic PDZ-interaction domain of ephrinB2 results in postnatal lethality, i.e., although the mutant pups exhibit a normal blood vasculature, they lack lymphatic valves, exhibit defective lymphatic remodeling, and have an abnormal partial accumulation of SMCs in their lymphatic capillaries [33]. The dermal lymphatic vasculature of these mutant mice fails to mature and resembles a primitive capillary network [33]. This result indicates that *ephrinB2* plays a cell-autonomous role during postnatal lymphatic remodeling [33]. On the other hand, conditional deletion of *ephrinB2* from SMCs leads to early postnatal death, and the mutant pups exhibit severe blood and lymphatic vasculature defects, abnormal migration of SMCs to the lymphatic vessels, and blood-filled lymphatics [16].

The ligand for the endothelial *Tie2* receptor tyrosine kinase, *Angiopoietin-2* (*Ang2* or *Agpt2*), also functions during postnatal lymphatic remodeling and maturation [17]. $Ang2^{-/-}$ newborn mice exhibit subcutaneous edema, chylous ascites, and leaky, abnormally organized lymphatic vessels [17]. $Ang2^{-/-}$ mice also lack proper SMC coverage in the larger collecting lymphatic vessels, and *Ang1* can rescue their lymphatic defects [17].

As discussed above, to prevent retrograde flow, lymphatic vessels contain primary valves at the junctions between LECs and secondary valves (modified endothelial structures) within the lymphatic lumen [53]. Although our knowledge about the genesis and positioning of lymphatic valves is very limited, recent data suggested that primary valves are associated with the discontinuous expression of some endothelial junction markers [38]. Currently, only a few genes essential for lymphatic valve formation and maturation have been identified: the forkhead transcription factor *Foxc2*, *ephrin B2*, and the class III ß-tubulin *Tuj1* [27, 33, 46].

Finally, recent work has determined that the differentiated LEC phenotype is a plastic, reprogrammable condition that depends on constant *Prox1* activity for its maintenance [25]. Conditional downregulation of Prox1 during embryonic, postnatal, or adult stages is sufficient to reprogram LECs into BECs [25].

2.5 Diseases Associated with a Defective Lymphatic Vasculature

There are numerous diseases associated with lymphatic defects. Congenital or acquired defects of the lymphatic vasculature can lead to lymphedema a disorder resulting from insufficient transportation of lymph owing to hypoplasia of the lymph vessels, impaired lymphatic function, or obstruction of lymph flow [64]. Typical features of lymphedema include disabling swelling of the limbs, tissue fibrosis, adipose degeneration of the connective tissue (Fig. 2.5), and susceptibility to infections. *Primary lymphedema* has a genetic origin and can be present at birth (Milroy disease) [37] or appear after puberty (Meige disease) [36]. In general, both of these diseases are characterized by dilated lymphatics and accumulation of lymph fluid. Heterozygous missense mutations in *VEGFR3* have been identified in several cases of Milroy disease [12, 15, 63]. Furthermore, heterozygous missense mutations in the *Vegfr3* gene have been identified in the *Chy*-mutant mice, which develop chylous ascites and lymphedematous limb swelling after birth [28].

Mutations in *FOXC2* have been identified in patients with another form of lymphedema, *lymphedema-distichiasis* [13]. This autosomal-dominant disorder is characterized by distichiasis (double row of eyelashes) at birth and bilateral lower limb lymphedema at puberty [31]. Mutations in the transcription factor *Sox18*

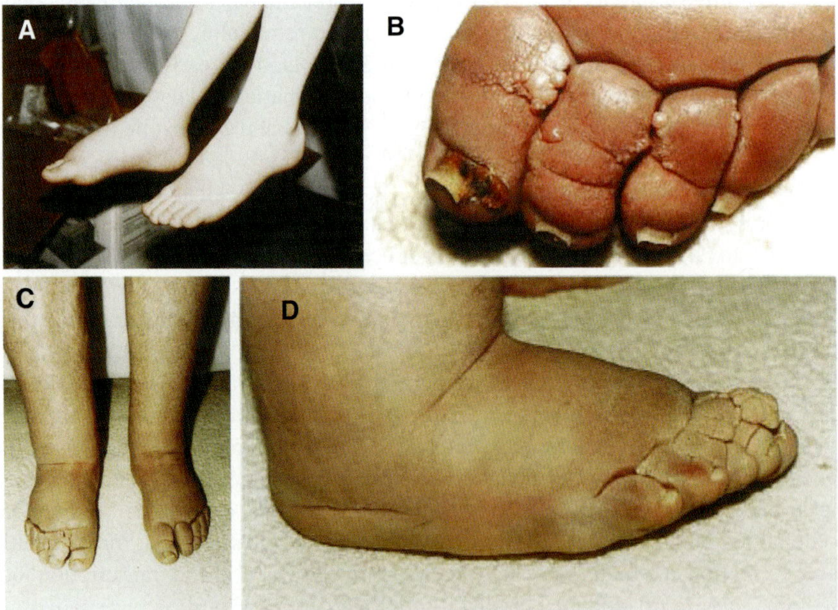

Fig. 2.5 Phenotypic consequences of lymphedema. Primary lymphedema occurs as a result of a congenital disorder of key genes controlling lymphatic development. The symptoms include but are not limited to **A** swelling of the extremities in childhood, **B** swelling with papillomatosis, and **C** bilateral edema with **D** papillomatosis. The figure was reproduced with permission from [12]

were identified in recessive and dominant forms of *hypotrichosis-lymphedema-telangiectasia* [24].

Secondary lymphedema occurs when the lymphatic system is damaged by surgery, infection (e.g., filariasis, which is caused by parasitic infection), radiation therapy (particularly after the treatment of breast cancer), or removal of lymph nodes [64]. As a result of radiation therapy, surgery, and/or removal of lymph nodes, secondary lymphedema develops in approximately 15–20% of women undergoing breast cancer treatment [59].

Another class of congenital disorder is *lymphangiectasia* (i.e., dilation of the lymphatic vessels). This condition is most often seen in the lung, intestine, and thoracic cavity [14]. *Congenital pulmonary lymphangiectasia* is a rare disorder of newborns and is often fatal. These children exhibit cyanosis, labored breathing, and often chylothorax or chylous effluence in the thoracic cavity [8]. *Intestinal lymphangiectasia* is characterized by highly dilated lymphatic capillaries in the intestinal villi. As in most cases of lymphangiectasia, hyperdilation of the lymphatic vessels does not permit the normal lymphatic response to interstitial pressure, and as a result, absorption by the intestine is compromised. The levels of *FOXC2* and *SOX18* transcription are significantly lower in these patients [22]. Less frequently, lymphangiectasia can occur in the kidney, and in even more rare cases, *secondary lymphangiectasia* can develop after tumor removal [4]. Treatment for these types of diseases is extremely limited. Manual drainage, massage, and dietary modification (i.e., limiting the consumption of long-chain fatty acids) comprise the current standards of care.

In addition to these relatively well-recognized disorders, *late-onset obesity* has recently been linked to leaky lymphatic vessels in an available mouse model of defective lymphatic vasculature [18]. *Prox1*$^{+/-}$ mice have mispatterned lymphatics with defective vessel integrity that allows chyle to leak from those vessels (Fig. 2.6)

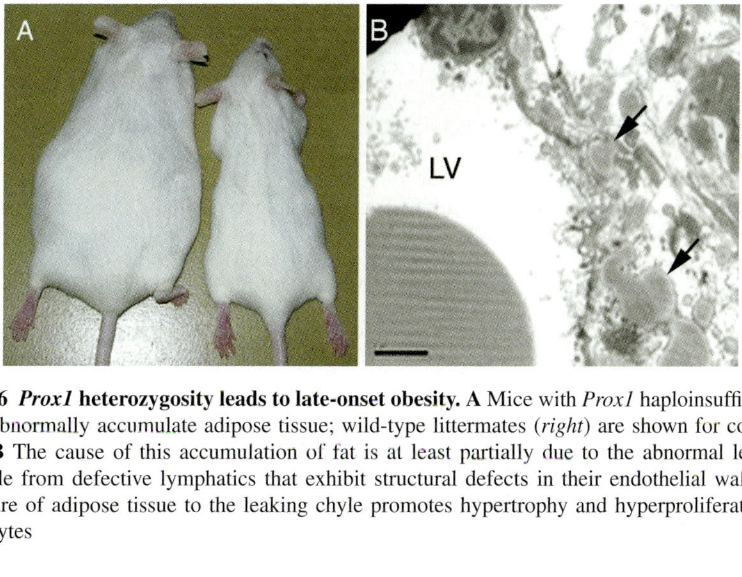

Fig. 2.6 *Prox1* heterozygosity leads to late-onset obesity. A Mice with *Prox1* haploinsufficiency (*left*) abnormally accumulate adipose tissue; wild-type littermates (*right*) are shown for comparison. **B** The cause of this accumulation of fat is at least partially due to the abnormal leakage of chyle from defective lymphatics that exhibit structural defects in their endothelial wall. The exposure of adipose tissue to the leaking chyle promotes hypertrophy and hyperproliferation of adipocytes

[18]. Chyle is adipogenic, and in vitro it can trigger immature preadipocytes to mature into lipid-accumulating adipocytes [18].

2.6 Future Perspectives

The mechanisms described above apply to developmental lymphangiogenesis. It is not yet clear how preexisting lymphatics regrow during tissue repair and inflammation. A similar statement can be made about the genes and mechanisms controlling lymphatic growth in and around tumors. Furthermore, the phenotypic differences between smaller, initial lymphatic capillaries and the larger, more complex collecting lymphatics suggest that at some point during their development and maturation, different regulatory mechanisms establish the final identities of these unique structures.

It is well accepted that similar to the blood vasculature, outgrowth from preexisting lymphatic vessels is at least one of the mechanisms of adult lymphangiogenesis; however, whether this process also involves other sources such the budding of LEC precursors from preexisting veins, transdifferentiating macrophages [34], or bone marrow-derived cells [30, 47, 51] is not yet known.

Finally, the finding that defective lymphatic vasculature can eventually lead to some forms of late-onset obesity [18] argues that other pathologic conditions may also be linked to a defective lymphatic vasculature. For example, individual differences in immune response could be linked to variations in the integrity and functionality of the lymphatic system.

Acknowledgments We want to thank Joshua. R. Stokes from the Biomedical Communications department for the generation of the piglet diagram shown in Fig. 2.3, Satish Srinivasan and Miriam Dillard for critical reading, and Angela McArthur for scientific editing.

References

1. Abtahian F, Guerriero A, Sebzda E, Lu MM, Zhou R, Mocsai A, Myers EE, Huang B, Jackson DG, Ferrari VA, et al. Regulation of blood and lymphatic vascular separation by signaling proteins SLP-76 and Syk. Science New York, NY 2003, 299: 247–51.
2. Ambrose C. Immunology's first priority dispute—an account of the 17th-century Rudbeck–Bartholin feud. Cell Immunol 2006, 242: 1.
3. Aselli G. De Lactibus Sive Lacteis Venis, Quarto Vasorum Mesaraicorum Genere, 1627.
4. Ashraf K, Raza SS, Ashraf O, Memon W, Memon A, Zubairi TA. Renal lymphangiectasia. Br J Radiol 2007, 80: e117–18.
5. Backhed F, Crawford PA, O'Donnell D, Gordon JI. Postnatal lymphatic partitioning from the blood vasculature in the small intestine requires fasting-induced adipose factor. Proc Natl Acad Sci USA 2007, 104: 606–11.
6. Baluk P, Fuxe J, Hashizume H, Romano T, Lashnits E, Butz S, Vestweber D, Corada M, Molendini C, Dejana, E, et al. Functionally specialized junctions between endothelial cells of lymphatic vessels. J Exp Med 2007, 205: 2349–62.
7. Beigneux AP, Davies BS, Gin P, Weinstein MM, Farber E, Qiao X, Peale F, Bunting S, Walzem RL, Wong JS, et al.. Glycosylphosphatidylinositol-anchored high-density lipoprotein-binding

protein 1 plays a critical role in the lipolytic processing of chylomicrons. Cell Met 2007, 5: 279–91.
8. Bellini C, Boccardo F, Campisi C, Bonioli E. Congenital pulmonary lymphangiectasia. Orp J Rare Dis 2006, 1: 43.
9. Black DD. Development and physiological regulation of intestinal lipid absorption. I. Development of intestinal lipid absorption: cellular events in chylomicron assembly and secretion. Am J Physiol Gastrointest Liver Physiol 2007, 293: G519–24.
10. Breiteneder-Geleff S, Soleiman A, Kowalski H, Horvat R, Amann G, Kriehuber E, Diem K, Weninger W, Tschachler E, Alitalo, K, et al.. Angiosarcomas express mixed endothelial phenotypes of blood and lymphatic capillaries: podoplanin as a specific marker for lymphatic endothelium. Am J Pathol 1999, 154: 385–94.
11. Dagenais SL, Hartsough RL, Erickson RP, Witte MH, Butler MG, Glover TW. Foxc2 is expressed in developing lymphatic vessels and other tissues associated with lymphedema-distichiasis syndrome. Gene Expr Patterns 2004, 4: 611–19.
12. Evans AL, Brice G, Sotirova V, Mortimer P, Beninson J, Burnand K, Rosbotham J, Child A, Sarfarazi M. Mapping of primary congenital lymphedema to the 5q35.3 region. Am J Hum Gen 1999, 64: 547–55.
13. Fang J, Dagenais SL, Erickson RP, Arlt MF, Glynn MW, Gorski JL, Seaver LH, Glover TW. Mutations in FOXC2 (MFH-1), a forkhead family transcription factor, are responsible for the hereditary lymphedema-distichiasis syndrome. Am J Hum Gen 2000, 67: 1382–8.
14. Faul JL, Berry GJ, Colby TV, Ruoss SJ, Walter MB, Rosen GD, Raffin TA. Thoracic lymphangiomas, lymphangiectasis, lymphangiomatosis, and lymphatic dysplasia syndrome. Am J Resp Crit Care Med 2000, 161: 1037–46.
15. Ferrell RE, Levinson KL, Esman JH, Kimak MA, Lawrence EC, Barmada MM, Finegold DN. Hereditary lymphedema: evidence for linkage and genetic heterogeneity. Hum Mol Gen 1998, 7: 2073–8.
16. Foo SS, Turner CJ, Adams S, Compagni A, Aubyn D, Kogata N, Lindblom P, Shani M, Zicha D, Adams RH. Ephrin-B2 controls cell motility and adhesion during blood-vessel-wall assembly. Cell 2006, 124: 161–73.
17. Gale NW, Thurston G, Hackett SF, Renard R, Wang Q, McClain J, Martin C, Witte C, Witte MH, Jackson, D, et al. Angiopoietin-2 is required for postnatal angiogenesis and lymphatic patterning, and only latter role is rescued by Angiopoietin-1. Dev Cell 2002, 3: 411–23.
18. Harvey NL, Srinivasan RS, Dillard ME, Johnson NC, Witte MH, Boyd K, Sleeman MW, Oliver G. Lymphatic vascular defects promoted by Prox1 haploinsufficiency cause adult-onset obesity. Nat Gen 2005, 37: 1072–81.
19. Havel RJ, Kane JP. Introduction: structure and metabolism of plasma lipoproteins. In The Metabolic and Molecular Bases of Inherited Diseases (Scriver CR, Beaudet AL, Sly WS, Valle D, Childs B, Kinzler KW, Vogelstein B, Eds.), pp. 2705–16. McGraw-Hill, New York, 2001.
20. Hewson W. Experimental Inquiries. Longman, London, 1774.
21. Hillman SS, Hedrick MS, Withers PC, Drewes RC. Lymph pools in the basement, sump pumps in the attic: the anuran dilemma for lymph movement. Physiol Biochem Zool 2004, 77: 161–73.
22. Hokari R, Kitagawa N, Watanabe C, Komoto S, Kurihara C, Okada Y, Kawaguchi A, Nagao S, Hibi T, Miura S. Changes in regulatory molecules for lymphangiogenesis in intestinal lymphangiectasia with enteric protein loss. J Gastroenterol Hepatol 2007, 23(7Pt2): e88–95.
23. Huntington GS, McClure CFW. The anatomy and development of the jugular lymph sac in the domestic cat. Am J Anat 1910, 10: 177–311.
24. Irrthum A, Devriendt K, Chitayat D, Matthijs G, Glade C, Steijlen PM, Fryns JP, Van Steensel MA, Vikkula M. Mutations in the transcription factor gene SOX18 underlie recessive and dominant forms of hypotrichosis-lymphedema-telangiectasia. Am J Hum Gen 2003, 72: 1470–8.

25. Johnson N, Dillard M, Baluk P, McDonald DM, Harvey NL, Frase SL, Oliver G. Lymphatic endothelial cell identity is reversible and its maintenance requires Prox1 activity. Genes Dev 2008, 22: 3282–91.
26. Kampmeier OF. Evolution and Comparative Morphology of the Lymphatic System. Thomas, Springfield, IL, 1969.
27. Kang J, Lee, I. TuJ1 (class III beta-tubulin) as phenotypic marker of lymphatic and venous valves. Cardiovasc Pathol 2006, 15: 218–21.
28. Karkkainen MJ, Saaristo A, Jussila L, Karila KA, Lawrence EC, Pajusola K, Bueler H, Eichmann A, Kauppinen R, Kettunen MI, et al. A model for gene therapy of human hereditary lymphedema. Proc Natl Acad Sci USA 2001, 98: 12677–82.
29. Karkkainen MJ, Haiko P, Sainio K, Partanen J, Taipale J, Petrova TV, Jeltsch M, Jackson DG, Talikka M, Rauvala, H, et al. Vascular endothelial growth factor C is required for sprouting of the first lymphatic vessels from embryonic veins. Nat Immunol 2004, 5: 74–80.
30. Kerjaschki D, Huttary N, Raab I, Regele H, Bojarski-Nagy K, Bartel G, Krober SM, Greinix H, Rosenmaier A, Karlhofer, F, et al. Lymphatic endothelial progenitor cells contribute to de novo lymphangiogenesis in human renal transplants. Nat Med 2006, 12: 230–4.
31. Kriederman BM, Myloyde TL, Witte MH, Dagenais SL, Witte CL, Rennels M, Bernas MJ, Lynch MT, Erickson RP, Caulder MS, et al. FOXC2 haploinsufficient mice are a model for human autosomal dominant lymphedema-distichiasis syndrome. Hum Mol Gen 2003, 12: 1179–85.
32. Lymboussaki A, Achen MG, Stacker SA, Alitalo K. Growth factors regulating lymphatic vessels. Cur Micro Immunol 2000, 251: 75–82.
33. Makinen T, Adams RH, Bailey J, Lu Q, Ziemiecki A, Alitalo K, Klein R, Wilkinson GA. PDZ interaction site in ephrinB2 is required for the remodeling of lymphatic vasculature. Genes Dev 2005, 19: 397–410.
34. Maruyama K, Ii M, Cursiefen C, Jackson DG, Keino H, Tomita M, Van Rooijen N, Takenaka H, D'Amore PA, Stein-Streilein, J, et al. Inflammation-induced lymphangiogenesis in the cornea arises from CD11b-positive macrophages. J Clin Invest 2005, 115: 2363–72.
35. McCray S, Parrish CR. When chyle leaks: nutrition management options. Prac Gast 2004, 17: 60–76.
36. Meige H. Dystophie oedematoeuse hereditaire. Presse Med 1898, 6: 341–3.
37. Milroy WF. An undescribed variety of hereditary oedema. NY Med J 1892, 56: 503.
38. Murfee WL, Rappleye JW, Ceballos M, Schmid-Schonbein GW. Discontinuous expression of endothelial cell adhesion molecules along initial lymphatic vessels in mesentery: the primary valve structure. Lymphat Res Biol 2007, 5: 81–90.
39. Nestel PJ, Havel RJ, Bezman A. Sites of initial removal of chylomicron triglyceride fatty acids from the blood. J Clin Invest 1962, 41: 1915–21.
40. Neufeld G, Kessler O, Herzog Y. The interaction of Neuropilin-1 and Neuropilin-2 with tyrosine-kinase receptors for VEGF. Adv Exp Med Biol 2002, 515: 81–90.
41. Ny A, Koch M, Schneider M, Neven E, Tong RT, Maity S, Fischer C, Plaisance S, Lambrechts D, Heligon, C, et al. A genetic Xenopus laevis tadpole model to study lymphangiogenesis. Nat Med 2005, 11: 998–1004.
42. Oliver G. Lymphatic vasculature development. Nat Rev Immunol 2004, 4: 35–45.
43. Oliver G, Harvey N. A stepwise model of the development of lymphatic vasculature. Ann NY Acad Sci 2002, 979: 159–65.
44. Oliver G, Alitalo K. The lymphatic vasculature: recent progress and paradigms. Annu Rev Cell Dev Biol 2005, 21: 457–83.
45. Pequet J. New anatomical experiments by which the hitherto unknown receptacle of the Chyle and transmission from thence to the Subclavian Veins by the now discovered Lacteal Channels of the Thorax is plainly made appear in Brutes. London, 1653.
46. Petrova TV, Karpanen T, Norrmen C, Mellor R, Tamakoshi T, Finegold D, Ferrell R, Kerjaschki D, Mortimer P, Yla-Herttuala, S, et al. Defective valves and abnormal mural cell recruitment underlie lymphatic vascular failure in lymphedema distichiasis. Nat Med 2004, 10: 974–81.

47. Religa P, Cao R, Bjorndahl M, Zhou Z, Zhu Z, Cao Y. Presence of bone marrow-derived circulating progenitor endothelial cells in the newly formed lymphatic vessels. Blood 2005, 106: 4184–90.
48. Rudbeck O. Nova Exercitatio Anatomica Exhibens Ductus Hepaticos Aquosus et Vasa Glandularum Serosa Lauringer, Vasteras, 1653.
49. Sabin F. On the origin of the lymphatics system from the veins and the development of the lymph hearts and the thoracic duct in the pig. Am J Anat 1902, 1: 367–91.
50. Sabin FR. The lymphatic system in human embryos, with a consideration of the system as a whole. Am J Anat 1909, 9: 43–91.
51. Salven P, Mustjoki S, Alitalo R, Alitalo K, Rafii S. VEGFR-3 and CD133 identify a population of CD34+ lymphatic/vascular endothelial precursor cells. Blood 2003, 101: 168–72.
52. Schacht V, Ramirez MI, Hong YK, Hirakawa S, Feng D, Harvey N, Williams M, Dvorak AM, Dvorak HF, Oliver, G, et al. T1alpha/podoplanin deficiency disrupts normal lymphatic vasculature formation and causes lymphedema. J EMBO 2003, 22: 3546–56.
53. Schmid-Schonbein GW. The second valve system in lymphatics. Lymphat Res Biol 2003, 1: 25–9; discussion 29–31.
54. Sleeman JP, Krishnan J, Kirkin V, Baumann P. Markers for the lymphatic endothelium: in search of the holy grail? Microsc Res Tech 2001, 55: 61–9.
55. Srinivasan RS, Dillard M, Lagutin O, Lin FJ, Tsai S, Tsai MJ, Samokhvalov IM, Oliver G. Lineage tracing demonstrates the venous origin of the mammalian lymphatic vasculature. Genes Dev 2007, 21: 2422–32.
56. Stacker SA, Achen MG, Jussila L, Baldwin ME, Alitalo K. Lymphangiogenesis and cancer metastasis. Nat Rev Cancer 2002, 2: 573–83.
57. Stadtfeld M, Graf T. Assessing the role of hematopoietic plasticity for endothelial and hepatocyte development by non-invasive lineage tracing. Development 2005, 132: 203–13.
58. van der Putte SC. The early development of the lymphatic system in mouse embryos. Acta Morphol Neerl Scand 1975, 13(4): 245–86.
59. Vignes S, Arrault M, Dupuy A. Factors associated with increased breast cancer-related lymphedema volume. Acta Oncol 2007, 46: 1138–42.
60. Wigle JT, Oliver G. Prox1 function is required for the development of the murine lymphatic system. Cell 1999, 98: 769–78.
61. Wigle JT, Harvey N, Detmar M, Lagutina I, Grosveld G, Gunn MD, Jackson DG, Oliver G. An essential role for Prox1 in the induction of the lymphatic endothelial cell phenotype. J EMBO 2002, 21: 1505–13.
62. Wilting J, Aref Y, Huang R, Tomarev SI, Schweigerer L, Christ B, Valasek P, Papoutsi M. Dual origin of avian lymphatics. Dev Biol 2006, 292: 165–73.
63. Witte MH, Erickson R, Bernas M, Andrade M, Reiser F, Conlon W, Hoyme HE, Witte CL. Phenotypic and genotypic heterogeneity in familial Milroy lymphedema. Lymphology 1998, 31: 145–55.
64. Witte MH, Bernas MJ, Martin CP, Witte CL. Lymphangiogenesis and lymphangiodysplasia: from molecular to clinical lymphology. Microsc Res Tech 2001, 55: 122–45.
65. Yaniv K, Isogai S, Castranova D, Dye L, Hitomi J, Weinstein BM. Live imaging of lymphatic development in the zebrafish. Nat Med 2006, 12: 711–16.
66. Yuan L, Moyon D, Pardanaud L, Breant C, Karkkainen MJ, Alitalo K, Eichmann A. Abnormal lymphatic vessel development in neuropilin 2 mutant mice. Development 2002, 129: 4797–806.

Chapter 3

NEW ANIMAL MODELS OF LYMPHANGIOGENESIS

Benjamin M. Hogan and Stefan Schulte-Merker
Hubrecht Institute (KNAW), University Medical Center Utrecht, Utrecht 3584 CT, The Netherlands, e-mail: s.schulte@niob.knaw.nl

Abstract: The discovery of new genes, genetic pathways and drug targets involved in the formation of lymphatic vessels and capillaries has long been hampered by the absence of a small model organism amenable to large scale genetic and pharmacological screening. The recent description of functional and conserved lymphatic vascular systems in the two small animal models *Xenopus laevis* (the African clawed frog) and *Danio rerio* (the zebrafish) now opens up the possibility to exploit these models for the study of developmental lymphangiogenesis. In this chapter we will describe the discovery and characterisation of the lymphatic vasculature in both frog and fish models. We will also describe and compare the available genetic tools for the study of lymphangiogenesis in these new model systems.

Key words: Lymphatic · Vasculature · Lymphangiogenesis · Zebrafish · *Xenopus*

3.1 Introduction

Lymphangiogenesis, the formation of lymphatic vessels from the pre-existing vasculature, is an important process in metabolism, immunity and pathological processes such as cancer metastasis. It is because of these important functions that lymphangiogenesis has received increased attention in recent years and the dramatic progress made in the field is described in other chapters of this book. In furthering our understanding of lymphangiogenesis, the use of sophisticated genetic analysis in mouse models has yielded much insight into the genetic control of embryonic and adult lymphangiogenesis (for review see [2]). Coupled with genetic analysis, cellular studies performed in both mouse and avian models have greatly improved our understanding of the origin and early development of the lymphatic vasculature [60, 70, 74, 84, 91, 92]. Furthermore, the molecular, morphological and functional analysis of cultured mammalian lymphatic endothelial cells has yielded

S.A. Stacker, M.G. Achen (eds.), *Lymphangiogenesis in Cancer Metastasis*, Cancer Metastasis – Biology and Treatment 13, DOI 10.1007/978-90-481-2247-9_3,

exciting new insights into both normal and pathological processes within the lymphatic vasculature.

As with any expanding field, new ideas are being explored and new ways to examine the object of interest are being developed. Despite the many advances, such as those highlighted above, a genetically tractable small animal model has not been available for unbiased, large scale gene and drug target discovery approaches until very recently. In this chapter we will describe the biology of lymphangiogenesis in the two newest model organisms to enter the lymphangiogenesis scene, namely the tadpole model from *Xenopus laevis* (the African clawed frog) and the larval model from the teleost *Danio rerio* (the zebrafish). Furthermore, as both frogs and fish offer a new spectrum of methodological approaches for the study of lymphangiogenesis, we will describe the available technologies in these model organisms and compare their amenability both to one another and to the mouse model.

3.2 Biology of Xenopus Laevis Lymphangiogenesis

It has long been held that anurans (frogs and toads), like fully terrestrial vertebrates, have a functional lymphatic vasculature comprising lymph hearts, vessels and measurable lymph flow as described in early studies, which examined the physiology and histology of their lymphatic vasculature [3, 8, 9]. As the initial studies were resticted to the lymphatic system of adult anurans, these organisms did not immediately present themselves as useful models for the study of lymphangiogenesis at the molecular and cellular levels. Recently, however, analysis of the development of the lymphatic system during the embryogenesis of *Xenopus laevis* tadpoles has opened up the opportunity to exploit this organism as a small animal model for detailed analysis of lymphatic vessel development [57].

3.2.1 Anatomical Localization of Early Lymphangiogenesis in the Xenopus Laevis Tadpole

The initial description of lymphatic development in *Xenopus laevis* tadpoles by Ny et al. [57] examined both the anatomy and the evolutionarily conserved molecular control of embryonic lymphangiogenesis. Ny et al. [57] used in situ hybridisation to examine the expression of the conserved lymphangiogenic transcription factor *prox1* as a marker for lymphangioblasts and lymphatic vessels. The earliest distinguishable structures of the lymphatic system that express *prox1* during development are the rostral lymph sac (RLS) and the lymph heart (LH), which develop from embryonic stages 28 (RLS) and 32 (LH) onwards. The expression of *prox1* was also seen in a population of cells likely to be lymphangioblasts in the region of the ventral caudal lymph vessel (VCLV) from stage 32 onwards. As has been observed for the initial sites of embryonic lymphangiogenesis in mammals [90], *prox1* expression in lymphangioblasts appears to be rapidly segregated from an origin initially overlapping with blood endothelial precursors derived from the lateral plate mesoderm.

In the RLS region, expression of *prox1* and *msr* (mesenchyme associated-serpentine receptor; a marker of blood vessels [12]), initially overlap in the lateral plate mesoderm adjacent to the cardiac field. However, *prox1* expression becomes progressively more lateral in localisation and eventually becomes restricted to the sub-ectodermal region lateral to the ventral aorta and heart tube [57]. In the LH region, *prox1* expression initially labels a population of cells which express *msr* and *fli* [7] located at the dorsal margins of the cardinal vein in the region where the vein fuses to the common cardinal vein to drain into the heart. Later, the expression of *prox1* suggests separation of lymphatic cells to a more dorso-lateral position where it labels the lymph hearts, from which the cephalic and lateral lymph ducts likely form [57].

In the posterior ventral region of the trunk at embryonic stage 32, *prox1* expression is observed in a region overlapping with *msr* expression within the developing posterior cardinal vein located immediately posterior to the rectal diverticle. Later, the expression is observed in developing lymphatic vessels of the VCLV found immediately ventral to the PCV [57]. Interestingly, expression of prox1 is also observed in cells immediately dorsal to the VCLV at stage 35/36 suggesting that a population of putative lymphangioblasts may actively migrate dorsally from an initial location in the VCLV. Expression is also observed at later stages in the dorsal longitudinal anastomosing vessel (DLAV), suggesting that a pool of VCLV precursor lymphangioblasts may give rise to the trunk lymphatic vasculature of the dorsal caudal lymphatic vessel and the ventral caudal lymphatic vessel by a process of dynamic cell migration [57] (see Fig. 3.1).

In summary, these analyses of *prox1* expression at different stages of embryogenesis, although not sufficient to show the physical movement of *prox1* expressing cells, are highly suggestive of dynamic migration. It seems likely that lymphangioblasts differentiate from an initial precursor pool of overlapping origin with the blood vasculature (most likely venous as observed in mammals [74]) and then migrate to populate defined regions of the developing tadpole. The process would involve at least three phases: differentiation of endothelial cells from a venous origin, active migration to distinct sites in the tadpole, and terminal differentiation including tubulogenesis to form a functional lymphatic vessel (Fig. 3.1).

In addition to the analysis of *prox1* expression, Ny et al. [57] also examined vessel morphology in detail, as well as structure and function of the trunk lymphatic vasculature. The VCLV fails to express Laminin indicating that it lacks a basement membrane, the absence of which is characteristic of lymphatic vessels (for a brief review of mammalian lymphatic vascular morphology and ultrastructure see [67]). In contrast, the adjacent blood vasculature (PCV) expresses detectable levels of Laminin. Examination of tadpole VCLVs using electron microscopy also indicates a high degree of ultrastructural similarity with mammalian lymphatic vessels. Tadpole VCLV cells are thin and irregular and lack well established junction formation with adjacent lymphatic endothelial cells. These are all characteristics observed in and typical of mammalian lymphatic endothelial cells [67]. Importantly, the introduction of fluorescent solutions (high molecular weight rhodamine dextran) directly into subcutaneous spaces using micro-injection, leads to the active drainage

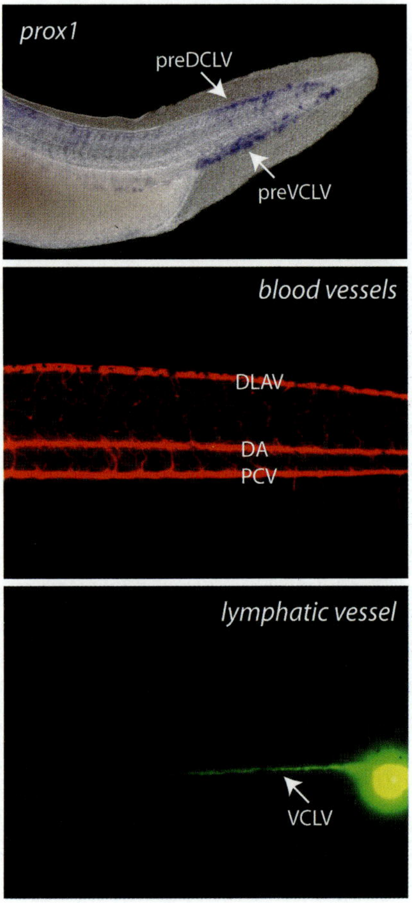

Fig. 3.1 *Xenopus laevis* embryonic lymphangiogenesis. *prox1* expression (*upper*) in the trunk of a stage 39 *Xenopus laevis* tadpole marks a population of putatively migrating lymphangioblasts which will give rise to the ventral caudal lymphatic vessel (VCLV) and the dorsal caudal lymphatic vessel (DCLV). The trunk blood vasculature (*middle*) highlighted by angiogram injection in a stage 45 *Xenopus laevis* tadpole. The dorsal aorta (DA), posterior cardinal vein (PCV) and dorsal longitudinal vessel (DLAV) are indicated. The ventral caudal lymphatic vessel (VCLV) in the trunk (*lower*) of a stage 45 *Xenopus laevis* tadpole as highlighted by lymphangiogram dye injection

of fluorescent dye into the dorsal caudal lymphatic vessel (DCLV) and VCLV. This drainage into lymphatic vessels occurs rapidly after micro-injection and indicates definitively that the vessels are not just morphologically similar, but the functional equivalent of mammalian lymphatic vessels (Fig. 3.1) [57]. In contrast the injection of fluorescent dye into the blood stream leads to specific accumulation in the blood vasculature indicating the physical separation of the two vascular systems (Fig. 3.1) [57].

3.2.2 Molecular Regulation of Developmental Lymphangiogenesis in Xenopus Laevis

As stated above, lymphatic vessels, lymphangioblasts and their precursors, the embryonic veins, all express the key marker of mammalian lymphatic identity, Prox1. In mammals Prox1 is both required for lymphangiogenesis and also sufficient to induce lymphangiogenic developmental programs in blood endothelial cells, identifying it as a master regulator of lymphangiogenesis [31, 91]. The examination of *prox1* loss-of-function in *Xenopus laevis* embryos also identified an essential role for this transcription factor in this species, thereby indicating an evolutionarily conserved molecular program for lymphangiogenesis [57]. Animals that were *prox1* depleted by injection of antisense morpholino oligomers (see section 4.1.1 for details) showed a reduction in the putative dorsal migration of lymphangioblasts from the VCLV as detected by examining *prox1* expression with in situ hybridisation. In addition to altered *prox1* expression, lymphatic function was reduced or absent in these animals as subcutaneously deposited tracer dyes no longer drained into the VCLV, indicating either a complete absence of the VCLV or severely impaired function. Morphologically, *prox1* depletion resulted in the development of massive edema, highly reminiscent of lymphedema in mammals [57].

In addition to requiring Prox1, mammalian lymphangiogenesis is dependent on signaling through Vegfr3 as initiated by binding to the ligand VegfC [35,37,44,85]. Depletion of *vegfC* in *Xenopus laevis*, again using morpholino oligomer microinjection, led to reduced dorsal migration of *prox1* expressing putative lymphangioblasts. Furthermore, the examination of lymphatic function revealed impaired drainage from interstitial spaces [57]. This indicates a critical, conserved role for VegfC and by extension also suggests a conserved role for Vegfr3 in *Xenopus laevis* embryonic lymphangiogenesis (Table 3.1). The conserved molecular regulation of lymphangiogenesis further highlights the utility of this model for examining the genetic pathways involved.

3.3 Biology of Lymphangiogenesis in the Zebrafish (Danio Rerio)

Unlike the lymphatic system in anurans, the existence of a functional lymphatic vasculature in teleost fishes was for a long time controversial (for review, see [75]). Although some early studies had reported the presence of 'atypical' vessels in adult fish and although it was known that bloodless vessels can be found in some fish species, teleosts were in many cases referred to as being devoid of a functional lymphatic vasculature. The simple argument was that fish, being an aquatic organism, have a reduced physiological need to regulate their fluid homeostasis, and that loss of water is a lesser problem for a water-borne organism. It was, therefore, not until the simultaneous reporting of the characterization of an embryonic lymphatic vascular system by both Küchler et al. [43] and Yaniv et al. [94] that the existence of a lymphatic vasculature was firmly established in zebrafish. Indeed, the presence of a functional and developmentally conserved lymphatic vascular system now opens up the possibility to exploit the zebrafish as a model for the study of developmental lymphangiogenesis.

Table 3.1 Conservation of the molecular regulation of embryonic lymphangiogenesis between *Xenopus*, Zebrafish and mouse

Gene	Xenopus *Expression*	*Phenotype*	Zebrafish *Expression*	*Phenotype*	Mouse *Expression*	*Phenotype*
VEGFC	–	Morpholino injection led to altered distribution of *prox1* positive lymphagioblasts with subsequent loss of lymphatic drainage and edema [57].	In hypochord at 18 somites (adjacent to the dorsal aorta). Later at 24 hours in the dorsal aorta [10].	Morpholino injection at high dose inhibited blood vessel development [10, 58], lower doses inhibited thoracic duct formation [43, 94].	Mesenchyme surrounding developing lymphatic vessels and smooth muscle cells surrounding embryonic arteries [37, 44].	Over-expression induced excess lymphangiogenesis [35]. Knock-out led to failure of sprouting of early lymphatic vessels [37].
VEGFR3	Expressed in lymphatic vessels [49].	–	Early embryonic veins [80].	Injection of soluble dominant negative VEGFR3 inhibited blood vessel formation at high doses [58] and inhibited thoracic duct formation at low doses [43, 94].	Early veins, developing and mature lymphatic vessels [4]. Adult expression in smooth muscle surrounded collecting lymphatic vessels down-regulated [86].	Over-activation induces excess lymphangiogenesis [85]. Knock-out mice were early lethal with cardiovascular defects [16]. Expression of soluble VEGFR3 inhibited lymphangiogenesis and induced regression of formed lymphatic vessels [38, 52].

(continued)

Table 3.1 (continued)

	Xenopus		Zebrafish		Mouse	
Gene	*Expression*	*Phenotype*	*Expression*	*Phenotype*	*Expression*	*Phenotype*
PROX1	Expressed in lymphangioblasts and lymphatic vessels [57].	Morpholino injection led to reduced putative migration of *prox1* positive lymphangioblasts and subsequently reduced lymphatic drainage [57].	Expressed in the parachordal region and thoracic duct [94].	Morpholino injection inhibited thoracic duct formation [94].	Early veins and progressively restricted to developing lymphatic vessels [90].	Knock-out mouse displayed a specific loss of lymphatic vessels [90, 91].
NRP2	–	–	Expressed in the thoracic duct [94].	–	Lymphatic vessels and at low levels in veins [95].	Knock-out mouse displayed transient hypoplasia of small lymphatic vessels and capillaries [95].

Table 3.1 (continued)

Gene	Xenopus Expression	Phenotype	Zebrafish Expression	Phenotype	Mouse Expression	Phenotype
TIE2	Embryonic heart and veins [32].	–	Early blood vessels and endocardium with some progressive restriction to veins [51].	–	Expressed in endothelial cells and precursors [14, 17]. Expressed in embryonic and adult lymphatic endothelial cells [55, 78].	Knock-out mice were embryonic lethal due to early cardiac defects [15, 66]. Loss of Ang2 induced signaling led to lymphatic hypoplasia [23]. Increased activation by Ang1 induced excess lymphangio-genesis [55, 78].
ANG2	–	–	Expression in adult lymphatic vessels and embryonic head mesenchyme [43, 94].	–	Aorta and major vessels [23].	Knock-out mouse displayed defective modeling and function of post-natal lymphatic vessels [23].

–=not yet examined

3.3.1 Anatomical Aspects of Early Zebrafish Lymphangiogenesis

The identification of lymphatic vessels in zebrafish larvae was made possible by the production of transgenic lines expressing GFP throughout the embryonic vasculature. The transgenic *tg[fli1:GFP]* line expresses GFP under the control of the *fli1* promoter, a gene expressed in all known endothelial cells in developing zebrafish embryos, larvae and adults [48]. In the trunk of *tg[fli1:GFP]* animals from 3.5 days post fertilisation (dpf) onwards an axial vessel is observable immediately ventral to the dorsal aorta and dorsal to the posterior cardinal vein (Fig. 3.2). By 4dpf the vessel runs the entire length of the trunk and is irregular in shape unlike the adjacent dorsal aorta or the posterior cardinal vein. This third major axial vessel lacks blood, as determined using angiogram micro-injection of fluorescent dye into the blood vasculature and it is for this reason that its existence had been missed in earlier angiography-based examinations of the zebrafish vasculature [33,43,94]. Detailed examination of the vessel, with electron microscopy, reveals that the vessel walls are far thinner than those of the adjacent blood vessels and that the cells display fibrous connections with the interstitial matrix, both characteristic of mammalian lymphatic vessels [25, 94]. Using sub-cutaneous micro-injections of fluorescent dyes both Yaniv et al. [94] and

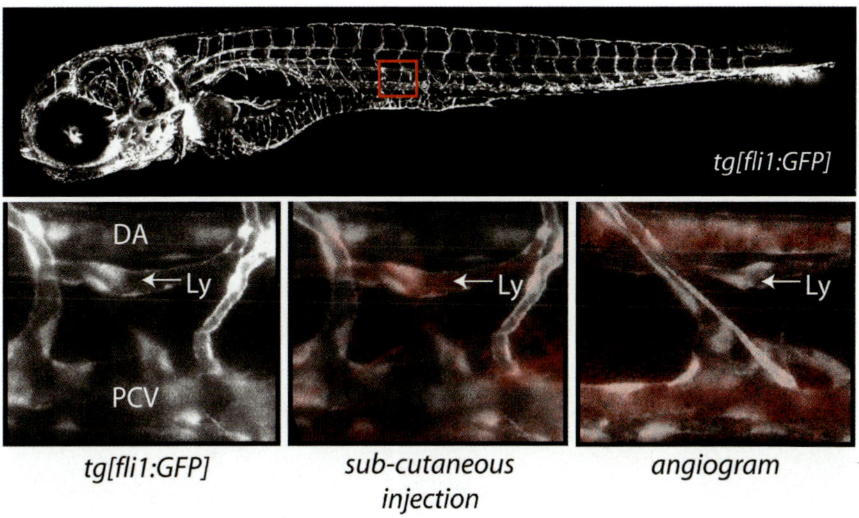

Fig. 3.2 Zebrafish embryonic lymphatic vessels.
The transgenic *tg[fli1:GFP]* strain expresses GFP in all endothelial cells of the embryo at 5 days post fertilisation (dpf) (*upper*).
The first observable lymphatic vessel (Ly), the thoracic duct, is seen as a long irregular vessel immediately ventral to the dorsal aorta (DA) and dorsal to the posterior cardinal vein (PCV) (*Lower left*).
Sub-cutaneously delivered fluorescein dextran dyes are specifically taken up into the thoracic duct (Ly) indicating a drainage function for the zebrafish lymphatic system (*lower middle*).
The blood vasculature and the lymphatic vasculature are physically separated as dye injected into the bloodstream in an angiogram assay is omitted from the lymphatic system (Ly) (*lower right*).
Lower images taken of region equivalent to the boxed region in the *upper panel*

Küchler et al. [43] found that this vessel is capable of taking up particles from interstitial spaces and eventually drains its contents into the blood vasculature. The vessel can also be labeled independent of the blood vasculature by direct lymphangiography [43, 94]. This axial lymphatic vessel hence displays many of the morphological and functional characteristics of a vertebrate thoracic duct and was the first *bone fide* lymphatic vessel identified in the developing zebrafish (Fig. 3.2).

Analysis of the lymphatic vasculature with Berlin blue dye injections has shown that in 5 week old zebrafish the anatomy of the anterior lymphatic vasculature is far more complex than at earlier stages. Older fish (5 week old) have facial, pectoral, lateral and jugular draining lymphatic vessels [94] indicating the increased branching complexity of the lymphatic vascular network in zebrafish, as is observed in other vertebrates. In adults, bloodless vessels have also been analysed using electron microscopy and antibody staining with an Angiopoietin2 antibody. Adult lymphatic vessels can be observed with anti-Ang2 immunohistochemistry, suggesting a conserved function for Angiopoietin/Tie2 signaling. In addition, they display a number of ultrastructural characteristics in common with mammalian lymphatic vessels [43]. The adult lymphatic capillaries of the skin can also be observed directly in adult zebrafish by taking advantage of a double transgenic line. Zebrafish carrying a *tg[fli1:GFP]* transgene as well as a *tg[gata1:GFP]* transgene (expressed in erythrocytes) display networks of both thin, regular, blood filled capillaries as well as irregular, bloodless capillaries in the skin (Fig. 3.3) (Hogan and Schulte-Merker, unpublished observation). Taken together these later stage analyses indicate that zebrafish form a complex lymphatic vascular network. It is, however, important to note that neither lymph nodes nor valves have been identified in the zebrafish, perhaps in-

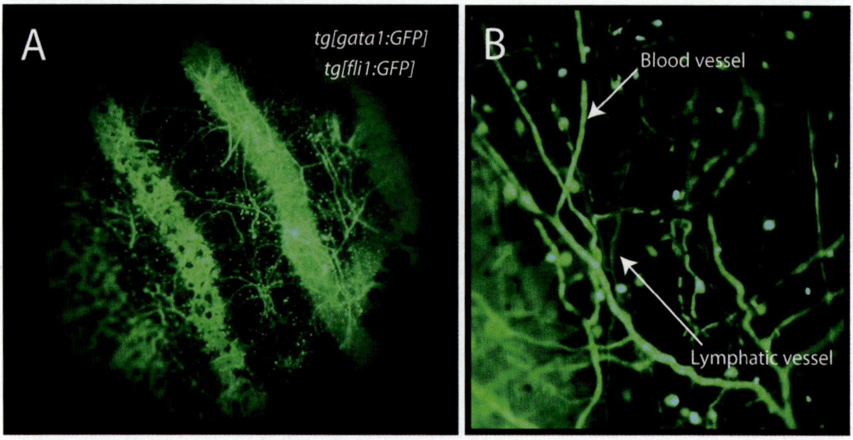

Fig. 3.3 Zebrafish adult lymphatic capillaries of the skin.
A. Low power fluorescence image of the skin of a double transgenic adult zebrafish carrying transgenes expressing GFP in erythrocytes (*tg[gata1:GFP]*) and the vasculature (*tg[fli1:GFP]*). Beds of capillaries are observable between the pigmented stripes of the zebrafish body. **B**. Capillary beds of the skin consist of regular blood filled vessels (visualised as solid bars due the presence of GFP in circulating erythrocytes) and irregular bloodless vessels, the putative lymphatic capillaries

dicating a level of functional diversification of the fully developed lymphatic system between teleosts and higher vertebrates.

3.3.2 Origin of the Zebrafish Thoracic Duct

The thoracic duct has been shown to form at discrete positions immediately ventral to the dorsal aorta [43, 94]. Using timelapse confocal microscopy, lymphatic endothelial cells are observed emerging at multiple dispersed sites in separate positions along the ventral dorsal aorta. These cells then actively migrate anteriorly or posteriorly towards the adjacent aggregates and make contact in between to form one long continuous and irregular vessel which will subsequently lumenise to form the thoracic duct [43, 94]. Although this analysis highlighted the morphogenesis of the thoracic duct, it failed to determine the earlier origins of the embryonic lymphatic endothelial cells (Fig. 3.3).

In examining the earlier development of the thoracic duct, another transgenic line was particularly informative. The *tg[fli1:NLSGFP]* line expresses a nuclear localised form of GFP in a pan-endothelial manner [64]. The origin of individual lymphatic endothelial cells of the thoracic duct was traced using live timelapse imaging of the developing trunk vasculature [94]. By back-tracking endothelial nuclei, using movies played in reverse, individual lymphatic endothelial cells could be fate-mapped back to their original locations in the vasculature at an earlier point in development (2dpf). Using this approach Yaniv et al. [94] were able to definitively demonstrate that lymphatic endothelial cells of the thoracic duct originate from the parachordal vessel (PAV) which in turn has a venous origin [33] (but see note added in proof). This elegant demonstration of a venous origin of the thoracic duct provided direct evidence for the 'centrifugal' sprouting theory generated in 1902 [65], and again seems to point towards evolutionarily conserved mechanisms of lymphangiogenesis. The finding of a venous origin of lymphatic endothelial cells has since been confirmed in the mouse model using cellular fate-mapping approaches [74].

The description of the zebrafish lymphatic system coupled with these cell tracing experiments suggests a similar developmental program for lymphangiogenesis in zebrafish to that described in *Xenopus laevis*, involving a dynamic process during which cells of venous origin are first specified to differentiate into lymphatic endothelial cells, before they then become migratory and eventually undergo terminal differentiation including tubulogenesis to form a patent lymphatic vessel.

3.3.3 Molecular Characteristics of the Zebrafish Lymphatic Vasculature

In the original descriptions of zebrafish embryonic lymphangiogenesis, the molecular control of the process was probed in order to determine whether mechanisms underpinning embryonic lymphangiogenesis were conserved between vertebrates

[43, 94]. As described above, Prox1 is considered a master regulator of lymphatic cell fate in mammals. In zebrafish, the expression of *prox1* is observed in the region of the thoracic duct using in situ hybridization and in the thoracic duct itself using immunohistochemistry at 7dpf [94]. The depletion of *prox1* with morpholinos was associated with several phenotypes including a deformed heart associated with decreased circulation, making the analysis of *prox1* loss-of-function phenotypes problematic [43]. However, in one study the injection of a *prox1* targeting morpholino led to a loss of the thoracic duct [94]. Likewise, the role of Vegfr3 and VegfC signaling has been shown to be required and conserved in the zebrafish. The depletion of VegfC with morpholinos led to a suppression of lymphangiogenesis in embryos which appear otherwise normal [43, 94]. The delivery of an mRNA encoding a soluble dominant negative form of the human VEGFR3 extracellular domain also caused a specific inhibition of lymphangiogenesis [43, 94]. These functional assays hence indicated that the molecular control of embryonic lymphangiogenesis is likely to be highly conserved between the zebrafish, *Xenopus* and other vertebrates (Table 3.1).

Finally, in addition to these functional analyses, the analysis of expression patterns conserved between fish and other vertebrates further suggests the activity of conserved pathways in zebrafish lymphangiogenesis (Table 3.1). Expression of *neuropilin2a*, a homologue of *Neuropilin2* (which is required for normal mammalian lymphatic development [95]) is detected in the thoracic duct in 5dpf zebrafish embryos by in situ hybridization [94]. Furthermore, Angiopoietin2 is expressed in adult lymphatics, as detected using immuno-histochemistry [43], suggesting a role for Angiopoietin/Tie2 signaling within the lymphatic endothelium, as has been shown previously in the mouse [23]. Although the expression of a number of classical markers of lymphatic endothelial cells (e.g. VEGFR3, Podoplanin and LYVE-1) has not yet been described in zebrafish, *vegfr3* transcripts are observed in the early embryonic veins, indicating conserved early expression of yet another key regulator of lymphangiogenesis [80] (Table 3.1).

3.4 Technologies for the Study of Lymphangiogenesis

With the description of these functionally and molecularly conserved vascular systems comes the opportunity to exploit both the tadpole and zebrafish as models to dissect the basic molecular mechanisms underpinning embryonic and possibly pathological lymphangiogenesis. To understand how to best take advantage of these research opportunities it is important to compare both the available technologies and the available methodologies which have been established in these organisms for other research purposes. Here we will outline the most useful methodologies available in these model organisms for the study of embryonic lymphangiogenesis and will directly compare their advantages and disadvantages with each other and with the existing murine model.

3.4.1 General Characteristics of the Xenopus Model

Historically, the accessibility of the large *Xenopus* oocyte, coupled with a large clutch size and ease of embryonic manipulation has made *Xenopus* an attractive model for the study of developmental biology. Well established embryonic fate maps, easily accessible extra-uterine development and highly robust embryos, coupled with the ability to perform cellular transplantations and 'animal cap' assays are a great strength of this model organism. More recently, a great deal of attention has been given to developing *Xenopus tropicalis* as a genetic model organism [6, 26], combining the advantages of *Xenopus laevis* embryonic accessibility with the concept of performing genetic screens. Additionally, the *Xenopus* model has proven an attractive tool for pharmacological, small molecule screens to identify potentially therapeutic small molecules [82].

3.4.1.1 The Xenopus Model for Genetic Studies

Unlike the genome of *Xenopus tropicalis*, *Xenopus laevis* experienced whole genome duplication and tetraploidization as recently as 40 million years ago [41, 83]. This has led to a degree of genome complexity and a level of functional redundancy that makes the (pseudo-) tetraploid *Xenopus laevis* undesirable for genetic manipulation. Consequently, the development of the diploid *Xenopus tropicalis* as a genetic model organism has been a major focus for the field and should greatly improve the utility of the frog as a model for the study of lymphangiogenesis [6].

Although mutant strains have been described in populations of wild *Xenopus laevis* [42], they occur at very low frequency due to the genetic constraints faced by tetra-ploidy. However, a recent report has described a focused forward genetic screen (firstly identifying a phenotype and then subsequently identifying the responsible gene) in *Xenopus tropicalis* [26] which yielded 77 stable developmental mutants (see Fig. 3.4 for screen overview). The identified mutants affected multiple lineages including the cardiovascular system, proving in principle that this model could be employed for focused forward genetic screens for lymphangiogenesis mutants. However, it is important to note that none of the isolated mutants has been genetically mapped or cloned and further development of community resources for genetic mapping are required to make the frog a viable option when compared with the zebrafish.

Reverse genetics (starting out with a given gene sequence and subsequently identifying the function) is the most widely used genetic approach in *Xenopus laevis*, particularly employing transient methods to inhibit gene function. Unlike in the mouse, there are no existing technologies for the creation of targeted gene knockouts or site directed knock-in animals. However, a large ENU induced TILLING (targeting induced local lesions in genomes) library, such as have been developed in plants [53,54] and zebrafish [88], has now been reported in *Xenopus tropicalis* [26]. The method is essentially the same as described in detail below for zebrafish [88] and is summarized in Fig. 3.4. To date, large numbers of stable genetic lesions in

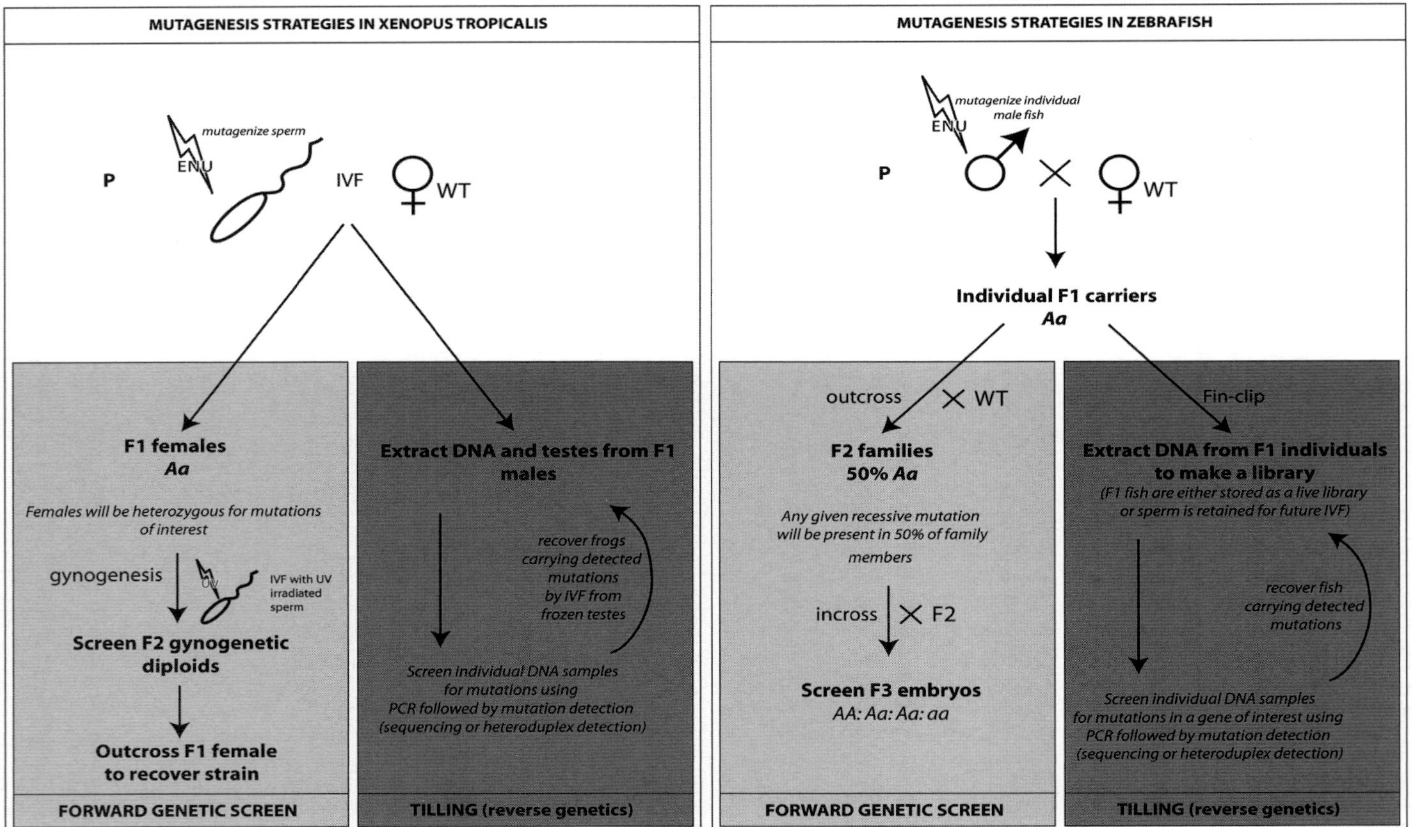

Fig. 3.4 ENU mutagenesis based genetic methodologies in *Xenopus* and zebrafish.
The approaches taken in forward genetic screens are depicted in the light grey boxes and the approaches taken for TILLING are depicted in the dark grey boxes for both *Xenopus tropicalis* and zebrafish. ENU= *N*-ethyl-*N*-nitrosourea, IVF=in vitro fertilisation, TILLING= targeting induced local lesions in genomes

genes of interest are yet to be described but seven mutants were isolated in the first description of the use of TILLING in *Xenopus tropicalis*, and more are likely in the future [26].

Transient genetic manipulation approaches involve the micro-injection of morpholino oligomers to create loss-of-function models or, alternatively, injection of DNA or mRNA to create loss- or gain-of-function models. Additionally, inhibiting gene function through forced expression or ablation of given cell types, can be carried out using tissue-specific promoter driven DNA constructs in a tissue of interest. Although the injection of mRNA or DNA constructs has not yet been used in the analysis of lymphangiogenesis in frogs, these are highly reliable methods and will undoubtedly provide novel insights in future analyses. A description of these basic methods can be found in [5,45].

Morpholino oligomers are antisense oligos in which standard basic residues (A, T, G or C) are connected by morpholine hexamer rings rather than ribose subunits. Each of these rings is linked by uncharged phosphodiamidates. This synthetic backbone stabilises the oligomers and renders them resistant to intracellular enzymatic degradation [76,77]. These highly stable oligomers can be targeted to inhibit mRNA translation or pre-mRNA splicing by direct binding to key regulatory sequences. Although inferior to stable genetic mutations because of possible non-specific or off-target effects, morpholino oligomers can be used to analyse specific loss-of-function phenotypes and protein depletion phenotypes for any given gene of interest. It is important to note that much like RNAi technologies in mammalian cells the use of morpholinos must be carefully controlled [19]. Typically, to indicate phenotypic specificity, multiple morpholinos are used to target different regions of a given gene, or a rescue of the phenotype (using exogenously introduced mRNA or DNA for the given gene) can be used if overexpression of the gene does not in itself cause a phenotype. When carefully controlled, morpholinos provide perhaps the most rapid approach with which to progress analyses from gene sequence to genetic phenotype. The analysis of Prox1 and VegfC function described above, provide two examples demonstrating the utility of morpholinos for the study of lymphangiogenesis in this model.

Another approach for the manipulation of protein function and the potential identification of therapeutically relevant molecules using this model is the use of small molecule inhibitors. The externally developing *Xenopus* zygote absorbs small molecules added to the culture medium, and this approach has been used either to inhibit specific signaling pathways or to screen for phenotypes induced by compounds that antagonize proteins [82] yet to be identified. This approach has already been used successfully to screen for specific chemical inhibitors of lymphangiogenesis (M. Detmar, personal communication).

3.4.1.2 The Xenopus Model for Cellular Studies

Standard immunohistochemistry and in situ techniques, such as the in situ hybidisation for *prox1* transcripts described above [57], are well established and can be used to examine blood and lymphatic vessels as well as lymphangioblasts directly. In addition, it is foreseeable that transgenic lines could also be developed to examine lymphangioblast migration in the future. Transgenic lines have not been widely used in either *Xenopus* species in the past, but there is no reason why this powerful in vivo visualization tool could not be employed in anuran tadpoles [73].

For the specific analysis of developing lymphatic vessels in *Xenopus*, the size and resilience of the embryos improve the ease of manipulation. There are two basic approaches to label lymphatic vessels: lymphangiogram and subcutaneous dye injections. Lymphangiograms involve arraying of anaesthetised embryos on agarose for injection and the subsequent delivery of a fluorescent dye (e.g. FITC-dextran) directly into the lymphatic vessels. This allows direct labeling of the lymphatic vessels in isolation from the blood vasculature. The other approach used, subcutaneous dye injections, involves the delivery of a bolus of dye immediately into the subcutaneous spaces by micro-injection. This approach tests the functionality of the lymphatic vessels, as functional vessels drain dye directly (for embryo alignment and injection see Fig. 3.5). In addition, the injection of dyes (e.g. rhodamine-dextran) into the blood stream can be used to examine the blood vasculature in isolation from

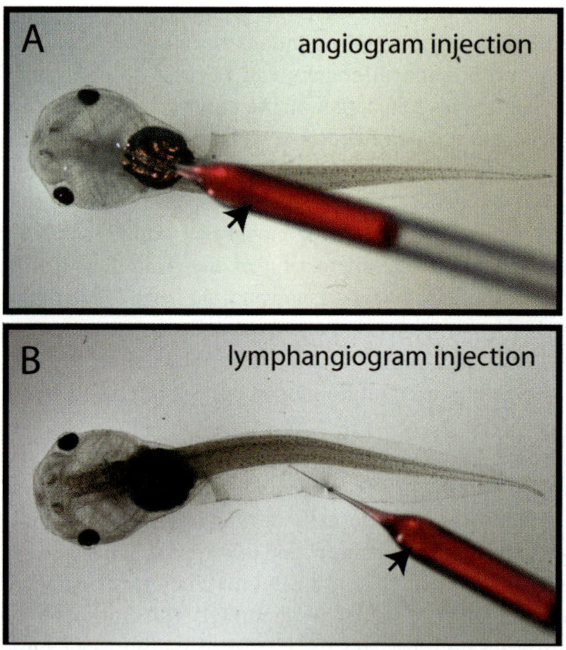

Fig. 3.5 Methodologies for examining *Xenopus* lymphatic vessels.
A. The procedure for angiogram dye injection involves the direct microinjection of fluorescent dye into the bloodstream in the region of the developing heart. Embryo viewed ventrally during injection.
B. The procedure for lymphangiogram dye injection involves the microinjection of dye directly into the trunk in the region of the developing lymphatic vessel of interest. Embryo viewed dorsally during injection in region of dorsal caudal lymphatic vessel (DCLV).
Arrowheads indicate the injection needle. Embryos are mounted on agarose

the developing lymphatic vessels and to examine the separation of the two systems (see Fig. 3.5).

3.4.2 General Characteristics of the Zebrafish Model

Transparent zebrafish embryos develop externally and acquire most of their major organs within 2–3 days post fertilization. Larvae do not need to be fed until day 8 of development and maintain synchronous development during this period. Zebrafish reach sexual maturity within 3 months depending on growth conditions, and can produce large numbers of progeny (up to 200 embryos per pair per week). This rapid external development and high fecundity coupled with a near complete genome sequence and the active development of genomic resources has, over the past decade, progressively increased the attractiveness of the zebrafish as a model for developmental genetics. Since the first reports of zebrafish developmental mutants, thousands of mutants have been identified [13, 18, 22, 28] and hundreds of them cloned (www.zfin.org). Transgenic lines have been developed to selectively label various tissues, making optimum use of the transparency of the early embryo. Both stable [88] and transient [56] reverse genetic techniques have been established, and small molecule screens have been successfully performed [62]. Each of these technologies has been reported in detail elsewhere and here we will simply provide an overview of those most useful for the study of lymphangiogenesis.

Interestingly, while zebrafish were initially introduced as a model for early embryogenesis, there are increasing numbers of reports using zebrafish to study cancer formation. Adult mutant and transgenic zebrafish models have now been described for various leukemias [47, 50], epithelial tumours [71], ocular tumors [20], rhabdomyosarcoma [50], malignant peripheral nerve sheath tumors [4] and melanoma [61]. These models and future zebrafish adult cancer models may add an extra facet to the use of zebrafish to study lymphangiogenesis once useful tools are developed and reported for the study of adult zebrafish lymphatic vessels. While this offers exciting prospects, we will focus here on the use of zebrafish to study early developmental lymphangiogenesis.

3.4.2.1 The Zebrafish Model for Genetic Studies

Forward Genetics in Zebrafish

One of the key events in establishing the zebrafish as a model for the study of developmental genetics came with the first description of large scale forward genetic screens for developmental defects performed in Tübingen (Germany) and Boston (USA), as reported in a 1996 December edition (Vol. 123) of Development. These initial reports proved the tractability of zebrafish for large scale detection of mutations affecting developmental processes [13, 22, 28]. These screens were followed by other screens, as well as by numerous reports of the genetic mapping and cloning of the underlying mutations.

In practical terms, the forward genetic approach involves (1) the mutagenesis of the male germ line by repeated exposure to the mutagen *N*-ethyl-*N*-nitrosourea (ENU), (2) the outcrossing of mutagenised males to create F1 individuals heterozygous for the induced mutations, and (3) the incrossing of unrelated F1 individuals to generate F2 families which can then (4) be set up for random brother-sister matings in order to (5) homozygose the respective mutations in F3 embryos. These F3 embryos are screened for mutant phenotypes (Fig. 3.4).

Large scale forward genetic screens performed over a short timeframe such as those described above require a large amount of space (thousands of fish tanks) for animal rearing and maintaining F2 families. Such screens can, in principle, allow for saturation screening of the genome whereby every gene is mutated at least once. However, it is important to note that even very large scale focused genetic screens reported to date have probably not yet reached saturation. Many research groups perform what are widely known as "rolling screens" whereby a smaller amount of space is required (for example a few hundred fish tanks) to raise F2 families. These families can be screened a few hundred at a time repetitively, thereby also allowing for the screening of large numbers of genomes, but over a considerably longer period.

Once mutants are identified in forward genetic screens, the genes responsible for the phenotypes must be identified. This involves the genetic mapping of mutations to a defined region, followed by the direct identification and validation of the gene responsible for the phenotype. Mapping involves the determination of genetic distance between polymorphic molecular markers and the mutation based on recombination frequency. A number of large scale community based projects have developed maps of known polymorphic markers which span the entire genome at variable density and allow the rapid identification of useful genetic mapping markers in any given region of the genome (www.zfin.org, http://zebrafish.mgh.harvard.edu/, http://cascad.niob.knaw.nl/snpview). The near complete genome sequence (www.ensembl.org) allows the tethering of physical genetic maps to real genomic space and makes it possible for researchers to identify the genes present in their region of interest. A full description of the use of genetic mapping technologies was reviewed by Geisler [24].

First generation screens were based on examining the morphology of live embryos (for example when scoring for heart function) or examining larvae with simple dye staining (e.g. alcian blue staining for cartilage). Second generation screens used in situ hybridization [30], antibody-based assays [72] or transgenic lines that allowed for identification of structures in live embryos [36, 59]. With the existence of the *tg[fli1:GFP]* transgenic line, such a screen could be carried out at present for mutants which specifically lack lymphatic vessels, as it allows for relatively easy assessment of the presence of lymphatic structures. Such unbiased screening approaches are expected to reveal mutants with deficiencies in lymphangiogenesis, or mutants with ectopic lymphangiogenesis, and will undoubtedly shed new light on the molecular regulation of developmental lymphangiogenesis in vertebrates (eg. Hogan et al., (2009) Nature Genetics, *in press,* see note added in proof). Similar screens are difficult to perform in the mouse, due to the relative inaccessibility of the embryos.

Reverse Genetics in Zebrafish

As with the frog, technologies allowing for site directed homologous recombination are yet to be established in the zebrafish. However, the development of TILLING and the creation of a community resource for the identification of targeted mutants has overcome the absence of homologous integration in allowing the creation of stable mutant lines and true loss-of-function models [88,89]. TILLING is a mutagenesis based approach for the identification of mutants in any given gene of interest. Male zebrafish are mutagenised and outcrossed to create an F1 generation carrying large numbers of heterozygous ENU induced mutations. DNA is then extracted from those F1 fish and used to create a DNA library representing the full complement of mutagenised F1 genomes. Typically, a library consists of DNA from several thousand fish. In order to identify mutations in a gene of interest, PCR is used to amplify a region in which mutations are desired (e.g. critical coding exons) and either direct sequencing of PCR products or heteroduplex detection against a wildtype PCR product is performed from each individual represented in the library. This allows the initial flagging of an individual that carries a mutation of interest. TILLING libraries can consist of either live fish from which DNA was extracted using fin clipping, or of cryo-preserved sperm samples from F1 males from which DNA was extracted. Recovery of detected mutations from a TILLING library therefore involves either direct crossing of a live fish confirmed as carrying the mutation of interest, or in vitro fertilization from frozen sperm samples (Fig. 3.4).

In addition to creating stable genetic lesions using TILLING, transient approaches, similar to those developed in the frog model, are available in zebrafish. As in the frog, this is made possible by the ability to directly inject nucleic acid constructs for direct assaying of gene function. In fact, the methods used to transiently manipulate zebrafish embryonic gene expression by direct microinjection were largely adapted from techniques developed in the frog. This approach is used for the expression of mRNA and DNA (e.g., constructs encoding dominant negative forms of a particular protein) as well as for morpholino oligomers.

The use of morpholinos has become perhaps the most widely employed tool for the creation of loss-of-function models in zebrafish. As described earlier in this chapter, morpholinos provide transient inhibition of normal transcript processing when designed against a splice donor or a splice acceptor site, or morpholinos inhibit translation, when directed against the first translated ATG or 5′ untranslated region of a given transcript. As described above for the frog model, in order to demonstrate the specificity of morpholino induced phenotypes and to avoid the publication of spurious data it is required that multiple and rigorous controls are used [19]. Hence, morpholinos can rapidly induce phenotypes of interest but the validation and interpretation of these phenotypes must be rigorous and is therefore usually the most time consuming aspect of this approach.

A complicating aspect in the use of morpholinos for the study of lymphangiogenesis is the temporal limitation of this transient approach. Morpholinos are not self replicative and after introduction their effective concentration within any given cell is therefore diluted at every cell division. Although morpholino oligos are highly efficient during the first few days of development, it is clear that by 4dpf the ability of some morpholinos to reduce mRNA processing is vastly reduced [93]. Given that the first observable lymphatic vessel, the thoracic duct, arises as late as 3.5dpf the use of morpholinos for lymphangiogenesis research may be limited. It is clear that *vegfC* targeting morpholinos used previously are capable of very specifically inhibiting embryonic lymphangiogenesis [43], but this is due to VegfC signaling acting very early during zebrafish development (Hogan et al., (2009) Nature Genetics, *in press,* see note added in proof), as has been observed in mammals. These morpholinos probably induce their phenotypic effect much earlier than the thoracic duct is observed (perhaps in the period of 2–3.5dpf). Hence, it is likely that whilst morpholinos will aid greatly the study of the earliest events of lymphangiogenesis and thoracic duct formation, they will be of limited use in dissecting later aspects such as lymphatic capillary formation and guidance, tubulogenesis or lymphatic function in late larvae or adults.

Transgenesis is an additional approach in zebrafish that can be used for cell labeling, manipulation of gene expression or for cell ablation. Constructs containing genomic promoter fragments fused to a coding sequence of a fluorescent protein (e.g. GFP) can be used to label specific cellular populations [48] and similarly by expressing cDNAs encoding dominant negative or dominant active proteins from a specific promoter can be used to interfere with normal gene expression [27]. In addition, approaches have been developed to express cellular toxins in a resticted manner, which then allows for inducible ablation of a cell or tissue type of interest [11]. A number of methods have been developed to optimize the efficiency of transgenesis in zebrafish including both I-*Sce*I mediated [79] and transposase mediated transgenesis (reviewed in [39]). Zebrafish transgenesis also affords a level of inducibility through the use of either heat shock inducible promoter (hsp70) driven transgenes [1,68,69] or Cre-mediated recombination driven transgenic lines [21,46,50,81].

3.4.2.2 The Zebrafish Model for Cellular Studies

In addition to its advantage as a genetic system, zebrafish offer a wonderful platform to study cellular behaviour in vivo. External development, transparency and high stress resilience inferred by experimental procedures have made this organism a popular system to study cellular behaviour and movements during development. Time lapse recording of larvae over extended periods allows the tracking of individual cells over time, and the combination of these features with transgenic lines that express GFP or another fluorophore in a cell specific manner has yielded important insights into the behavior of particular cell types in vivo. In addition to standard techniques such as in situ hybridization and immunohistochemistry, approaches such as cellular transplantation can be used for the direct labeling of cell types of interest [87] and for the analysis of gene function in mosaic embryos [29]. This approach

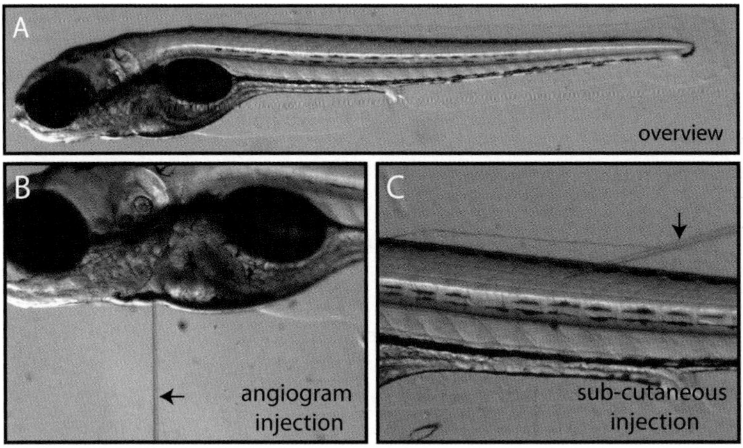

Fig. 3.6 Methodologies for examining zebrafish lymphatic vessels.
A. 5 day post fertilisation zebrafish larvae prepared in agarose for angiogram or lymphangiogram assays. **B**. Injection of dye immediately posterior to the heart in an angiogram assay. **C**. Injection of dye immediately beneath the skin in a sub-cutaneous injection to test lymphatic drainage and to label lymphatic vessels. *Arrows* indicate the injection needle in B and C

has been used to demonstrate the existence of a hemangioblast in zebrafish [87], and may also be useful in examining the ontogeny of lymphatic precursor cells in vivo. Looking forward, the development of transgenic lines to directly label lymphatic vessels and to test their function will greatly assist experimental and genetic approaches in this model.

For the specific examination of lymphangiogenesis, the most important approach in this section is the use of direct dye injection to label embryonic lymphatic vessels (Fig. 3.6). This method was originally used to demonstrate that the lymphatic vasculature has a function conserved with the mammalian lymphatic vasculature, in the drainage of interstitial fluids [43, 94]. In practice, embryos are aligned under mild anaesthaesia in agarose and a bolus of dye (usually a large molecular weight rhodamine dextran) is then delivered immediately under the skin by microinjection. One important note is that, given the small size of the embryos (even at one week of age) the accurate delivery of a bolus specifically into the interstitial spaces is challenging. An alternative approach is the direct injection into lymphatic vessels. One can inject directly into the thoracic duct in the trunk or into the jugular lymphatic vessel targeting immediately dorsal to the pectoral fin [43, 94]. As with sub-cutaneous injections this approach is a technical challenge and the use of large numbers of embryos followed by immediate observation post-delivery is suggested. Direct injection of dye into the embryonic lymphatic vasculature is best used to label lymphatic vessels in the absence of the blood vasculature, whereas subcutaneous dye injections test the functionality of lymphatic vessels by highlighting vessels upon drainage from subcutaneous spaces (Fig. 3.6).

3.5 Two New Tools for the Lymphangiogenesis Toolbox

The recent discovery in frogs and fish of a functional, evolutionarily conserved lymphatic vascular system, which develops via conserved molecular pathways during embryogenesis, now greatly broadens the approaches that can be used to examine lymphangiogenesis. At the moment, both models have been developed for the study of embryonic/larval lymphangiogenesis, which in itself serves as a model of adult and tumour lymphangiogenesis. Whilst both models have varying strengths and weaknesses, both complement the use of each other and the mouse.

The mouse remains unassailable in terms of the model's utility for obtaining knock-out and knock-in animals, the creation of adult cancer models and the panels of antibodies and molecular markers available. However, carrying out embryonic mutant screens in the mouse has not been feasible, and as a consequence an unbiased, genome-wide approach to discover genes essential for mammalian lymphangiogenesis has not been carried out. Rather, the identification of lymphatic gene function had sometimes to rely on fortuitous findings. Zebrafish have an excellent track record of identifying novel genes or novel gene functions on the basis of forward genetic screens. Are screens for lymphatic phenotypes feasible in fish? Even given the current tools they are, and all that would be required is to carry out a mutagenesis in the *tg[fli1:GFP]* transgenic line that has been used to detect thoracic duct formation in zebrafish [43, 94]. Screening a few hundred genomes would give an indication of how many vertebrate genes are essential to govern the early steps of lymphangiogenesis. Developing improved imaging tools (e.g. transgenic lines that better highlight specific aspects of the lymphatic system, such as jugular lymphatics versus the thoracic duct) will make such screens even more informative and rewarding.

In addition to these classical screens there are other exciting scenarios entailing sensitized screening set-ups: for example, creating a *vegfr3* mutant in zebrafish via TILLING would pave the way for carrying out a genetic screen in a Vegfr3-sensitized background. The mutagenesis would have to be done in males heterozygous for a *vegfr3* mutation, and F2 males [28] would need to be genotyped to identify heterozygous carriers, but even though the latter step involves some additional work, this concept offers the opportunity to identify factors interacting with Vegfr3.

A variation of this scheme is the use of compounds such as MAZ51 [40], a chemical inhibitor which preferentially inhibits VEGFR-3. Using this or other inhibitors at sub-critical levels would again achieve a sensitization of the Vegfr3 signaling pathway, and in a forward genetic screen or a morpholino based screen this is an elegant way to probe for genes which on their own will not mutate to yield a phenotype, but will show an effect in conjunction with reduced Vegfr3 signaling.

While the above mentioned screens require some time and considerable infrastructure, it should be clearly stated that genetic interactions and gene functions can be analyzed very quickly in fish and frogs on a gene-by-gene basis. The equivalent of a "double mutant" by combining a mutant with a chemical inhibitor, or a chemical inhibitor with a morpholino, is an extremely efficient and fast way to analyze the interaction of different genes.

Therefore, forward genetics, the use of morpholinos, and the application of chemicals in the zebrafish and in *Xenopus* can be used to identify novel players in embryonic lymphangiogenesis and can be proceeded by analysis in mouse knockout models, allowing for analysis of early developmental function in fish and frog as well as physiological, pathological and biochemical functions in adult mammals. Using the relative advantages of the respective systems in a combinatorial manner will lead to new and exciting insights into the genetic and cellular control of lymphangiogenesis.

Note Added in Proof

During the proofing for publication of this chapter, we have found that, in zebrafish, the initial budding of endothelial precursors from the posterior cardinal vein is regulated by vegfc. We have also found that the cells previously thought to constitute the "parachordal vessel" are actually lymphatic endothelial precursor cells at the level of the horizontal myoseptum. These cells are individual, migratory cells which sprout from the PCV, migrate along the horizontal myoseptum, and eventually migrate ventrally to form the thoracic duct, or dorsally to contribute to other aspects of the mature lymphatic system (Hogan BM, Bos FL, Bussmann J, Witte M, Chi NC, Duckers HJ and Schulte-Merker S. *ccbe1 is required for embryonic lymphangiogenesis and venous sprouting* (2009) Nature genetics, *in press*).

Acknowledgments The authors would like to thank Annelii Ny and Peter Carmeliet for providing the *Xenopus laevis* images, Michael Detmar for communicating unpublished results as well as Terhi Karpanen and Chrissy Hammond for critical reading of the manuscript.

Abbreviations

RLS	Rostral lymph sac
LH	Lymph heart
VCLV	Ventral caudal lymph vessel
DCLV	Dorsal caudal lymph vessel
DLAV	Dorsal longitudinal anastomosing vessel
PCV	Posterior cardinal vein
DA	Dorsal aorta
PAV	Parachordal vessel
GFP	Green flourescent protein
Tg	transgenic
Dfp	Days post fertilisation
NLS	Nuclear localisation signal

VEGFR3	Vascular endothelial growth factor receptor 3
VEGFC	Vascular endothelial growth factor C
LYVE	lymphatic vessel endothelial HA receptor
ENU	*N*-ethyl-*N*-nitrosourea
TILLING	targeting induced local lesions in genomes

References

1. Adam A, Bartfai R, Lele Z, Krone PH, Orban L. Heat-inducible expression of a reporter gene detected by transient assay in zebrafish. Exp Cell Res 2000, 256: 282–90.
2. Alitalo K, Tammela T, Petrova TV. Lymphangiogenesis in development and human disease. Nature 2005, 438: 946–53.
3. Baldwin AL, Ferrer P, Rozum JS, Gore RW. Regulation of water balance between blood and lymph in the frog, Rana pipiens. Lymphology 1993, 26: 4–18.
4. Berghmans S, Murphey RD, Wienholds E, Neuberg D, Kutok JL, Fletcher CD, Morris JP, Liu TX, Schulte-Merker S, Kanki JP, Plasterk R, Zon LI, Look AT. tp53 mutant zebrafish develop malignant peripheral nerve sheath tumors. Proc Natl Acad Sci USA 2005, 102: 407–12.
5. Bossi E, Fabbrini MS, Ceriotti A. Exogenous protein expression in Xenopus oocytes: basic procedures. Methods Mol Biol 2007, 375: 107–31.
6. Carruthers S, Stemple DL. Genetic and genomic prospects for Xenopus tropicalis research. Semin Cell Dev Biol 2006, 17: 146–53.
7. Cleaver O, Tonissen KF, Saha MS, Krieg PA. Neovascularization of the Xenopus embryo. Dev Dyn 1997, 210: 66–77.
8. Cooper E. Lympho-myeloid organs of Amphibia. I. Appearance during larval and adult stages of Rana catesbeiana. J Morphol 1967a, 122: 381–97.
9. Cooper E. Lympho-myeloid organs of Amphibia. II. Vasculature in larval and adult Rana catesbeiana. J Morphol 1967b, 123: 463–79.
10. Covassin LD, Villefranc JA, Kacergis MC, Weinstein BM, Lawson ND. Distinct genetic interactions between multiple Vegf receptors are required for development of different blood vessel types in zebrafish. Proc Natl Acad Sci USA 2006, 103: 6554–9.
11. Curado S, Anderson RM, Jungblut B, Mumm J, Schroeter E, Stainier DY. Conditional targeted cell ablation in zebrafish: a new tool for regeneration studies. Dev Dyn 2007, 236: 1025–35.
12. Devic E, Paquereau L, Vernier P, Knibiehler B, Audigier Y. Expression of a new G protein-coupled receptor X-msr is associated with an endothelial lineage in Xenopus laevis. Mech Dev 1996, 59: 129–40.
13. Driever W, Solnica-Krezel L, Schier AF, Neuhauss SC, Malicki J, Stemple DL, Stainier DY, Zwartkruis F, Abdelilah S, Rangini Z, Belak J, Boggs C. A genetic screen for mutations affecting embryogenesis in zebrafish. Development 1996, 123: 37–46.
14. Dumont DJ, Gradwohl GJ, Fong GH, Auerbach R, Breitman ML. The endothelial-specific receptor tyrosine kinase, tek, is a member of a new subfamily of receptors. Oncogene 1993, 8: 1293–301.
15. Dumont DJ, Gradwohl G, Fong GH, Puri MC, Gertsenstein M, Auerbach A, Breitman ML. Dominant-negative and targeted null mutations in the endothelial receptor tyrosine kinase, tek, reveal a critical role in vasculogenesis of the embryo. Genes Dev 1994, 8: 1897–909.
16. Dumont DJ, Jussila L, Taipale J, Lymboussaki A, Mustonen T, Pajusola K, Breitman M, Alitalo K. Cardiovascular failure in mouse embryos deficient in VEGF receptor-3. Science 1998, 282: 946–9.
17. Dumont DJ, Yamaguchi TP, Conlon RA, Rossant J, Breitman ML. tek, a novel tyrosine kinase gene located on mouse chromosome 4, is expressed in endothelial cells and their presumptive precursors. Oncogene 1992, 7: 1471–80.
18. Eisen JS. Zebrafish make a big splash. Cell 1996, 87: 969–77.

19. Eisen JS, Smith JC Controlling morpholino experiments: don't stop making antisense. Development 2008, 135: 1735–43.
20. Faucherre A, Taylor GS, Overvoorde J, Dixon JE, Hertog JD Zebrafish pten genes have overlapping and non-redundant functions in tumorigenesis and embryonic development. Oncogene 2008, 27: 1079–86.
21. Feng H, Langenau DM, Madge JA, Quinkertz A, Gutierrez A, Neuberg DS, Kanki JP, Look AT. Heat-shock induction of T-cell lymphoma/leukaemia in conditional Cre/lox-regulated transgenic zebrafish. Br J Haematol 2007, 138: 169–75.
22. Frohnhofer HG. Table of zebrafish mutations. In Zebrafish: Practical approach (Dahm R. Nusslein-Volhard C, Eds.), pp. 237–81. Oxford University Press, New York, 2002.
23. Gale NW, Thurston G, Hackett SF, Renard R, Wang Q, McClain J, Martin C, Witte C, Witte MH, Jackson D, Suri C, Campochiaro PA, Wiegand SJ, Yancopoulos GD. Angiopoietin-2 is required for postnatal angiogenesis and lymphatic patterning, and only the latter role is rescued by Angiopoietin-1. Dev Cell 2002, 3: 411–23.
24. Geisler R. Mapping and cloning. In Zebrafish: Practical approach (Nusslein-Volhard C, Dahm R, Eds.), pp 175–212. Oxford University Press, New York, 2002
25. Gerli R, Ibba L, Fruschelli C. Ultrastructural cytochemistry of anchoring filaments of human lymphatic capillaries and their relation to elastic fibers. Lymphology 1991, 24: 105–12.
26. Goda T, Abu-Daya A, Carruthers S, Clark MD, Stemple DL, Zimmerman LB. Genetic screens for mutations affecting development of Xenopus tropicalis. PLoS Genet 2006, 2: e91.
27. Gupta S, Zhu H, Zon LI, Evans T. BMP signaling restricts hemato-vascular development from lateral mesoderm during somitogenesis. Development 2006, 133: 2177–87.
28. Haffter P, Granato M, Brand M, Mullins MC, Hammerschmidt M, Kane DA, Odenthal J, van Eeden FJ, Jiang YJ, Heisenberg CP, Kelsh RN, Furutani-Seiki M, Vogelsang E, Beuchle D, Schach U, Fabian C, Nusslein-Volhard C. The identification of genes with unique and essential functions in the development of the zebrafish, Danio rerio. Development 1996, 123: 1–36.
29. Ho RK, Kane DA. Cell-autonomous action of zebrafish spt-1 mutation in specific mesodermal precursors. Nature 1990, 348: 728–30.
30. Hogan BM, Layton JE, Pyati UJ, Nutt SL, Hayman JW, Varma S, Heath JK, Kimelman D, Lieschke GJ. Specification of the primitive myeloid precursor pool requires signaling through Alk8 in zebrafish. Curr Biol 2006, 16: 506–11.
31. Hong YK, Harvey N, Noh YH, Schacht V, Hirakawa S, Detmar M, Oliver G. Prox1 is a master control gene in the program specifying lymphatic endothelial cell fate. Dev Dyn 2002, 225: 351–7.
32. Iraha F, Saito Y, Yoshida K, Kawakami M, Izutsu Y, Daar IO, Maeno M. Common and distinct signals specify the distribution of blood and vascular cell lineages in Xenopus laevis embryos. Dev Growth Differ 2002, 44: 395–407.
33. Isogai S, Horiguchi M, Weinstein BM. The vascular anatomy of the developing zebrafish: an atlas of embryonic and early larval development. Dev Biol 2001, 230: 278–301.
34. Isogai S, Lawson ND, Torrealday S, Horiguchi M, Weinstein BM. Angiogenic network formation in the developing vertebrate trunk. Development 2003, 130: 5281–90.
35. Jeltsch M, Kaipainen A, Joukov V, Meng X, Lakso M, Rauvala H, Swartz M, Fukumura D, Jain RK, Alitalo K. Hyperplasia of lymphatic vessels in VEGF-C transgenic mice. Science 1997, 276: 1423–5.
36. Jin SW, Herzog W, Santoro MM, Mitchell TS, Frantsve J, Jungblut B, Beis D, Scott IC, DAmico LA, Ober EA, Verkade H, Field HA, Chi NC, Wehman AM, Baier H, Stainier DY. A transgene-assisted genetic screen identifies essential regulators of vascular development in vertebrate embryos. Dev Biol 2007, 307: 29–42.
37. Karkkainen MJ, Haiko P, Sainio K, Partanen J, Taipale J, Petrova TV, Jeltsch M, Jackson DG, Talikka M, Rauvala H, Betsholtz C, Alitalo K. Vascular endothelial growth factor C is required for sprouting of the first lymphatic vessels from embryonic veins. Nat Immunol 2004, 5: 74–80.

38. Karpanen T, Wirzenius M, Makinen T, Veikkola T, Haisma HJ, Achen MG, Stacker SA, Pytowski B, Yla-Herttuala S, Alitalo K. Lymphangiogenic growth factor responsiveness is modulated by postnatal lymphatic vessel maturation. Am J Pathol 2006, 169: 708–18.
39. Kawakami K. Transposon tools and methods in zebrafish. Dev Dyn 2005, 234: 244–54.
40. Kirkin V, Mazitschek R, Krishnan J, Steffen A, Waltenberger J, Pepper MS, Giannis A, Sleeman JP. Characterization of indolinones which preferentially inhibit VEGF-C- and VEGF-D-induced activation of VEGFR-3 rather than VEGFR-2. Eur J Biochem 2001, 268: 5530–40.
41. Kobel HR. Allopolyploid speciation. In The Biology of Xenopus. (Tinsley RC, Kobel HR, Eds.), pp. 391–401. Clarendon Press, Oxford., 1996
42. Krotoski DM, Reinschmidt DC, Tompkins R. Developmental mutants isolated from wild-caught Xenopus laevis by gynogenesis and inbreeding. J Exp Zool 1985, 233: 443–9.
43. Küchler AM, Gjini E, Peterson-Maduro J, Cancilla B, Wolburg H, Schulte-Merker S. Development of the zebrafish lymphatic system requires VEGFC signaling. Curr Biol 2006, 16: 1244–8.
44. Kukk E, Lymboussaki A, Taira S, Kaipainen A, Jeltsch M, Joukov V, Alitalo K. VEGF-C receptor binding and pattern of expression with VEGFR-3 suggests a role in lymphatic vascular development. Development 1996, 122: 3829–37.
45. Lane CD. The fate of genes, messengers, and proteins introduced into Xenopus oocytes. Curr Top Dev Biol 1983, 18: 89–116.
46. Langenau DM, Feng H, Berghmans S, Kanki JP, Kutok JL, Look AT. Cre/lox-regulated transgenic zebrafish model with conditional myc-induced T cell acute lymphoblastic leukemia. Proc Natl Acad Sci USA 2005, 102: 6068–73.
47. Langenau DM, Traver D, Ferrando AA, Kutok JL, Aster JC, Kanki JP, Lin S, Prochownik E, Trede NS, Zon LI, Look AT. Myc-induced T cell leukemia in transgenic zebrafish. Science 2003, 299: 887–90.
48. Lawson ND, Weinstein BM. In vivo imaging of embryonic vascular development using transgenic zebrafish. Dev Biol 2002, 248: 307–18.
49. Le Bras B, Barallobre MJ, Homman-Ludiye J, Ny A, Wyns S, Tammela T, Haiko P, Karkkainen MJ, Yuan L, Muriel MP, Chatzopoulou E, Breant C, Zalc B, Carmeliet P, Alitalo K, Eichmann A, Thomas JL. VEGF-C is a trophic factor for neural progenitors in the vertebrate embryonic brain. Nat Neurosci 2006, 9: 340–8.
50. Le X, Langenau DM, Keefe MD, Kutok JL, Neuberg DS, Zon LI. Heat shock-inducible Cre/Lox approaches to induce diverse types of tumors and hyperplasia in transgenic zebrafish. Proc Natl Acad Sci USA 2007, 104: 9410–5.
51. Lyons MS, Bell B, Stainier D, Peters KG. Isolation of the zebrafish homologues for the tie-1 and tie-2 endothelium-specific receptor tyrosine kinases. Dev Dyn 1998, 212: 133–40.
52. Makinen T, Jussila L, Veikkola T, Karpanen T, Kettunen MI, Pulkkanen KJ, Kauppinen R, Jackson DG, Kubo H, Nishikawa S, Yla-Herttuala S, Alitalo K. Inhibition of lymphangiogenesis with resulting lymphedema in transgenic mice expressing soluble VEGF receptor-3. Nat Med 2001, 7: 199–205.
53. McCallum CM, Comai L, Greene EA, Henikoff S. Targeted screening for induced mutations. Nat Biotechnol 2000a, 18: 455–7.
54. McCallum CM, Comai L, Greene EA, Henikoff S. Targeting induced local lesions IN genomes (TILLING) for plant functional genomics. Plant Physiol 2000b, 123: 439–42.
55. Morisada T, Oike Y, Yamada Y, Urano T, Akao M, Kubota Y, Maekawa H, Kimura Y, Ohmura M, Miyamoto T, Nozawa S, Koh GY, Alitalo K, Suda T. Angiopoietin-1 promotes LYVE-1-positive lymphatic vessel formation. Blood 2005, 105: 4649–56.
56. Nasevicius A, Ekker SC. Effective targeted gene 'knockdown' in zebrafish. Nat Genet 2000, 26: 216–20.
57. Ny A, Koch M, Schneider M, Neven E, Tong RT, Maity S, Fischer C, Plaisance S, Lambrechts D, Heligon C, Terclavers S, Ciesiolka M, Kalin R, Man WY, Senn I, Wyns S, Lupu F, Brandli A, Vleminckx K, Collen D, Dewerchin M, Conway EM, Moons L, Jain RK, Carmeliet P. A genetic Xenopus laevis tadpole model to study lymphangiogenesis. Nat Med 2005, 11: 998–1004.

58. Ober EA, Olofsson B, Makinen T, Jin SW, Shoji W, Koh GY, Alitalo K, Stainier DY. Vegfc is required for vascular development and endoderm morphogenesis in zebrafish. EMBO Rep 2004, 5: 78–84.
59. Ober EA, Verkade H, Field HA, Stainier DY. Mesodermal Wnt2b signalling positively regulates liver specification. Nature 2006, 442: 688–91.
60. Papoutsi M, Tomarev SI, Eichmann A, Prols F, Christ B, Wilting J. Endogenous origin of the lymphatics in the avian chorioallantoic membrane. Dev Dyn 2001, 222: 238–51.
61. Patton EE, Widlund HR, Kutok JL, Kopani KR, Amatruda JF, Murphey RD, Berghmans S, Mayhall EA, Traver D, Fletcher CD, Aster JC, Granter SR, Look AT, Lee C, Fisher DE, Zon LI. BRAF mutations are sufficient to promote nevi formation and cooperate with p53 in the genesis of melanoma. Curr Biol 2005, 15: 249–54.
62. Peterson RT, Shaw SY, Peterson TA, Milan DJ, Zhong TP, Schreiber SL, MacRae CA, Fishman MC. Chemical suppression of a genetic mutation in a zebrafish model of aortic coarctation. Nat Biotechnol 2004, 22: 595–9.
63. Pham VN, Roman BL, Weinstein BM. Isolation and expression analysis of three zebrafish angiopoietin genes. Dev Dyn 2001, 221: 470–4.
64. Roman BL, Pham VN, Lawson ND, Kulik M, Childs S, Lekven AC, Garrity DM, Moon RT, Fishman MC, Lechleider RJ, Weinstein BM. Disruption of acvrl1 increases endothelial cell number in zebrafish cranial vessels. Development 2002, 129: 3009–19.
65. Sabin FR. On the origin of the lymphatic system from the veins and the development of the lymph hearts and thoraci duct in the pig. Am J Anat 1902, 1: 367–89.
66. Sato TN, Tozawa Y, Deutsch U, Wolburg-Buchholz K, Fujiwara Y, Gendron-Maguire M, Gridley T, Wolburg H, Risau W, Qin Y. Distinct roles of the receptor tyrosine kinases Tie-1 and Tie-2 in blood vessel formation. Nature 1995, 376: 70–4.
67. Scavelli C, Weber E, Agliano M, Cirulli T, Nico B, Vacca A, Ribatti D. Lymphatics at the crossroads of angiogenesis and lymphangiogenesis. J Anat 2004, 204: 433–49.
68. Scheer N, Groth A, Hans S, Campos-Ortega JA. An instructive function for Notch in promoting gliogenesis in the zebrafish retina. Development 2001, 128: 1099–107.
69. Scheer N, Riedl I, Warren JT, Kuwada JY, Campos-Ortega JA. A quantitative analysis of the kinetics of Gal4 activator and effector gene expression in the zebrafish. Mech Dev 2002, 112: 9–14.
70. Schneider M, Othman-Hassan K, Christ B, Wilting J. Lymphangioblasts in the avian wing bud. Dev Dyn 1999, 216: 311–9.
71. Shepard JL, Amatruda JF, Finkelstein D, Ziai J, Finley KR, Stern HM, Chiang K, Hersey C, Barut B, Freeman JL, Lee C, Glickman JN, Kutok JL, Aster JC, Zon LI. A mutation in separase causes genome instability and increased susceptibility to epithelial cancer. Genes Dev 2007, 21: 55–9.
72. Shepard JL, Amatruda JF, Stern HM, Subramanian A, Finkelstein D, Ziai J, Finley KR, Pfaff KL, Hersey C, Zhou Y, Barut B, Freedman M, Lee C, Spitsbergen J, Neuberg D, Weber G, Golub TR, Glickman JN, Kutok JL, Aster JC, Zon LI. A zebrafish bmyb mutation causes genome instability and increased cancer susceptibility. Proc Natl Acad Sci USA 2005, 102: 13194–9.
73. Sparrow DB, Latinkic B, Mohun TJ. A simplified method of generating transgenic Xenopus. Nucleic Acids Res 2000, 28: E12.
74. Srinivasan RS, Dillard ME, Lagutin OV, Lin FJ, Tsai S, Tsai MJ, Samokhvalov IM, Oliver G. Lineage tracing demonstrates the venous origin of the mammalian lymphatic vasculature. Genes Dev 2007, 21: 2422–32.
75. Steffensen JF, Lomholt JP. The secondary vascular system. In Fish Physiology (Hoar WS, Randall DJ, Farrell AP, Eds.), Vol. XII, pp. 185–217. Academic Press, San Diego, 1992.
76. Summerton J. Morpholino antisense oligomers: the case for an RNase H-independent structural type. Biochem Biophys Acta 1999, 1489: 141–58.
77. Summerton J, Weller D. Morpholino antisense oligomers: design, preparation, and properties. Antisense Nucleic Acid Drug Dev 1997, 7: 187–95.

78. Tammela T, Saaristo A, Lohela M, Morisada T, Tornberg J, Norrmen C, Oike Y, Pajusola K, Thurston G, Suda T, Yla-Herttuala S, Alitalo K. Angiopoietin-1 promotes lymphatic sprouting and hyperplasia. Blood 2005, 105: 4642–8.
79. Thermes V, Grabher C, Ristoratore F, Bourrat F, Choulika A, Wittbrodt J, Joly JS. I-SceI meganuclease mediates highly efficient transgenesis in fish. Mech Dev 2002, 118: 91–8.
80. Thompson MA, Ransom DG, Pratt SJ, MacLennan H, Kieran MW, Detrich HW, 3rd, Vail B, Huber TL, Paw B, Brownlie AJ, Oates AC, Fritz A, Gates MA, Amores A, Bahary N, Talbot WS, Her H, Beier DR, Postlethwait JH, Zon LI. The cloche and spadetail genes differentially affect hematopoiesis and vasculogenesis. Dev Biol 1998, 197: 248–69.
81. Thummel R, Burket CT, Brewer JL, Sarras MP, Jr., Li L, Perry M, McDermott JP, Sauer B, Hyde DR, Godwin AR. Cre-mediated site-specific recombination in zebrafish embryos. Dev Dyn 2005, 233: 1366–77.
82. Tomlinson ML, Field RA, Wheeler GN. Xenopus as a model organism in developmental chemical genetic screens. Mol Biosyst 2005, 1: 223–8.
83. Tymowska L. Polyploidy and cytogenetic variation in frogs of the genus Xenopus. In Amphibian Cytogenetics and Evolution (Green DS, Sessions SK, Eds.), pp. 259–97. Academic Press, San Diego, 1991.
84. Valasek P, Macharia R, Neuhuber WL, Wilting J, Becker DL, Patel K. Lymph heart in chick–somitic origin, development and embryonic oedema. Development 2007, 134: 4427–36.
85. Veikkola T, Jussila L, Makinen T, Karpanen T, Jeltsch M, Petrova TV, Kubo H, Thurston G, McDonald DM, Achen MG, Stacker SA, Alitalo K. Signalling via vascular endothelial growth factor receptor-3 is sufficient for lymphangiogenesis in transgenic mice. Embo J 2001, 20: 1223–31.
86. Veikkola T, Lohela M, Ikenberg K, Makinen T, Korff T, Saaristo A, Petrova T, Jeltsch M, Augustin HG, Alitalo K. Intrinsic versus microenvironmental regulation of lymphatic endothelial cell phenotype and function. Faseb J 2003, 17: 2006–13.
87. Vogeli KM, Jin SW, Martin GR, Stainier DY. A common progenitor for haematopoietic and endothelial lineages in the zebrafish gastrula. Nature 2006, 443: 337–9.
88. Wienholds E, Schulte-Merker S, Walderich B, Plasterk RH. Target-selected inactivation of the zebrafish rag1 gene. Science 2002, 297: 99–102.
89. Wienholds E, van Eeden F, Kosters M, Mudde J, Plasterk RH, Cuppen E. Efficient target-selected mutagenesis in zebrafish. Genome Res 2003, 13: 2700–7.
90. Wigle JT, Oliver G. Prox1 function is required for the development of the murine lymphatic system. Cell 1999, 98: 769–78.
91. Wigle JT, Harvey N, Detmar M, Lagutina I, Grosveld G, Gunn MD, Jackson DG, Oliver G. An essential role for Prox1 in the induction of the lymphatic endothelial cell phenotype. Embo J 2002, 21: 1505–13.
92. Wilting J, Aref Y, Huang R, Tomarev SI, Schweigerer L, Christ B, Valasek P, Papoutsi M. Dual origin of avian lymphatics. Dev Biol 2006, 292: 165–73.
93. Yan YL, Miller CT, Nissen RM, Singer A, Liu D, Kirn A, Draper B, Willoughby J, Morcos PA, Amsterdam A, Chung BC, Westerfield M, Haffter P, Hopkins N, Kimmel C, Postlethwait JH, Nissen R. A zebrafish sox9 gene required for cartilage morphogenesis. Development 2002, 129: 5065–79.
94. Yaniv K, Isogai S, Castranova D, Dye L, Hitomi J, Weinstein BM. Live imaging of lymphatic development in the zebrafish. Nat Med 2006, 12: 711–6.
95. Yuan L, Moyon D, Pardanaud L, Breant C, Karkkainen MJ, Alitalo K, Eichmann A. Abnormal lymphatic vessel development in neuropilin 2 mutant mice. Development 2002, 129: 4797–806.

Chapter 4

MOLECULAR MECHANISMS OF LYMPH NODE METASTASIS

Matthias Rinderknecht and Michael Detmar
Institute of Pharmaceutical Sciences, Swiss Federal Institute of Technology, ETH Zurich, CH-8093 Zurich, Switzerland, e-mail: michael.detmar@pharma.ethz.ch

Abstract: Lymph node metastasis represents the first step of spread for most human cancers and serves as an important prognostic indicator. Recent studies have revealed that tumors can actively induce the growth of tumor-associated lymphatic vessels (lymphangiogenesis), thus promoting their metastasis to draining (sentinel) lymph nodes. Several tumor lymphangiogenesis factors have been identified, including vascular endothelial growth factor (VEGF)-C, VEGF-D, VEGF-A and hepatocyte growth factor. A large number of clinical studies have confirmed the correlation of tumor lymphangiogenesis and metastasis of different types of human cancers. Importantly, recent results indicate that tumors can also induce lymphatic vessel growth in sentinel lymph nodes, even before the onset of metastasis, and that lymph node lymphangiogenesis further promotes cancer spread to distant lymph nodes and to distant organs. Thus, in addition to its prognostic importance, lymphangiogenesis has become a new target for the prevention, treatment and imaging of lymph node metastases.

Key words: Cancer · Lymphangiogenesis · Lymphatic metastasis · VEGF-A · VEGF-C · VEGFR-3

4.1 Introduction to the Lymphatic System

The lymphatic vascular system was first described in the 17th century by the Italian surgeon and anatomist Gasparo Aselli [6]. He described the lacteal lymphatic vessels, conveying milky-white chyle, as "milky veins" in the intestine of a dog. Later, it was found that these lacteals contain lipids and the fat-soluble vitamins A, D, E and K that are absorbed from the intestine. In higher vertebrates, the lymphatic system is the secondary circulatory system (for a comprehensive review see [35]).

S.A. Stacker, M.G. Achen (eds.), *Lymphangiogenesis in Cancer Metastasis*,
Cancer Metastasis – Biology and Treatment 13, DOI 10.1007/978-90-481-2247-9_4,

Unlike the cardiovascular system, where the blood is pumped in a closed circuit by the heart, the lymphatic vascular system drains interstitial fluid in a one-way, open-ended fashion without a central driving force. The interstitial fluid, also called lymph, is a protein-rich exsudate from the blood capillaries. In the periphery, lymph is passively taken up by blind-ended lymphatic capillaries, drained to larger collecting lymphatic vessels and returned to the blood circulation via the thoracic duct, which drains into the inferior vena cava. Further constituents of the lymphatic system are the lymphoid organs including lymph nodes, tonsils, Peyer's patches, spleen and thymus. These structures are part of the immune system and play an important in the immune surveillance of the body. Immune cells including lymphocytes and antigen-presenting dendritic cells travel via the lymphatic vessels from the skin to regional lymph nodes, where specific immune responses are mounted.

Histologically, lymphatic capillary vessels differ from blood vessels by the lack of a basement membrane, of surrounding smooth muscle cells and of pericytes [3]. Because LECs are anchored to the extracellular matrix by elastic anchoring filaments, increased hydrostatic pressure in the tissue causes lymphatic vessels to expand rather than to collapse [30]. These physiological features render the lymphatic capillaries more permeable than blood vessels and facilitate cellular invasion by immune and also by metastasizing cancer cells.

Over the last few years, several growth factors promoting lymphatic vessel growth have been identified. The first lymphangiogenic factors discovered were vascular endothelial growth factor (VEGF)-C and VEGF-D. Both growth factors act via their receptors VEGF receptor-3 (VEGF-R3), which is specifically expressed on LECs in normal tissues, and VEGF-R2. The role of VEGF-R2 in lymphangiogenesis is supported by findings that VEGF-A – which binds to VEGF-R2 but not to VEGF-R3 – promotes lymphatic endothelial cell growth in vitro and in vivo [42,46,73,88]. Hepatocyte growth factor has been found to also act as a potent lymphangiogenic factor in vitro and in vivo, [55], and angiopoietin-1, a ligand for the endothelial

Table 4.1 Lymphatic growth factors and their receptors

Growth factor	Cognate receptor on lymphatic endothelial cells	Ref
VEGF-A	VEGF-R2, Nrp2	[88]
VEGF-C	VEGF-R3, VEGF-R2, Nrp2	[48,57]
VEGF-D	VEGF-R3, VEGF-R2, Nrp2	[57,130]
HGF	HGF-R	[14,55]
IGF1/2	IGF-1R	[9]
PDGF-BB	PDGFRα/β	[13]
Angiopoietin-1	Tie-2	[60,92]
Adrenomedullin	CRLR/RAMP2	[25]
FGF-2	FGF-R3	[15,112]

CRLR, calcitonin receptor-like receptor; FGF(-R), fibroblast growth factor (receptor); HGF(-R), hepatocyte growth factor (receptor); IGF(-R), insulin-like growth factor (receptor); Nrp2, neuropilin-2; PDGF-BB, platelet-derived growth factor-BB; PDGFR, platelet-derived growth factor receptor; RAMP2, receptor activity modifying protein 2; Tie-2, tunica internal endothelial cell kinase; VEGF(-R), vascular endothelial growth factor (receptor)

receptor Tie2 [21], also induces lymphangiogenesis after viral or transgenic delivery to the skin of mice [86, 123]. In addition, fibroblast growth factor-2 [15, 72, 112], platelet-derived growth factors [13] and insulin-like growth factors [9] have been found to promote lymphangiogenesis, as well as adrenomedullin, acting via calcitonin receptor-like receptor (CRLR) and receptor activity-modifying protein-2 (RAMP2) [25]. An overview of the currently known lymphangiogenic growth factors and their respective receptors is provided in Table 4.1.

4.2 Experimental Evidence for an Active Role of Tumor Lymphangiogenesis in Promoting Lymph Node Metastasis

For several decades, lymphatic metastasis was considered as a rather passive process, where tissue-invading cancer cells happen upon preexisting lymphatic vessels and are then taken up and drained to lymph nodes. However, the recent progress in the identification of lymphatic vessel-specific markers, as well as the identification of lymphatic growth factors, has enabled experimental studies in rodents that have provided compelling evidence for an active role of tumor-induced lymphangiogenesis in the promotion of lymph node metastasis. Overexpression of VEGF-C or VEGF-D by cancer cells significantly promoted tumor lymphangiogenesis and lymph node metastasis in mice [80, 115, 118, 132]. In a chemically induced, multistep skin cancer model, transgenic overexpression of VEGF-A or VEGF-C by epidermal keratinocytes enhanced both tumor-associated lymphatic vessel growth and metastasis to sentinel lymph nodes [43, 44].

Inhibition of the VEGF-C/VEGF-D/VEGF-R3 axis has now been shown in several studies to reduce the incidence of lymph node metastasis in different cancer models. Inhibition of VEGF-R3 signaling by systemic treatment with VEGF-R3 blocking antibodies reduced the incidence of lymph node and organ metastasis in a mouse breast carcinoma model [104]. In an orthotopic model of gastric cancer, therapy with VEGF-R3 blocking antibodies reduced the lymphatic vessel density in the primary tumors and also inhibited regional lymph node metastasis [111]. Neutralizing antibodies against VEGF-D have been shown to inhibit VEGF-D induced lymphatic tumor spread to lymph nodes in mice [117]. Virally delivered soluble VEGF-R3 fusion protein ("VEGF-C/D trap") inhibited the formation of tumor-associated lymphatic vessels and suppressed lymph node metastasis in mouse models of melanoma, gastric and prostate cancer [76], and overexpression of soluble VEGF-R3 in lung cancer cells inhibited tumor lymphangiogenesis and reduced the incidence of metastases in draining lymph nodes [39]. Moreover, soluble VEGF-R3 also inhibited VEGF-C-induced tumor lymphangiogenesis and metastatic spread in a breast cancer xenotransplant model [56], and overexpression of soluble VEGF-R3 in a rat mammary cancer model inhibited metastasis formation in lymph nodes and lungs [70]. Finally, in a murine mammary cancer model, inhibition of tumor cell VEGF-C expression by stably transfected siRNA inhibited tumor lymphangiogenesis, lymph node and lung metastasis [16], and nanoparticle-delivered

VEGF-C siRNA suppressed tumor lymphangiogenesis and lymph node metastasis in subcutaneous xenografts of gastric cancer in mice [38]. Overall, these results indicate that tumor lymphangiogenesis actively contributes to cancer dissemination, that blockade of lymphatic vessel growth might inhibit tumor metastasis to lymph nodes, and that the extent of tumor-associated lymphangiogenesis represents a prognostic parameter for the metastatic risk of cancers.

4.3 The Role of Intratumoral Versus Peritumoral Lymphangiogenesis

In several of the experimental cancer metastasis models discussed above, lymphatic vessel growth and proliferation was not only observed at the tumor border, but also inside of tumors. Hence, a heavily discussed topic has been the relative importance of intratumoral versus peritumoral lymphatics for cancer metastasis to the draining lymph nodes. In particular, it has been proposed that intratumoral lymphatic vessels might be compressed and non-functional. In mouse models of orthotopically implanted fibrosarcomas or malignant melanoma cells overexpressing VEGF-C, increased tumor-associated lymphangiogenesis was observed, based on stains for the lymphatic-specific hyaluronan receptor LYVE-1 (lymphatic vascular endothelium receptor-1) and the blood vascular marker MECA-32 [100]. However, intratumoral lymphatic vessels were often found to be collapsed or filled with tumor cells that occluded the lumen, while peritumoral lymphatics apparently exhibited a more normal morphology. Microlymphangiography with ferritin injected deeply into the tumor did not revealed colocalization of ferritin with LYVE-1 positive structures, suggesting that no functional intratumoral lymphatics were present. In contrast, about 85% of ferritin staining in the tumor margin and the normal tissue colocalized with LYVE-1 [100]. However, the capacity of lymphatics for fluid transport (as measured by ferritin injection) does not necessarily correlate with their role in transporting tumor cells. Moreover, the single use of LYVE-1 as a lymphatic marker might possibly to an over- or underestimation of the lymphatic vessel density since tumor-associated macrophages have been shown to express LYVE-1 [109] and since lymphatic LYVE-1 expression might be downregulated under certain conditions [51]. Thus, combined stains for additional lymphatic markers such as the mucin-type glycoprotein podoplanin [10, 106, 107] or the transcription factor Prox1 [96] might be beneficial for such studies. Another study in experimental prostate cancer in mice indicated that peritumoral lymphatic vessels might play a more important role for cancer metastasis to lymph nodes than intratumoral lymphatics, since specific inhibition of intratumoral lymphatics did not prevent lymph node metastasis [135].

It has been suggested that differential expression levels of VEGF-C or VEGF-D might result in differences of intra- versus peritumoral lymphangiogenesis. In human pancreatic adenocarcinomas, VEGF-C and VEGF-D protein levels as estimated by IHC were higher at the margin (within 2 mm of the external invasive edge) than in the center of the tumor, and only the expression levels at the margins but not in the tumor center correlated with lymph node metastasis [74]. Similarly, another study in human prostate adenocarcinomas indicated that peritumoral, but not intratumoral lymphangiogenesis correlated with lymph node metastasis, although the number of investigated cases was rather small (n = 14) [105].

On the other hand, a number of studies have revealed proliferation of intratumoral lymphatics in mouse tumor models, and a significant correlation between intratumoral lymphatic vessel density and lymph node metastasis has been found in different types of human cancer. In head and neck squamous cell carcinomas, the extent of intratumoral but not peritumoral lymphatics was associated with lymph node metastasis and poor prognosis [82]. Intratumoral lymphatic vessel density was also associated with lymph node metastasis in oropharyngeal carcinomas, but not in oral cavity or laryngeal carcinoma [8]. In malignant melanomas of the skin, the incidence of intratumoral lymphatics was significantly higher in metastatic melanomas and correlated with poor disease-free survival [19]. One has to keep in mind that most studies performed thus far only investigated static parameters (i.e., the number or size of lymphatic vessels) but not the functional properties of tumor-associated endothelium, and that activated lymphatic vessels might promote metastasis regardless of lymphatic hyperplasia. There is increasing evidence that lymphatic vessels can actively recruit certain types of tumor cells. As an example, it has been shown that lymphatic endothelium releases the chemokine CCL21 that binds to its receptor CCR7 on some types of tumor cells, and that this mechanism actively promotes the metastasis of CCR7-expressing melanoma cells to lymph nodes [134]. Moreover, detection of lymphovascular invasion by tumor cells on histological sections – stained for lymphatic-specific markers such as podoplanin – might represent a more sensitive parameter than the quantitation of lymphatic vessel size and number.

4.4 Lymph Node Lymphangiogenesis: A Tumor-Induced (Pre)Metastatic Niche Promoting Distant Cancer Metastasis

Surprisingly, we recently found – in a multistep skin carcinogenesis model in transgenic mice with skin-specific overexpression of VEGF-A – that metastatic tumor cells within draining lymph nodes continued to induce lymphangiogenesis involving active proliferation of lymphatic endothelium [43]. Importantly, lymph node lymphangiogenesis was induced in sentinel lymph nodes draining cutaneous squamous cell carcinomas even before these tumors had metastasized. This was also observed in a carcinogenesis model applied to transgenic mice with skin-specific overexpression of VEGF-C [44]. These findings give a new twist to the seed-and-soil hypothesis, proposed more than a century ago, which suggested that tumor cells (the "seeds") can only metastasize to organs that provide a fertile soil [101]. Our recent results indicate that tumors – via release of VEGF-A and/or VEGF-C that are drained via lymphatic vessels to the sentinel lymph node – induce lymphatic vessel

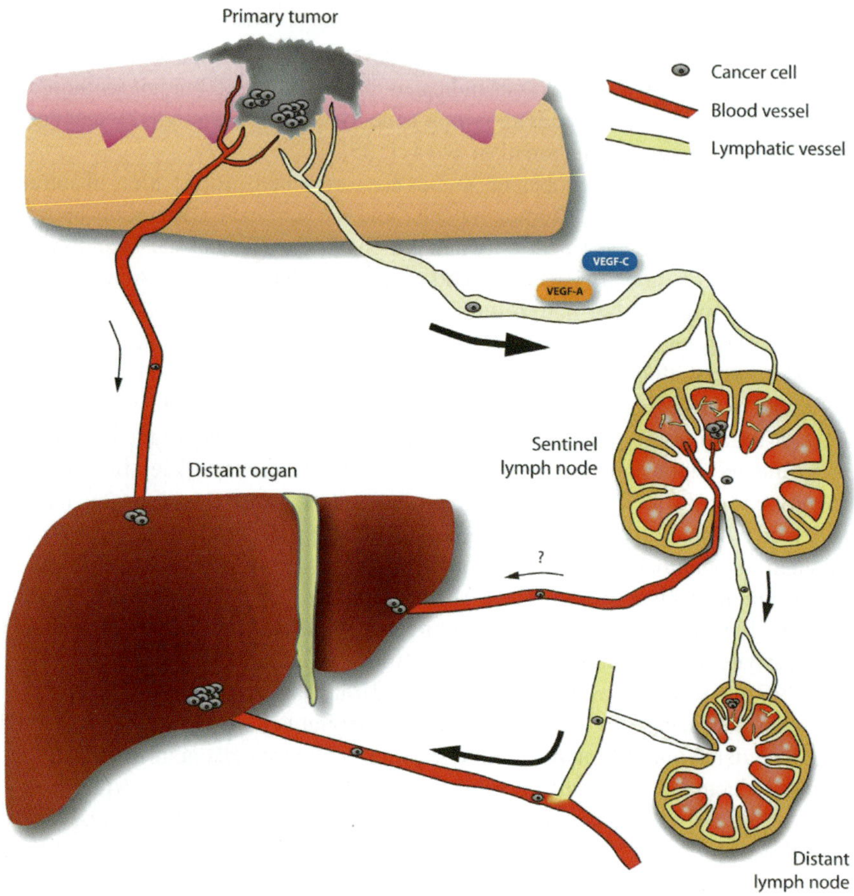

Fig. 4.1 Proposed model of cancer metastasis to sentinel lymph nodes and beyond. VEGF-A and/or VEGF–C secreted from the primary tumor promote the formation of peritumoral lymphatic vessels. Drainage of VEGF-A and/or VEGF–C to the sentinel lymph node via newly formed and preexisting lymphatic vessels around the tumor leads to lymphangiogenesis in the sentinel lymph node. These premetastatic niches are subsequently colonized by metastatic cancer cells from the primary tumor, which continue to secrete lymphangiogenic growth factors to induce lymphangiogenesis in distant lymph nodes, promoting their metastatic spread to other lymph nodes and eventually to distant organs

expansion in the draining lymph node (Fig. 4.1), thereby preparing a premetastatic niche. Upon arrival of metastatic tumor cells in the sentinel lymph node, there is an even more pronounced induction of lymphangiogenesis. Importantly, we recently found that induction of lymph node lymphangiogenesis by VEGF-C secreting tumor cells promoted further metastasis to distant lymph nodes and to organs [44]. Thus, expansion of the lymphatic network within lymph nodes actively contributes to metastatic cancer spread and, therefore, represents a novel target for therapies aimed at preventing or treating cancer metastasis.

This concept has since then been experimentally confirmed by other groups in mouse models of metastatic nasopharyngeal carcinoma [103] and of malignant melanoma [37]. Moreover, we also detected the occurrence of lymph node lymphangiogenesis in metastatic human malignant melanomas [20]. Very recently, the metastasis-enhancing action of lymph node lymphangiogenesis was confirmed in breast cancer [128]. Increased sentinel lymph node lymphangiogenesis was found to be associated with nonsentinel axillary lymph node metastasis in breast cancer patients with a positive sentinel node. These findings further indicate that inhibition of lymph node lymphangiogenesis might prevent lymph node metastasis and further metastatic spread of certain types of cancer.

4.5 Correlation of Lymphangiogenic Growth Factor Expression, Tumor Lymphangiogenesis and Metastasis in Human Cancers

A rapidly increasing number of studies have investigated the link between lymphatic growth factor expression, the extent of tumor lymphangiogenesis, the presence of lymphatic vessel invasion, the presence of sentinel lymph node metastasis and the clinical prognosis in many different types of human cancers (Table 4.2). Among the most intensely investigated cancers are breast cancer, gastric cancer, colorectal carcinoma and malignant melanoma. Since VEGF-C and -D were the first identified lymphangiogenic factors, a large number of studies have investigated their possible role in cancer lymphangiogenesis and metastasis.

In epithelial cancers, the correlation between the expression of lymphangiogenic growth factors and lymph node metastasis has been most intensely studied for VEGF-C. More than 70 studies have investigated the influence of VEGF-C expression on cancer metastasses, either at the mRNA level using (quantitative) RT-PCR or RNase protection assays, or – most frequently – at the protein level using immunohistochemistry (IHC) or ELISA assays (Table 4.2). VEGF-C mRNA and/or protein levels have often been found to be upregulated in tumor samples and/or serum of cancer patients, as compared to normal controls [12, 29, 36, 77, 84, 95, 113, 122]. Over all studies and techniques used, 56 out of 73 studies (77%) found a significant positive correlation between VEGF-C expression levels and lymph node metastasis. Assessment via IHC most often detected a positive correlation (41 out of 49 studies; 83.7%), as compared with mRNA based techniques (12 out of 18 studies; 66.7%) and ELISA detection (3 out of 6 studies; 50%). Some tumors might be better suited for prediction of lymph node metastases by measuring VEGF-C expression levels. For instance, among the three most frequently studied cancer types, VEGF-C expression correlated with lymph node metastasis in 14 out of 17 studies (82.3%) on gastric carcinoma, as compared with colorectal carcinoma (9 out of 13 studies; 69.2%) and breast cancer (5/9, 55.6%).

The association of VEGF-D expression with lymph node metastasis has been investigated in a total of 32 studies; 14 of these (43.8%) reported a positive correlation

Table 4.2 Publications on correlation of LNM and VEGF-A/C/D expression

Marker	Detection method	Correlation of marker expression with LVD	LVI	LNM	Poor prognosis	Ref
Malignant melanoma						
VEGF-C	IHC, Zymed 18–2255	+	nd	+	nd	[20]
	IHC, SC-7133, R&D goat pab	nd	nd	+	nd	[108]
	ELISA (IBL), serum ΔNΔC-VC	nd	nd	+	nd	[131]
VEGF-D	IHC, Achen mab	nd	nd	–	nd	[20]
	IHC, SC-7602	nd	nd	–	nd	[108]
Breast cancer						
VEGF-A	IHC, LabVision RM-9128-R7	+	+	+	+	[85]
VEGF-C	IHC, Zymed 18-2255	+	+	+	+	[146]
	IHC, Zymed 18-2255	+	–	+	+	[85]
	IHC, SC goat pab	nd	+	–	+	[61]
	IHC, SC-9047	nd	–	–	nd	[45]
	IHC, R&D AF752	nd	nd	+	+uv (–mv)	[90]
	ELISA, intratumoral ΔNΔC-VC	nd	nd	–	Inv. correl. (uv/mv)	[7]
	qRT-PCR	+	nd	+	nd	[91]
	qRT-PCR	nd	–	–(+[a])	nd	[69]
	RNase protection assay	nd	nd	–	nd	[34]
VEGF-D	IHC, R&D Mab286	nd	nd	+	+(uv/mv)	[89]
	IHC, R&D Mab286	–	–	–	–	[85]
	IHC, SC-7602	nd	nd	–	–	[17]
	qRT-PCR	nd	Inv. correl.	–	nd	[69]

(continued)

Table 4.2 (continued)

Marker	Detection method	Correlation of marker expression with LVD	LVI	LNM	Poor prognosis	Ref
Colorectal carcinoma						
VEGF-A	ELISA (R&D)/RT-PCR	nd/nd	nd/nd	+/+	nd/nd	[29]
	RNase protection assay	nd	nd	–	nd	[36]
VEGF-C	IHC, SC-9047	+	+	+	+uv (–mv)	[116]
	IHC, Zymed 18-2255	+	+	+	nd	[26]
	IHC, SC-7133 Tumor invasive edge & surface	nd	+(inv. edge) – (surface)	+(uv/mv) (inv. edge)	+(uv/mv)	[97]
	IHC, SC rabbit pab	nd	nd	+	nd	[50]
	IHC, SC rabbit pab	nd	nd	+	nd	[79]
	IHC, SC-1881	nd	+	+	+(mv)	[28]
	IHC, SC-7133/qRT-PCR	nd/nd	+/ + (uv)	+/+	nd/nd	[59]
	RT-PCR	nd	nd	–	nd	[29]
	qRT-PCR	nd	nd	–	nd	[122]
	RNase protection assay	nd	nd	–	nd	[36]
VEGF-D	IHC, SC-7603 Tumor invasive edge & surface	nd	– (inv. edge & surface)	+uv, –mv (inv. edge)	+uv (–mv)	[97]
	IHC, R&D Mab622	nd	–	+(uv/mv)	–	[27]
	IHC, SC goat pab	nd	nd	+	+	[133]
	RT-PCR	nd	nd	–	nd	[29]
	RNase protection assay	nd	nd	–	nd	[36]

Table 4.2 (continued)

Marker	Detection method	Correlation of marker expression with LVD	LVI	LNM	Poor prognosis	Ref
Gastric cancer						
VEGF-A	IHC, SC rabbit pab	nd	–	–	nd	[54]
VEGF-A/C	IHC, VA: IBL mab/VC: SC-1881	nd	–/ – /+[b]	–/ – /+[b]	nd	[67]
VEGF-C	IHC, Zhongshan ZA-0266	+	+	–	+	[18]
	IHC, SC-7133	nd	+	+(uv/mv)	nd	[4]
	IHC, Zymed 18-2255	nd	nd	+	nd	[145]
	IHC, SC-9047	nd	+	+	nd	[98]
	IHC, Zymed 18-2255	nd	+	+	nd	[136]
	IHC, Zymed 18-2255/SC-7133	nd	–	+	–	[53]
	IHC, Zymed 18-2255	nd	+	+	nd	[137]
	IHC, SC goat pab	nd	+	–	nd	[54]
	IHC, IBL goat pab	nd	+	+/–[c]	nd	[47]
	IHC, SC-7133	nd	+	+uv (–mv)	+	[121]
	IHC, SC-1881/RT-PCR	nd/nd	+/nd	+/+	+(uv/mv)/ nd	[140]
	RT-PCR	nd	+	+	–	[110]
	RT-PCR	nd	nd	+	nd	[65]
	RT-PCR	nd	nd	+	nd	[77]
	qRT-PCR	nd	nd	+	nd	[142]
VEGF-D	IHC, SC-7603	nd	–	–	nd	[98]
	IHC, R&D goat pab	nd	+	+/–[c]	nd	[47]
	IHC, R&D AF286, SC-7602	nd	+	+	+	[53]
	RT-PCR	nd	nd	–	nd	[65]
	RT-PCR	nd	+	–	–	[110]
	qRT-PCR	–	nd	–	nd	[142]

(continued)

Table 4.2 (continued)

Marker	Detection method	Correlation of marker expression with LVD	LVI	LNM	Poor prognosis	Ref
Bladder transitional cell carcinoma						
VEGF-C	IHC, SC-1881	nd	+	+	+	[120]
	IHC, SC goat pab	nd	nd	+	+	[147]
Cervical cancer						
VEGF-A	ELISA (IBL), serum VA	nd	nd	−	−	[84]
VEGF-C	IHC, Zymed 18-2255	+	+	+	+uv (−mv)	[31]
	IHC, Van Trappen rabbit pabs	nd	−	−	nd	[129]
VEGF-C	IHC, SC goat pabs	nd	nd	+	+	[125]
	ELISA (IBL), serum ΔNΔC-VC	nd	nd	−	+	[84]
VEGF-D	IHC, Kirkin rabbit serum 4292	nd	−	−	nd	[129]
Endometrial cancer						
VEGF-A	IHC, SC-7269	nd	+	−	+uv (−mv)	[41]
VEGF-C	IHC, SC-1881	nd	+	+	+(uv/mv)	[41]
VEGF-D	IHC, R&D AF286 Tumor and stroma cells	nd	+[d] (vessel invasion)	+[d]	+uv/ − mv (tumor) +uv/mv (stroma)	[139]
Esophageal cancer						
VEGF-A	ELISA (IBL), serum VA	nd	nd	−	nd	[71]
	qRT-PCR	nd	nd	−	nd	[78]
VEGF-C	IHC, SC-7133	nd	+[d]	+[d]	nd	[64]
	IHC, SC goat pab	nd	nd	+	nd	[12]
	ELISA (IBL), serum ΔNΔC-VC	nd	nd	+	nd	[71]
	qRT-PCR	nd	nd	+	nd	[78]

Table 4.2 (continued)

Marker	Detection method	Correlation of marker expression with LVD	LVI	LNM	Poor prognosis	Ref
VEGF-D	qRT-PCR	nd	nd	–	nd	[78]
Laryngopharyngeal squamous cell carcinoma						
VEGF-C	IHC, Zymed 18-2255	– (it)	nd	+	+	[40]
Lung adenocarcinoma						
VEGF-A	qRT-PCR	nd	–	+[e]	nd	[93]
VEGF-C	qRT-PCR	nd	+[f]	+[f]	nd	[93]
VEGF-D	qRT-PCR	nd	Inv.correl.	–	nd	[93]
Neck squamous cell carcinoma						
VEGF-A	RT-PCR	nd	nd	+(uv/mv)	nd	[95]
VEGF-C	RT-PCR	nd	nd	+(uv/mv)	nd	[95]
VEGF-D	RT-PCR	nd	nd	–	nd	[95]
Neuroblastoma						
VEGF-A	qRT-PCR	nd	nd	–	nd	[66]
VEGF-C	qRT-PCR	nd	nd	–	nd	[66]
Non-small cell lung cancer						
VEGF-A	ELISA (R&D), serum VA	nd	+	+	nd	[124]
VEGF-C	IHC, IBL 18415	nd	nd	–	+uv (–mv)	[5]
	IHC, Zymed 18-2255	+	+	+	+	[75]
	ELISA (IBL), serum ΔNΔC-VC	nd	+	+	nd	[124]

(continued)

Table 4.2 (continued)

Marker	Detection method	Correlation of marker expression with LVD	LVI	LNM	Poor prognosis	Ref
Oral squamous cell carcinoma						
VEGF-A	IHC, SC goat pab	nd	nd	–	nd	[113]
VEGF-C	IHC, IBL mouse mab	nd	nd	+	nd	[113]
	IHC, IBL 18415	nd	–	+T1/2; –T3/4	+uv (–mv)	[63]
VEGF-D	IHC, SC goat pab	nd	nd	+	nd	[113]
Ovarian cancer						
VEGF-A	IHC, SC ab	nd	nd	–	–	[94]
VEGF-C	IHC, Zymed 18-2255	nd	nd	+	+uv	[94]
	IHC, SC goat pab	+(it)	nd	+	+	[126]
Pancreatic adenocarcinoma						
VEGF-C	IHC, R&D AF752	–(it/pt)	nd	–	–	[114]
	IHC, SC-1881 At margin & center of tumor	nd	+ at margin – at center	+ at margin – at center	–+[g]	[74]
VEGF-D	qRT-PCR	–	nd	–	–	[114]
	IHC, SC rabbit pab At margin & center of tumor	nd	– at margin & center	+ at margin – at center	–+[g]	[74]
Prostate cancer						
VEGF-A	qRT-PCR WB, SC-507 ELISA (R&D), plasma VA	nd	nd	tumor protein & mRNA: inv. correl. + plasma VEGF-A	nd	[58]

Table 4.2 (continued)

Marker	Detection method	Correlation of marker expression with: LVD	LVI	LNM	Poor prognosis	Ref
						[58]
						[58]
VEGF-C	IHC, R&D AF752	nd	nd	+	nd	[49]
	RNase protection assay	nd	nd	–	nd	[119]
VEGF-D	qRT-PCR WB, SC-13085 ELISA, R&D, plasma VD	nd	nd	– for mRNA + for tumor and plasma VEGF-D	nd	[58]
						[58]
						[58]
	RNase protection assay	nd	nd	+	nd	[119]
Thyroid cancer						
VEGF-A	RT-PCR	nd	nd	–	nd	[11]
VEGF-C	RT-PCR	nd	nd	+	nd	[11]
VEGF-C	IHC, R&D goat pab	nd	+	+	nd	[141]
VEGF-D	IHC, R&D ab/qRT-PCR	+(it/pt)/nd	nd/nd	+/+	nd/nd	[138]

[a] for VEGF-C/D ratio; [b] + for VEGF-A and -C combined, – for VEGF-A or -C alone; [c] + in undifferentiated tumors, – in differentiated tumors; [d] for VEGF-C or -D from tumor and stromal cells; [e] for VEGF-A/VEGF-D ratio; [f] for VEGF-C/VEGF-D ratio; [g] for VEGF-C and -D combined; it, intratumoral; LNM, lymph-node metastasis; LVD, lymphatic vessel density; LVI, lymphatic vessel invasion; mab, monoclonal antibody; mv, multivariate; nd, not determined; pab, polyclonal antibody; pt, peritumoral; uv, univariate; VA, VEGF-A; VC, VEGF-C; VD, VEGF-D; +, statistically significant correlation ($p \leq 0.05$); –, no significant correlati

(Table 4.2). Similar to the studies on VEGF-C, assessment by IHC found the highest correlation (11 out of 20 studies, 55%), whereas VEGF-D mRNA expression was correlated with lymph node metastasis in only 2 out of 10 studies (20%), and ELISA assays detected a positive correlation in 1 of 2 studies (50%). Unlike the findings with VEGF-C, however, VEGF-D mRNA and protein levels were sometimes found to be decreased in tumor samples [36, 59, 69, 95] or metastatic lymph nodes [95] of cancer patients, whereas some studies found elevated levels [113, 119, 139]. Interestingly, two studies [69, 93] found a positive correlation between lymph node metastasis and a decreased VEGF-D to VEGF-C mRNA ratio. In these cases, VEGF-C or VEGF-D levels themselves did not correlate with lymphatic metastasis. The reasons for these findings are at present unclear – it remains to be investigated whether decreased VEGF-D levels might allow more VEGF-C to be bound to VEGF-R3 and VEGF-R2 and whether this might lead to a more potent stimulation of lymphatic endothelium. In 8 out of 19 studies (42.1%), VEGF-A mRNA or protein expression was positively correlated to LNM (Table 4.2). Detection by IHC was positively correlated in 1 of 6 studies (17%), as compared to mRNA expression (4 out of 8 studies; 50%) or ELISA detection (2 out of 4 studies; 50%). It is of interest that in a large number of studies, in particular in gastric cancer, elevated levels of VEGF-C expression were correlated with increased detection of lymphatic vessel invasion (LVI; Table 4.2), indicating that VEGF-C might directly influence the interaction between tumor cells and lymphatic endothelium.

In human malignant melanomas of the skin, lymphatic vessels represent the major route of metastatic dissemination [83], and sentinel lymph node biopsies are performed to assess the metastatic potential of the primary melanoma, with important implications for the choice of treatment and for prognosis [68, 87]. Two recent studies indicate that the extent of lymphangiogenesis in the primary cutaneous melanoma can predict the presence of sentinel lymph node metastases at the time of surgery [20, 81]. Tumor lymphangiogenesis was the most significant independent prognostic indicator for metastasis [20]. VEGF-C expression levels were significantly correlated with lymphatic vessel density in primary melanomas [20] and with lymph node metastasis [20, 108, 131]. Primary melanomas were reported to express more VEGF-C in the vertical growth phase than in the horizontal growth phase [32]. No correlation was found between the expression levels of VEGF-D and the incidence of lymph node metastasis in two studies [20, 108] but VEGF-D has been detected in melanomas and has been associated with tumorangiogenesis [1]. The contribution of other recently identified lymphangiogenesis factors towards melanoma lymphangiogenesis and lymphatic metastasis remains unclear at this time.

Overall, these studies indicate VEGF-C as the most important lymphangiogenic factor in the majority of human cancers. The observed discrepancies between different studies might be explained, at least in part, by the lack of standardized criteria for the evaluation and quantification of tumor lymphangiogenesis, as well as for the determination of expression levels of lymphangiogenic mediators.

4.6 Evaluation of the Different Detection Methods Used for Assessment of Expression Levels of Lymphangiogenic Growth Factors in Human Cancers

It is obvious that a thorough interpretation of the studies investigating the correlation between growth factor expression, tumor lymphangiogenesis and lymph node metastasis is hampered by the lack of a direct comparability between these studies. This is due to the use of several different, non-standardized methods for assessing lymphatic growth factor expression and for reporting clinicopathological features. Expression of molecules involved in lymphatic vessel growth has been measured by (quantitative) RT-PCR, ELISA or by visual scoring of IHC stains. What is the most appropriate technique to perform this task? What parameters need to be considered when applying a given technique? In the following paragraphs, we intend to review the potential advantages and drawbacks of each method.

4.6.1 mRNA vs Protein Expression

Whereas ELISA and IHC evaluate the expression of protein, (quantitative) RT-PCR measures the expression of mRNA. Due to possible posttranscriptional and posttranslational regulation, mRNA levels do not necessarily reflect protein levels [33]. It is therefore unclear whether detection of VEGF-C mRNA levels by quantitative RT-PCR truly reflects the levels of the respective proteins. In fact, at least two studies failed to detect a correlation between VEGF-C mRNA and protein levels, or encountered cases where mRNA overexpression did not translate into increased VEGF-C protein expression [59, 61]. In contrast, a significant correlation between VEGF-C mRNA levels and mature VEGF-C protein levels was reported in cells isolated from patients with acute myeloid leukemia [22], and mRNA and protein levels of VEGF-D were also positively correlated in tumor samples of colorectal cancer [29]. One also has to keep in mind that tissue preservation crucially impacts the detection of growth factors at the mRNA and protein level. This is of particular relevance to the large number of studies that have been performed on tissues routinely embedded in paraffin. Moreover, the parameters for RT-PCR detection and normalization have not been well defined in a number of studies. Overall, more studies are needed to evaluate the best-suited approach for the quantitative measurement of lymphatic growth factors in tissues. Based on the previously published studies, detection at the protein level might have some advantages over mRNA-based methods, and fresh-frozen material might be better suited than paraffin-embedded samples.

4.6.2 Sampling Area and Cellular Resolution

A major problem when using (quantitative) RT-PCR and ELISA assays of tumor samples arises from the area from where the samples were obtained. Most often, biopsies are taken from some undefined part of the tumor tissue. For analysis, tissue samples are then homogenized, which results in a loss of resolution and a masking of focal lymphangiogenic growth factor levels. Moreover, since tumor lymphangiogenesis predominantly occurs in the peritumoral area, samples taken from the tumor center may not appropriately represent VEGF-C or VEGF-D levels at the hot spots of lymphangiogenic activity. Indeed, several studies found that intratumoral VEGF-C and VEGF-D levels are lower than peritumoral levels [31, 74, 97, 113, 144].

Immunohistochemistry of whole tumor samples has the advantage of preserving the histological environment of the assessed tissue sample, thereby enabling a more detailed picture of lymphangiogenic growth factor expression. Samples encompassing intra- and peritumoral structures can be investigated at the same time, and hot spots of VEGF-C/VEGF-D expression can be easily visualized. Moreover, the cell type-specific expression can often be distinguished by IHC, whereas RT-PCR or ELISA cannot differentiate between tumor and stromal cells. On the other hand, a rather large number of different antibodies has been used for IHC studies, and their relative specificity has not been comprehensively evaluated.

4.6.3 Systemic Measurement of Lymphangiogenic Growth Factors

In a number of studies, circulating VEGF-C [2, 24, 71, 84, 124, 131], VEGF-D [24, 29, 58] or VEGF-A [58, 84, 124] levels were measured in the plasma or serum of cancer patients. Whereas this method would be the preferred way of assessing VEGF-C and other growth factor levels, due to its minimal invasiveness and the possibility for repeat measurements of biomarkers over time, its validity remains unclear. For instance, VEGF-C is produced locally in rather low amounts whereas serum VEGF-C levels are typically in the range of 1–2 ng/ml with a significant variation between individuals [124]. While local VEGF-C levels might be highly increased at the tumor edge, an increase in systemic VEGF-C levels might be delayed and obscured when VEGF-C is drained via the lymphatics and the thoracic duct into the blood circulation. Nevertheless, there are a few studies which reported a significant correlation of serum VEGF-A, -C or -D levels and lymph node metastasis [58, 71, 124, 131].

4.6.4 Quantitative vs Qualitative Assessment Methods

To determine the risk for metastasis of primary cancers, it would be advantageous to be able to define clear thresholds for the levels of lymphangiogenic growth factor in or around the primary tumor or in the circulation when measured with quantitative methods such as ELISA or qRT-PCR. Although retrospective cohort studies found a significant increase of VEGF-C protein in patients with lymph node metastasis [71, 124, 131], this marker might not be applicable for prospective diagnosis of lymph node metastasis in individuals, since VEGF-C serum levels vary significantly between individual patients. High serum levels of VEGF-C in patients with non-metastatic tumors are well within the range of VEGF-C serum levels of patients bearing metastatic tumors [124]. Without corresponding negative controls from the pre-disease state, circulating growth factor levels are difficult to interpret.

By contrast, qualitative (semiquantitative) assessment of lymphangiogenic growth factor expression by IHC allows comparison with healthy tissue from the same patient and subsequent determination of the extent of tumor-driven lymphangiogenesis.

4.6.5 Choice of Antibodies for ELISA and IHC

Since VEGF-C and D undergo extensive proteolytic processing before and after secretion, it is important to consider which isoform to detect. Antibodies against the VEGF homology domain (VHD), which is contained in the fully processed, N- and C-terminally cleaved forms of VEGF-C and D, appear to be suited best since they recognize the mature and most potent protein as well as all precursor proteins containing the VHD. Moreover, the completely processed forms of VEGF-C and VEGF-D (ΔNΔC-VEGF-C/-D) activate VEGF-R2 much more potently than the unprocessed forms, thereby inducing enhanced vascular permeability in lymphatics and blood vessels [23, 52, 98], which might lead to facilitated lymphatic vessel invasion by tumor cells. However, one of the most frequently used antibodies for detecting VEGF-C, the rabbit polyclonal Zymed anti-VEGF-C antibody (cat. no 18-2255) was raised against the C-terminal part of unprocessed VEGF-C and therefore does not recognize mature VEGF-C (Table 4.3).

In conclusion, there are, at present, no established protocols for the standardized assessment of lymphangiogenic growth factor expression (as well as of tumor lymphangiogenesis) in tumors, tumor metastases and/or serum samples. Several of the methods used are time-consuming, expensive and difficult to quantify. Thus, because of the emerging importance of tumor (and lymph node) lymphangiogenesis as a prognostic marker and as a therapeutic and diagnostic target, there is an urgent need for large, multi-center studies to comparatively evaluate the prognostic/diagnostic value of growth factor expression, including some of the newly identified lymphangiogenic factors. At present, evaluation of VEGF-A, -C and -D expression by visual evaluation of histological samples – that include the peritumoral area of primary tumors or a representative area of the draining lymph nodes – by immunohistochemistry with antibodies against the mature forms of these growth factors might represent the most promising approach.

Table 4.3 Overview of antibodies used for detection of VEGF-C and VEGF-D in tissue sections

Manufacturer	Cat. No.	Species	Peptide used for immunization	Reactive against	Ref.
Anti-VEGF-C					
Immuno-Biological Labs (IBL)	18415	rabbit	aa 105–118	VHD	M
R&D Systems	AF752	goat	aa 104–330	VHD	M, [108, 143]
Santa Cruz	SC-7133	goat	aa 1–100	N-terminus	M
	SC-1881	goat	aa 369–419	C-terminus	M
	SC-9047	rabbit	aa 230–419	C-terminus	M
Van Trappen et al.	N/A	rabbit	aa 112–227	VHD	[129]
Zhongshan*	ZA-0266	rabbit	C-terminus	C-terminus	M
Zymed/Invitrogen	18-2255	rabbit	C-terminus	C-terminus	M
Anti-VEGF-D					
Achen et al.	N/A	rabbit	aa 93–201	VHD	[1, 20]
Kirkin et al.	Serum 4292	rabbit	aa 92–218	VHD	[62, 129]
R&D Systems	Mab286	mouse	rhVEGF-D	VHD	[102, 143]
	AF286	goat	rhVEGF-D	VHD	M, [127]
	Mab622	mouse	rhVEGF-D	VHD	[27]
Santa Cruz	SC-7602	goat	aa 301–400	C-terminus	M
	SC-7603	goat	aa 1–89	N-terminus	M
	SC-13085	rabbit	aa 211–354	C-terminus	M

aa, amino acid; M, manufacturer; N/A, not available; VHD, VEGF homology domain; *, repacked Zymed 18-2255

4.7 Perspectives

There is now compelling evidence, stemming from experimental tumor metastasis studies in mice and from clinicopathological studies in a large number of different human cancer types, that tumor-associated lymphangiogenesis promotes cancer metastasis and represents a novel and powerful prognostic indicator for the risk of metastasis. Moreover, the newly identified process of lymph node lymphangiogenesis – induced even before cancer metastasis – indicates that lymphangiogenesis represents a novel target for the prevention, treatment, and early detection of cancer metastases. While VEGF-C appears at present to be the most important tumor lymphangiogenesis factors in the majority of human cancers, also VEGF-A and VEGF-D have potent lymphangiogenic activity. Moreover, there are a number of newly identified lymphangiogenic growth factors whose relative importance for cancer metastasis needs to be clarified. Therapies aimed at inhibiting the VEGF-C/VEGF-D/VEGF-R3 axis need to take into consideration that VEGF-R3 is also expressed by several non-endothelial cell types, including cells in the bone marrow, and that e.g. VEGF-D is also expressed in osteoblasts [99]. Thus, it remains to be seen whether or not therapies blocking this axis will also lead to adverse effects on the recruitment of bone marrow-derived precursor cells, hematopoiesis or bone growth and repair. Importantly, considerable efforts are needed to standardize the qualitative and quantitative evaluation of tumor lymphangiogenesis and of the expression levels of lymphangiogenic growth factors in tumors and metastases.

Acknowledgments This work was supported by National Institutes of Health grant CA69184, Swiss National Fund grant 3100A0-108207, Austrian Science Foundation grant S9408-B11, Cancer League Zurich and Commission of the European Communities grant LSHC-CT-2005-518178 (M.D.).

References

1. Achen MG, Williams RA, Minekus MP, Thornton GE, Stenvers K, Rogers PA, Lederman F, Roufail S, Stacker SA. Localization of vascular endothelial growth factor-D in malignant melanoma suggests a role in tumour angiogenesis. J Pathol 2001, 193: 147–54.
2. Al-Mowallad A, Kirwan C, Byrne G, McDowell G, Li C, Stewart A, Al-Qouzi A, Kumar S. Vascular endothelial growth factor-C in patients with breast cancer. In Vivo 2007, 21: 549–51.
3. Alitalo K, Tammela T, Petrova TV. Lymphangiogenesis in development and human disease. Nature 2005, 438: 946–53.
4. Amioka T, Kitadai Y, Tanaka S, Haruma K, Yoshihara M, Yasui W, Chayama K. Vascular endothelial growth factor-C expression predicts lymph node metastasis of human gastric carcinomas invading the submucosa. Eur J Cancer 2002, 38: 1413–9.
5. Arinaga M, Noguchi T, Takeno S, Chujo M, Miura T, Uchida Y. Clinical significance of vascular endothelial growth factor C and vascular endothelial growth factor receptor 3 in patients with nonsmall cell lung carcinoma. Cancer 2003, 97: 457–64.
6. Aselli G. De Lactibus Sive Lacteis Venis. Dissertatio Mediolan 1627.
7. Bando H, Weich HA, Horiguchi S, Funata N, Ogawa T, Toi M. The association between vascular endothelial growth factor-C, its corresponding receptor, VEGFR-3, and prognosis in primary breast cancer: a study with 193 cases. Oncol Rep 2006, 15: 653–9.
8. Beasley NJ, Prevo R, Banerji S, Leek RD, Moore J, van Trappen P, Cox G, Harris AL, Jackson DG. Intratumoral lymphangiogenesis and lymph node metastasis in head and neck cancer. Cancer Res 2002, 62: 1315–20.
9. Bjorndahl M, Cao R, Nissen LJ, Clasper S, Johnson LA, Xue Y, Zhou Z, Jackson D, Hansen AJ, et al. Insulin-like growth factors 1 and 2 induce lymphangiogenesis in vivo. Proc Natl Acad Sci USA 2005, 102: 15593–8.
10. Breiteneder-Geleff S, Soleiman A, Kowalski H, Horvat R, Amann G, Kriehuber E, Diem K, Weninger W, Tschachler E, et al. Angiosarcomas express mixed endothelial phenotypes of blood and lymphatic capillaries: podoplanin as a specific marker for lymphatic endothelium. Am J Pathol 1999, 154: 385–94.
11. Bunone G, Vigneri P, Mariani L, Buto S, Collini P, Pilotti S, Pierotti MA, Bongarzone I. Expression of angiogenesis stimulators and inhibitors in human thyroid tumors and correlation with clinical pathological features. Am J Pathol 1999, 155: 1967–76.
12. Byeon JS, Jung HY, Lee YJ, Lee D, Lee GH, Myung SJ, Yang SK, Hong WS, Kim JH, et al. Clinicopathological significance of vascular endothelial growth factor-C and cyclooxygenase-2 in esophageal squamous cell carcinoma. J Gastroenterol Hepatol 2004, 19: 648–54.
13. Cao R, Bjorndahl MA, Religa P, Clasper S, Garvin S, Galter D, Meister B, Ikomi F, Tritsaris K, et al. PDGF-BB induces intratumoral lymphangiogenesis and promotes lymphatic metastasis. Cancer Cell 2004, 6: 333–45.

14. Cao R, Bjorndahl MA, Gallego MI, Chen S, Religa P, Hansen AJ, Cao Y. Hepatocyte growth factor is a lymphangiogenic factor with an indirect mechanism of action. Blood 2006, 107: 3531–6.
15. Chang LK, Garcia-Cardena G, Farnebo F, Fannon M, Chen EJ, Butterfield C, Moses MA, Mulligan RC, Folkman J, et al. Dose-dependent response of FGF-2 for lymphangiogenesis. Proc Natl Acad Sci USA 2004, 101: 11658–63.
16. Chen Z, Varney ML, Backora MW, Cowan K, Solheim JC, Talmadge JE, Singh RK. Down-regulation of vascular endothelial cell growth factor-C expression using small interfering RNA vectors in mammary tumors inhibits tumor lymphangiogenesis and spontaneous metastasis and enhances survival. Cancer Res 2005, 65: 9004–11.
17. Currie MJ, Hanrahan V, Gunningham SP, Morrin HR, Frampton C, Han C, Robinson BA, Fox SB. Expression of vascular endothelial growth factor D is associated with hypoxia inducible factor (HIF-1alpha) and the HIF-1alpha target gene DEC1, but not lymph node metastasis in primary human breast carcinomas. J Clin Pathol 2004, 57: 829–34.
18. Da MX, Wu XT, Wang J, Guo TK, Zhao ZG, Luo T, Zhang MM, Qian K. Expression of cyclooxygenase-2 and vascular endothelial growth factor-C correlates with lymphangiogenesis and lymphatic invasion in human gastric cancer. Arch Med Res 2008, 39: 92–9.
19. Dadras SS, Paul T, Bertoncini J, Brown LF, Muzikansky A, Jackson DG, Ellwanger U, Garbe C, Mihm MC, et al. Tumor lymphangiogenesis: a novel prognostic indicator for cutaneous melanoma metastasis and survival. Am J Pathol 2003, 162: 1951–60.
20. Dadras SS, Lange-Asschenfeldt B, Velasco P, Nguyen L, Vora A, Muzikansky A, Jahnke K, Hauschild A, Hirakawa S, et al. Tumor lymphangiogenesis predicts melanoma metastasis to sentinel lymph nodes. Mod Pathol 2005, 18: 1232–42.
21. Davis S, Aldrich TH, Jones PF, Acheson A, Compton DL, Jain V, Ryan TE, Bruno J, Radziejewski C, et al. Isolation of angiopoietin-1, a ligand for the Tie2 receptor, by secretion-trap expression cloning. Cell 1996, 87: 1161–9.
22. de Jonge HJ, Weidenaar AC, Ter Elst A, Boezen HM, Scherpen FJ, Bouma-Ter Steege JC, Kaspers GJ, Goemans BF, Creutzig U, et al. Endogenous vascular endothelial growth factor-C expression is associated with decreased drug responsiveness in childhood acute myeloid leukemia. Clin Cancer Res 2008, 14: 924–30.
23. Dixelius J, Makinen T, Wirzenius M, Karkkainen MJ, Wernstedt C, Alitalo K, Claesson-Welsh L. Ligand-induced vascular endothelial growth factor receptor-3 (VEGFR-3) heterodimerization with VEGFR-2 in primary lymphatic endothelial cells regulates tyrosine phosphorylation sites. J Biol Chem 2003, 278: 40973–9.
24. Duff SE, Saunders M, McCredie V, Kumar S, O'Dwyer ST, Jayson GC. Pre-operative plasma levels of vascular endothelial growth factor A, C and D in patients with colorectal cancer. Clin Oncol (R Coll Radiol) 2005, 17: 367–71.
25. Fritz-Six KL, Dunworth WP, Li M, Caron KM. Adrenomedullin signaling is necessary for murine lymphatic vascular development. J Clin Invest 2008, 118: 40–50.
26. Fukunaga S, Maeda K, Noda E, Inoue T, Wada K, Hirakawa K. Association between expression of vascular endothelial growth factor C, chemokine receptor CXCR4 and lymph node metastasis in colorectal cancer. Oncology 2006, 71: 204–11.
27. Funaki H, Nishimura G, Harada S, Ninomiya I, Terada I, Fushida S, Tani T, Fujimura T, Kayahara M, et al. Expression of vascular endothelial growth factor D is associated with lymph node metastasis in human colorectal carcinoma. Oncology 2003, 64: 416–22.
28. Furudoi A, Tanaka S, Haruma K, Kitadai Y, Yoshihara M, Chayama K, Shimamoto F. Clinical significance of vascular endothelial growth factor C expression and angiogenesis at the deepest invasive site of advanced colorectal carcinoma. Oncology 2002, 62: 157–66.
29. George ML, Tutton MG, Janssen F, Arnaout A, Abulafi AM, Eccles SA, Swift RI. VEGF-A, VEGF-C, and VEGF-D in colorectal cancer progression. Neoplasia 2001, 3: 420–7.
30. Gerli R, Solito R, Weber E, Agliano M. Specific adhesion molecules bind anchoring filaments and endothelial cells in human skin initial lymphatics. Lymphology 2000, 33: 148–57.

31. Gombos Z, Xu X, Chu CS, Zhang PJ, Acs G. Peritumoral lymphatic vessel density and vascular endothelial growth factor C expression in early-stage squamous cell carcinoma of the uterine cervix. Clin Cancer Res 2005, 11: 8364–71.
32. Goydos JS, Gorski DH. Vascular endothelial growth factor C mRNA expression correlates with stage of progression in patients with melanoma. Clin Cancer Res 2003, 9: 5962–7.
33. Gu X, Lundqvist EN, Coates PJ, Thurfjell N, Wettersand E, Nylander K. Dysregulation of TAp63 mRNA and protein levels in psoriasis. J Invest Dermatol 2006, 126: 137–41.
34. Gunningham SP, Currie MJ, Han C, Robinson BA, Scott PA, Harris AL, Fox SB. The short form of the alternatively spliced flt-4 but not its ligand vascular endothelial growth factor C is related to lymph node metastasis in human breast cancers. Clin Cancer Res 2000, 6: 4278–86.
35. Guyton A, Hall, J. Textbook of Medical Physiology, 9th ed. W.B. Saunder Company, Philadelphia, 1996.
36. Hanrahan V, Currie MJ, Gunningham SP, Morrin HR, Scott PA, Robinson BA, Fox SB. The angiogenic switch for vascular endothelial growth factor (VEGF)-A, VEGF-B, VEGF-C, and VEGF-D in the adenoma-carcinoma sequence during colorectal cancer progression. J Pathol 2003, 200: 183–94.
37. Harrell MI, Iritani BM, Ruddell A. Tumor-induced sentinel lymph node lymphangiogenesis and increased lymph flow precede melanoma metastasis. Am J Pathol 2007, 170: 774–86.
38. He XW, Liu T, Chen YX, Cheng DJ, Li XR, Xiao Y, Feng YL. Calcium carbonate nanoparticle delivering vascular endothelial growth factor-C siRNA effectively inhibits lymphangiogenesis and growth of gastric cancer in vivo. Cancer Gene Ther 2008, 15: 193–202.
39. He Y, Kozaki K, Karpanen T, Koshikawa K, Yla-Herttuala S, Takahashi T, Alitalo K. Suppression of tumor lymphangiogenesis and lymph node metastasis by blocking vascular endothelial growth factor receptor 3 signaling. J Natl Cancer Inst 2002, 94: 819–25.
40. Hinojar-Gutierrez A, Fernandez-Contreras ME, Gonzalez-Gonzalez R, Fernandez-Luque MJ, Hinojar-Arzadun A, Quintanilla M, Gamallo C. Intratumoral lymphatic vessels and VEGF-C expression are predictive factors of lymph node relapse in T1-T4 N0 laryngopharyngeal squamous cell carcinoma. Ann Surg Oncol 2007, 14: 248–57.
41. Hirai M, Nakagawara A, Oosaki T, Hayashi Y, Hirono M, Yoshihara T. Expression of vascular endothelial growth factors (VEGF-A/VEGF-1 and VEGF-C/VEGF-2) in postmenopausal uterine endometrial carcinoma. Gynecol Oncol 2001, 80: 181–8.
42. Hirakawa S, Hong YK, Harvey N, Schacht V, Matsuda K, Libermann T, Detmar M. Identification of vascular lineage-specific genes by transcriptional profiling of isolated blood vascular and lymphatic endothelial cells. Am J Pathol 2003, 162: 575–86.
43. Hirakawa S, Kodama S, Kunstfeld R, Kajiya K, Brown LF, Detmar M. VEGF-A induces tumor and sentinel lymph node lymphangiogenesis and promotes lymphatic metastasis. J Exp Med 2005, 201: 1089–99.
44. Hirakawa S, Brown LF, Kodama S, Paavonen K, Alitalo K, Detmar M. VEGF-C-induced lymphangiogenesis in sentinel lymph nodes promotes tumor metastasis to distant sites. Blood 2007, 109: 1010–7.
45. Hoar FJ, Chaudhri S, Wadley MS, Stonelake PS. Co-expression of vascular endothelial growth factor C (VEGF-C) and c-erbB2 in human breast carcinoma. Eur J Cancer 2003, 39: 1698–703.
46. Hong YK, Lange-Asschenfeldt B, Velasco P, Hirakawa S, Kunstfeld R, Brown LF, Bohlen P, Senger DR, Detmar M. VEGF-A promotes tissue repair-associated lymphatic vessel formation via VEGFR-2 and the alpha1beta1 and alpha2beta1 integrins. Faseb J 2004, 18: 1111–3.
47. Ishikawa M, Kitayama J, Kazama S, Nagawa H. Expression of vascular endothelial growth factor C and D (VEGF-C and -D) is an important risk factor for lymphatic metastasis in undifferentiated early gastric carcinoma. Jpn J Clin Oncol 2003, 33: 21–7.
48. Jeltsch M, Kaipainen A, Joukov V, Meng X, Lakso M, Rauvala H, Swartz M, Fukumura D, Jain RK, et al. Hyperplasia of lymphatic vessels in VEGF-C transgenic mice. Science 1997, 276: 1423–5.

49. Jennbacken K, Vallbo C, Wang W, Damber JE. Expression of vascular endothelial growth factor C (VEGF-C) and VEGF receptor-3 in human prostate cancer is associated with regional lymph node metastasis. Prostate 2005, 65: 110–6.
50. Jia YT, Li ZX, He YT, Liang W, Yang HC, Ma HJ. Expression of vascular endothelial growth factor-C and the relationship between lymphangiogenesis and lymphatic metastasis in colorectal cancer. World J Gastroenterol 2004, 10: 3261–3.
51. Johnson LA, Prevo R, Clasper S, Jackson DG. Inflammation-induced uptake and degradation of the lymphatic endothelial hyaluronan receptor LYVE-1. J Biol Chem 2007, 282: 33671–80.
52. Joukov V, Sorsa T, Kumar V, Jeltsch M, Claesson-Welsh L, Cao Y, Saksela O, Kalkkinen N, Alitalo K. Proteolytic processing regulates receptor specificity and activity of VEGF-C. Embo J 1997, 16: 3898–911.
53. Juttner S, Wissmann C, Jons T, Vieth M, Hertel J, Gretschel S, Schlag PM, Kemmner W, Hocker M. Vascular endothelial growth factor-D and its receptor VEGFR-3: two novel independent prognostic markers in gastric adenocarcinoma. J Clin Oncol 2006, 24: 228–40.
54. Kabashima A, Maehara Y, Kakeji Y, Sugimachi K. Overexpression of vascular endothelial growth factor C is related to lymphogenous metastasis in early gastric carcinoma. Oncology 2001, 60: 146–50.
55. Kajiya K, Hirakawa S, Ma B, Drinnenberg I, Detmar M. Hepatocyte growth factor promotes lymphatic vessel formation and function. Embo J 2005, 24: 2885–95.
56. Karpanen T, Egeblad M, Karkkainen MJ, Kubo H, Yla-Herttuala S, Jaattela M, Alitalo K. Vascular endothelial growth factor C promotes tumor lymphangiogenesis and intralymphatic tumor growth. Cancer Res 2001, 61: 1786–90.
57. Karpanen T, Heckman CA, Keskitalo S, Jeltsch M, Ollila H, Neufeld G, Tamagnone L, Alitalo K. Functional interaction of VEGF-C and VEGF-D with neuropilin receptors. Faseb J 2006, 20: 1462–72.
58. Kaushal V, Mukunyadzi P, Dennis RA, Siegel ER, Johnson DE, Kohli M. Stage-specific characterization of the vascular endothelial growth factor axis in prostate cancer: expression of lymphangiogenic markers is associated with advanced-stage disease. Clin Cancer Res 2005, 11: 584–93.
59. Kawakami M, Furuhata T, Kimura Y, Yamaguchi K, Hata F, Sasaki K, Hirata K. Quantification of vascular endothelial growth factor-C and its receptor-3 messenger RNA with real-time quantitative polymerase chain reaction as a predictor of lymph node metastasis in human colorectal cancer. Surgery 2003, 133: 300–8.
60. Kim KE, Cho CH, Kim HZ, Baluk P, McDonald DM, Koh GY. In vivo actions of angiopoietins on quiescent and remodeling blood and lymphatic vessels in mouse airways and skin. Arterioscler Thromb Vasc Biol 2007, 27: 564–70.
61. Kinoshita J, Kitamura K, Kabashima A, Saeki H, Tanaka S, Sugimachi K. Clinical significance of vascular endothelial growth factor-C (VEGF-C) in breast cancer. Breast Cancer Res Treat 2001, 66: 159–64.
62. Kirkin V, Mazitschek R, Krishnan J, Steffen A, Waltenberger J, Pepper MS, Giannis A, Sleeman JP. Characterization of indolinones which preferentially inhibit VEGF-C- and VEGF-D-induced activation of VEGFR-3 rather than VEGFR-2. Eur J Biochem 2001, 268: 5530–40.
63. Kishimoto K, Sasaki A, Yoshihama Y, Mese H, Tsukamoto G, Matsumura T. Expression of vascular endothelial growth factor-C predicts regional lymph node metastasis in early oral squamous cell carcinoma. Oral Oncol 2003, 39: 391–6.
64. Kitadai Y, Amioka T, Haruma K, Tanaka S, Yoshihara M, Sumii K, Matsutani N, Yasui W, Chayama K. Clinicopathological significance of vascular endothelial growth factor (VEGF)-C in human esophageal squamous cell carcinomas. Int J Cancer 2001, 93: 662–6.
65. Kitadai Y, Kodama M, Cho S, Kuroda T, Ochiumi T, Kimura S, Tanaka S, Matsumura S, Yasui W, et al. Quantitative analysis of lymphangiogenic markers for predicting metastasis of human gastric carcinoma to lymph nodes. Int J Cancer 2005, 115: 388–92.

66. Komuro H, Kaneko S, Kaneko M, Nakanishi Y. Expression of angiogenic factors and tumor progression in human neuroblastoma. J Cancer Res Clin Oncol 2001, 127: 739–43.
67. Kondo K, Kaneko T, Baba M, Konno H. VEGF-C and VEGF-A synergistically enhance lymph node metastasis of gastric cancer. Biol Pharm Bull 2007, 30: 633–7.
68. Koskivuo I, Talve L, Vihinen P, Maki M, Vahlberg T, Suominen E. Sentinel lymph node biopsy in cutaneous melanoma: a case-control study. Ann Surg Oncol 2007, 14: 3566–74.
69. Koyama Y, Kaneko K, Akazawa K, Kanbayashi C, Kanda T, Hatakeyama K. Vascular endothelial growth factor-C and vascular endothelial growth factor-d messenger RNA expression in breast cancer: association with lymph node metastasis. Clin Breast Cancer 2003, 4: 354–60.
70. Krishnan J, Kirkin V, Steffen A, Hegen M, Weih D, Tomarev S, Wilting J, Sleeman JP. Differential in vivo and in vitro expression of vascular endothelial growth factor (VEGF)-C and VEGF-D in tumors and its relationship to lymphatic metastasis in immunocompetent rats. Cancer Res 2003, 63: 713–22.
71. Krzystek-Korpacka M, Matusiewicz M, Diakowska D, Grabowski K, Blachut K, Banas T. Up-regulation of VEGF-C secreted by cancer cells and not VEGF-A correlates with clinical evaluation of lymph node metastasis in esophageal squamous cell carcinoma (ESCC). Cancer Lett 2007, 249: 171–7.
72. Kubo H, Cao R, Brakenhielm E, Makinen T, Cao Y, Alitalo K. Blockade of vascular endothelial growth factor receptor-3 signaling inhibits fibroblast growth factor-2-induced lymphangiogenesis in mouse cornea. Proc Natl Acad Sci USA 2002, 99: 8868–73.
73. Kunstfeld R, Hirakawa S, Hong YK, Schacht V, Lange-Asschenfeldt B, Velasco P, Lin C, Fiebiger E, Wei X, et al. Induction of cutaneous delayed-type hypersensitivity reactions in VEGF-A transgenic mice results in chronic skin inflammation associated with persistent lymphatic hyperplasia. Blood 2004, 104: 1048–57.
74. Kurahara H, Takao S, Maemura K, Shinchi H, Natsugoe S, Aikou T. Impact of vascular endothelial growth factor-C and -D expression in human pancreatic cancer: its relationship to lymph node metastasis. Clin Cancer Res 2004, 10: 8413–20.
75. Li Q, Dong X, Gu W, Qiu X, Wang E. Clinical significance of co-expression of VEGF-C and VEGFR-3 in non-small cell lung cancer. Chin Med J (Engl) 2003, 116: 727–30.
76. Lin J, Lalani AS, Harding TC, Gonzalez M, Wu WW, Luan B, Tu GH, Koprivnikar K, VanRoey MJ, et al. Inhibition of lymphogenous metastasis using adeno-associated virus-mediated gene transfer of a soluble VEGFR-3 decoy receptor. Cancer Res 2005, 65: 6901–9.
77. Liu XE, Sun XD, Wu JM. Expression and significance of VEGF-C and FLT-4 in gastric cancer. World J Gastroenterol 2004, 10: 352–5.
78. Loges S, Clausen H, Reichelt U, Bubenheim M, Erbersdobler A, Schurr P, Yekebas E, Schuch G, Izbicki J, et al. Determination of microvessel density by quantitative real-time PCR in esophageal cancer: correlation with histologic methods, angiogenic growth factor expression, and lymph node metastasis. Clin Cancer Res 2007, 13: 76–80.
79. Maeda K, Yashiro M, Nishihara T, Nishiguchi Y, Sawai M, Uchima K, Onoda N, Ohira M, Ishikawa T, et al. Correlation between vascular endothelial growth factor C expression and lymph node metastasis in T1 carcinoma of the colon and rectum. Surg Today 2003, 33: 736–9.
80. Mandriota SJ, Jussila L, Jeltsch M, Compagni A, Baetens D, Prevo R, Banerji S, Huarte J, Montesano R, et al. Vascular endothelial growth factor-C-mediated lymphangiogenesis promotes tumour metastasis. Embo J 2001, 20: 672–82.
81. Massi D, Puig S, Franchi A, Malvehy J, Vidal-Sicart S, Gonzalez-Cao M, Baroni G, Ketabchi S, Palou J, et al. Tumour lymphangiogenesis is a possible predictor of sentinel lymph node status in cutaneous melanoma: a case-control study. J Clin Pathol 2006, 59: 166–73.
82. Maula SM, Luukkaa M, Grenman R, Jackson D, Jalkanen S, Ristamaki R. Intratumoral lymphatics are essential for the metastatic spread and prognosis in squamous cell carcinomas of the head and neck region. Cancer Res 2003, 63: 1920–6.

83. Meier F, Will S, Ellwanger U, Schlagenhauff B, Schittek B, Rassner G, Garbe C. Metastatic pathways and time courses in the orderly progression of cutaneous melanoma. Br J Dermatol 2002, 147: 62–70.
84. Mitsuhashi A, Suzuka K, Yamazawa K, Matsui H, Seki K, Sekiya S. Serum vascular endothelial growth factor (VEGF) and VEGF-C levels as tumor markers in patients with cervical carcinoma. Cancer 2005, 103: 724–30.
85. Mohammed RA, Green A, El-Shikh S, Paish EC, Ellis IO, Martin SG. Prognostic significance of vascular endothelial cell growth factors -A, -C and -D in breast cancer and their relationship with angio- and lymphangiogenesis. Br J Cancer 2007, 96: 1092–100.
86. Morisada T, Oike Y, Yamada Y, Urano T, Akao M, Kubota Y, Maekawa H, Kimura Y, Ohmura M, et al. Angiopoietin-1 promotes LYVE-1-positive lymphatic vessel formation. Blood 2005, 105: 4649–56.
87. Morton DL, Thompson JF, Cochran AJ, Mozzillo N, Elashoff R, Essner R, Nieweg OE, Roses DF, Hoekstra HJ, et al. Sentinel-node biopsy or nodal observation in melanoma. N Engl J Med 2006, 355: 1307–17.
88. Nagy JA, Vasile E, Feng D, Sundberg C, Brown LF, Detmar MJ, Lawitts JA, Benjamin L, Tan X, et al. Vascular permeability factor/vascular endothelial growth factor induces lymphangiogenesis as well as angiogenesis. J Exp Med 2002, 196: 1497–506.
89. Nakamura Y, Yasuoka H, Tsujimoto M, Yang Q, Imabun S, Nakahara M, Nakao K, Nakamura M, Mori I, et al. Prognostic significance of vascular endothelial growth factor D in breast carcinoma with long-term follow-up. Clin Cancer Res 2003, 9: 716–21.
90. Nakamura Y, Yasuoka H, Tsujimoto M, Yang Q, Tsukiyama A, Imabun S, Nakahara M, Nakao K, Nakamura M, et al. Clinicopathological significance of vascular endothelial growth factor-C in breast carcinoma with long-term follow-up. Mod Pathol 2003, 16: 309–14.
91. Nakamura Y, Yasuoka H, Tsujimoto M, Imabun S, Nakahara M, Nakao K, Nakamura M, Mori I, Kakudo K. Lymph vessel density correlates with nodal status, VEGF-C expression, and prognosis in breast cancer. Breast Cancer Res Treat 2005, 91: 125–32.
92. Nguyen VP, Chen SH, Trinh J, Kim H, Coomber BL, Dumont DJ. Differential response of lymphatic, venous and arterial endothelial cells to angiopoietin-1 and angiopoietin-2. BMC Cell Biol 2007, 8: 10.
93. Niki T, Iba S, Tokunou M, Yamada T, Matsuno Y, Hirohashi S. Expression of vascular endothelial growth factors A, B, C, and D and their relationships to lymph node status in lung adenocarcinoma. Clin Cancer Res 2000, 6: 2431–9.
94. Nishida N, Yano H, Komai K, Nishida T, Kamura T, Kojiro M. Vascular endothelial growth factor C and vascular endothelial growth factor receptor 2 are related closely to the prognosis of patients with ovarian carcinoma. Cancer 2004, 101: 1364–74.
95. O-charoenrat P, Rhys-Evans P, Eccles SA. Expression of vascular endothelial growth factor family members in head and neck squamous cell carcinoma correlates with lymph node metastasis. Cancer 2001, 92: 556–68.
96. Oliver G, Detmar M. The rediscovery of the lymphatic system: old and new insights into the development and biological function of the lymphatic vasculature. Genes Dev 2002, 16: 773–83.
97. Onogawa S, Kitadai Y, Tanaka S, Kuwai T, Kimura S, Chayama K. Expression of VEGF-C and VEGF-D at the invasive edge correlates with lymph node metastasis and prognosis of patients with colorectal carcinoma. Cancer Sci 2004, 95: 32–9.
98. Onogawa S, Kitadai Y, Amioka T, Kodama M, Cho S, Kuroda T, Ochiumi T, Kimura S, Kuwai T, et al. Expression of vascular endothelial growth factor (VEGF)-C and VEGF-D in early gastric carcinoma: correlation with clinicopathological parameters. Cancer Lett 2005, 226: 85–90.
99. Orlandini M, Spreafico A, Bardelli M, Rocchigiani M, Salameh A, Nucciotti S, Capperucci C, Frediani B, Oliviero S. Vascular endothelial growth factor-D activates VEGFR-3 expressed in osteoblasts inducing their differentiation. J Biol Chem 2006, 281: 17961–7.

100. Padera TP, Kadambi A, di Tomaso E, Carreira CM, Brown EB, Boucher Y, Choi NC, Mathisen D, Wain J, et al. Lymphatic metastasis in the absence of functional intratumor lymphatics. Science 2002, 296: 1883–6.
101. Paget S. The distribution of secondary growths in cancer of the breast. Lancet 1889, 1: 571–573.
102. Partanen TA, Arola J, Saaristo A, Jussila L, Ora A, Miettinen M, Stacker SA, Achen MG, Alitalo K. VEGF-C and VEGF-D expression in neuroendocrine cells and their receptor, VEGFR-3, in fenestrated blood vessels in human tissues. Faseb J 2000, 14: 2087–96.
103. Qian CN, Berghuis B, Tsarfaty G, Bruch M, Kort EJ, Ditlev J, Tsarfaty I, Hudson E, Jackson DG, et al. Preparing the "soil": the primary tumor induces vasculature reorganization in the sentinel lymph node before the arrival of metastatic cancer cells. Cancer Res 2006, 66: 10365–76.
104. Roberts N, Kloos B, Cassella M, Podgrabinska S, Persaud K, Wu Y, Pytowski B, Skobe M. Inhibition of VEGFR-3 activation with the antagonistic antibody more potently suppresses lymph node and distant metastases than inactivation of VEGFR-2. Cancer Res 2006, 66: 2650–7.
105. Roma AA, Magi-Galluzzi C, Kral MA, Jin TT, Klein EA, Zhou M. Peritumoral lymphatic invasion is associated with regional lymph node metastases in prostate adenocarcinoma. Mod Pathol 2006, 19: 392–8.
106. Schacht V, Ramirez MI, Hong YK, Hirakawa S, Feng D, Harvey N, Williams M, Dvorak AM, Dvorak HF, et al. T1alpha/podoplanin deficiency disrupts normal lymphatic vasculature formation and causes lymphedema. Embo J 2003, 22: 3546–56.
107. Schacht V, Dadras SS, Johnson LA, Jackson DG, Hong YK, Detmar M. Up-regulation of the lymphatic marker podoplanin, a mucin-type transmembrane glycoprotein, in human squamous cell carcinomas and germ cell tumors. Am J Pathol 2005, 166: 913–21.
108. Schietroma C, Cianfarani F, Lacal PM, Odorisio T, Orecchia A, Kanitakis J, D'Atri S, Failla CM, Zambruno G. Vascular endothelial growth factor-C expression correlates with lymph node localization of human melanoma metastases. Cancer 2003, 98: 789–97.
109. Schledzewski K, Falkowski M, Moldenhauer G, Metharom P, Kzhyshkowska J, Ganss R, Demory A, Falkowska-Hansen B, Kurzen H, et al. Lymphatic endothelium-specific hyaluronan receptor LYVE-1 is expressed by stabilin-1+, F4/80+, CD11b+ macrophages in malignant tumours and wound healing tissue in vivo and in bone marrow cultures in vitro: implications for the assessment of lymphangiogenesis. J Pathol 2006, 209: 67–77.
110. Shida A, Fujioka S, Kobayashi K, Ishibashi Y, Nimura H, Mitsumori N, Yanaga K. Expression of vascular endothelial growth factor (VEGF)-C and -D in gastric carcinoma. Int J Clin Oncol 2006, 11: 38–43.
111. Shimizu K, Kubo H, Yamaguchi K, Kawashima K, Ueda Y, Matsuo K, Awane M, Shimahara Y, Takabayashi A, et al. Suppression of VEGFR-3 signaling inhibits lymph node metastasis in gastric cancer. Cancer Sci 2004, 95: 328–33.
112. Shin JW, Min M, Larrieu-Lahargue F, Canron X, Kunstfeld R, Nguyen L, Henderson JE, Bikfalvi A, Detmar M, et al. Prox1 promotes lineage-specific expression of fibroblast growth factor (FGF) receptor-3 in lymphatic endothelium: a role for FGF signaling in lymphangiogenesis. Mol Biol Cell 2006, 17: 576–84.
113. Shintani S, Li C, Ishikawa T, Mihara M, Nakashiro K, Hamakawa H. Expression of vascular endothelial growth factor A, B, C, and D in oral squamous cell carcinoma. Oral Oncol 2004, 40: 13–20.
114. Sipos B, Kojima M, Tiemann K, Klapper W, Kruse ML, Kalthoff H, Schniewind B, Tepel J, Weich H, et al. Lymphatic spread of ductal pancreatic adenocarcinoma is independent of lymphangiogenesis. J Pathol 2005, 207: 301–12.
115. Skobe M, Hawighorst T, Jackson DG, Prevo R, Janes L, Velasco P, Riccardi L, Alitalo K, Claffey K, et al. Induction of tumor lymphangiogenesis by VEGF-C promotes breast cancer metastasis. Nat Med 2001, 7: 192–8.

116. Soumaoro LT, Uetake H, Takagi Y, Iida S, Higuchi T, Yasuno M, Enomoto M, Sugihara K. Coexpression of VEGF-C and Cox-2 in human colorectal cancer and its association with lymph node metastasis. Dis Colon Rectum 2006, 49: 392–8.
117. Stacker SA, Caesar C, Baldwin ME, Thornton GE, Williams RA, Prevo R, Jackson DG, Nishikawa S, Kubo H, et al. VEGF-D promotes the metastatic spread of tumor cells via the lymphatics. Nat Med 2001, 7: 186–91.
118. Stacker SA, Baldwin ME, Achen MG. The role of tumor lymphangiogenesis in metastatic spread. Faseb J 2002, 16: 922–34.
119. Stearns ME, Wang M, Hu Y, Kim G, Garcia FU. Expression of a flt-4 (VEGFR3) splicing variant in primary human prostate tumors. VEGF D and flt-4t(Delta773–1081) overexpression is diagnostic for sentinel lymph node metastasis. Lab Invest 2004, 84: 785–95.
120. Suzuki K, Morita T, Tokue A. Vascular endothelial growth factor-C (VEGF-C) expression predicts lymph node metastasis of transitional cell carcinoma of the bladder. Int J Urol 2005, 12: 152–8.
121. Takahashi A, Kono K, Itakura J, Amemiya H, Feng Tang R, Iizuka H, Fujii H, Matsumoto Y. Correlation of vascular endothelial growth factor-C expression with tumor-infiltrating dendritic cells in gastric cancer. Oncology 2002, 62: 121–7.
122. Takeuchi H, Bilchik A, Saha S, Turner R, Wiese D, Tanaka M, Kuo C, Wang HJ, Hoon DS. c-MET expression level in primary colon cancer: a predictor of tumor invasion and lymph node metastases. Clin Cancer Res 2003, 9: 1480–8.
123. Tammela T, Saaristo A, Lohela M, Morisada T, Tornberg J, Norrmen C, Oike Y, Pajusola K, Thurston G, et al. Angiopoietin-1 promotes lymphatic sprouting and hyperplasia. Blood 2005, 105: 4642–8.
124. Tamura M, Ohta Y. Serum vascular endothelial growth factor-C level in patients with primary nonsmall cell lung carcinoma: a possible diagnostic tool for lymph node metastasis. Cancer 2003, 98: 1217–22.
125. Ueda M, Terai Y, Yamashita Y, Kumagai K, Ueki K, Yamaguchi H, Akise D, Hung YC, Ueki M. Correlation between vascular endothelial growth factor-C expression and invasion phenotype in cervical carcinomas. Int J Cancer 2002, 98: 335–43.
126. Ueda M, Hung YC, Terai Y, Kanda K, Kanemura M, Futakuchi H, Yamaguchi H, Akise D, Yasuda M, et al. Vascular endothelial growth factor-C expression and invasive phenotype in ovarian carcinomas. Clin Cancer Res 2005, 11: 3225–32.
127. Vacca A, Ria R, Ribatti D, Semeraro F, Djonov V, Di Raimondo F, Dammacco F. A paracrine loop in the vascular endothelial growth factor pathway triggers tumor angiogenesis and growth in multiple myeloma. Haematologica 2003, 88: 176–85.
128. Van den Eynden GG, Vandenberghe MK, van Dam PJ, Colpaert CG, van Dam P, Dirix LY, Vermeulen PB, Van Marck EA. Increased sentinel lymph node lymphangiogenesis is associated with nonsentinel axillary lymph node involvement in breast cancer patients with a positive sentinel node. Clin Cancer Res 2007, 13: 5391–7.
129. Van Trappen PO, Steele D, Lowe DG, Baithun S, Beasley N, Thiele W, Weich H, Krishnan J, Shepherd JH, et al. Expression of vascular endothelial growth factor (VEGF)-C and VEGF-D, and their receptor VEGFR-3, during different stages of cervical carcinogenesis. J Pathol 2003, 201: 544–54.
130. Veikkola T, Jussila L, Makinen T, Karpanen T, Jeltsch M, Petrova TV, Kubo H, Thurston G, McDonald DM, et al. Signalling via vascular endothelial growth factor receptor-3 is sufficient for lymphangiogenesis in transgenic mice. Embo J 2001, 20: 1223–31.
131. Vihinen PP, Hilli J, Vuoristo MS, Syrjanen KJ, Kahari VM, Pyrhonen SO. Serum VEGF-C is associated with metastatic site in patients with malignant melanoma. Acta Oncol 2007, 46: 678–84.
132. Von Marschall Z, Scholz A, Stacker SA, Achen MG, Jackson DG, Alves F, Schirner M, Haberey M, Thierauch KH, et al. Vascular endothelial growth factor-D induces lymphangiogenesis and lymphatic metastasis in models of ductal pancreatic cancer. Int J Oncol 2005, 27: 669–79.

133. White JD, Hewett PW, Kosuge D, McCulloch T, Enholm BC, Carmichael J, Murray JC. Vascular endothelial growth factor-D expression is an independent prognostic marker for survival in colorectal carcinoma. Cancer Res 2002, 62: 1669–75.
134. Wiley HE, Gonzalez EB, Maki W, Wu MT, Hwang ST. Expression of CC chemokine receptor-7 and regional lymph node metastasis of B16 murine melanoma. J Natl Cancer Inst 2001, 93: 1638–43.
135. Wong SY, Haack H, Crowley D, Barry M, Bronson RT, Hynes RO. Tumor-secreted vascular endothelial growth factor-C is necessary for prostate cancer lymphangiogenesis, but lymphangiogenesis is unnecessary for lymph node metastasis. Cancer Res 2005, 65: 978–98.
136. Yan C, Zhang J, Ji J, Yu Y, Zhang Y, Ji Y, Liu B, Zhu Z, Yin H, et al. Vascular Endothelial Growth Factor C Expression in Gastric Carcinoma and Its Relationship with Lymph Node Metastasis. Chin Ger J Clin Oncol 2004, 3: 74–77.
137. Yan C, Zhu ZG, Yu YY, Ji J, Zhang Y, Ji YB, Yan M, Chen J, Liu BY, et al. Expression of vascular endothelial growth factor C and chemokine receptor CCR7 in gastric carcinoma and their values in predicting lymph node metastasis. World J Gastroenterol 2004, 10: 783–90.
138. Yasuoka H, Nakamura Y, Zuo H, Tang W, Takamura Y, Miyauchi A, Nakamura M, Mori I, Kakudo K. VEGF-D expression and lymph vessels play an important role for lymph node metastasis in papillary thyroid carcinoma. Mod Pathol 2005, 18: 1127–33.
139. Yokoyama Y, Charnock-Jones DS, Licence D, Yanaihara A, Hastings JM, Holland CM, Emoto M, Sakamoto A, Sakamoto T, et al. Expression of vascular endothelial growth factor (VEGF)-D and its receptor, VEGF receptor 3, as a prognostic factor in endometrial carcinoma. Clin Cancer Res 2003, 9: 1361–9.
140. Yonemura Y, Endo Y, Fujita H, Fushida S, Ninomiya I, Bandou E, Taniguchi K, Miwa K, Ohoyama S, et al. Role of vascular endothelial growth factor C expression in the development of lymph node metastasis in gastric cancer. Clin Cancer Res 1999, 5: 1823–9.
141. Yu XM, Lo CY, Chan WF, Lam KY, Leung P, Luk JM. Increased expression of vascular endothelial growth factor C in papillary thyroid carcinoma correlates with cervical lymph node metastases. Clin Cancer Res 2005, 11: 8063–9.
142. Yuanming L, Feng G, Lei T, Ying W. Quantitative analysis of lymphangiogenic markers in human gastroenteric tumor. Arch Med Res 2007, 38: 106–12.
143. Zeng Y, Opeskin K, Baldwin ME, Horvath LG, Achen MG, Stacker SA, Sutherland RL, Williams ED. Expression of vascular endothelial growth factor receptor-3 by lymphatic endothelial cells is associated with lymph node metastasis in prostate cancer. Clin Cancer Res 2004, 10: 5137–44.
144. Zhang B, Zhao WH, Zhou WY, Yu WS, Yu JM, Li S. Expression of vascular endothelial growth factors-C and -D correlate with evidence of lymphangiogenesis and angiogenesis in pancreatic adenocarcinoma. Cancer Detect Prev 2007, 31: 436–42.
145. Zhang J, Ji J, Yuan F, Zhu L, Yan C, Yu YY, Liu BY, Zhu ZG, Lin YZ. Cyclooxygenase-2 expression is associated with VEGF-C and lymph node metastases in gastric cancer patients. Biomed Pharmacother 2005, 59 Suppl 2: S285–8.
146. Zhang XH, Huang DP, Guo GL, Chen GR, Zhang HX, Wan L, Chen SY. Coexpression of VEGF-C and COX-2 and its association with lymphangiogenesis in human breast cancer. BMC Cancer 2008, 8: 4.
147. Zu X, Tang Z, Li Y, Gao N, Ding J, Qi L. Vascular endothelial growth factor-C expression in bladder transitional cell cancer and its relationship to lymph node metastasis. BJU Int 2006, 98: 1090–3.

Chapter 5

THE CLINICAL SIGNIFICANCE OF LYMPH-NODE METASTASIS

Mark B. Faries and Donald L. Morton
Division of Surgical Oncology, The Roy E. Coats Research Laboratories, John Wayne Cancer Institute, Saint John's Health Center, Santa Monica, CA 90404, USA, e-mail: fariesm@jwci.org

Abstract: Lymph nodes are the initial site of metastasis in many cancers. For patients with clinically localized tumors, the pathologic status of regional nodes is the most important prognostic variable, and techniques to evaluate these nodes radiographically and surgically are critical components in management. Advances in nodal evaluation, particularly sentinel node biopsy, have enabled more accurate nodal evaluation with less morbidity. The therapeutic impact of lymph node dissection remains a subject of controversy in several solid tumors, but growing evidence suggests that early removal of microscopically involved lymph nodes improves the long-term outcome of patients. Nodal metastasis should remain a central focus of clinical research to build on these recent discoveries.

Key words: Lymph nodes metastases · Prognosis · Lymph node staging · Lymph node imaging · Lymph node dissection

Cancer cannot progress to a fatal metastatic phenotype unless it escapes its site of origin and travels to a new anatomic site where it successfully establishes a focus of malignant growth. Because this new site is often the regional lymph nodes, evaluation of these nodes becomes pivotal for prognostic assessment and treatment planning.

As might be expected, the frequency of nodal metastases at the time of presentation varies considerably among tumor types but tends to be lower when the anatomic site or screening technique facilitates early detection. (Fig. 5.1) Perhaps more surprising is the absence of a uniform correlation between frequency and prognostic significance of nodal metastases. (Fig. 5.2) For example, only 12% of melanoma patients present with regional metastasis, but they have a 33% absolute decrease in survival, whereas the 34% of thyroid cancer patients with nodal involvement have only a 2.8% decrease in survival.

S.A. Stacker, M.G. Achen (eds.), *Lymphangiogenesis in Cancer Metastasis*, Cancer Metastasis – Biology and Treatment 13, DOI 10.1007/978-90-481-2247-9_5,

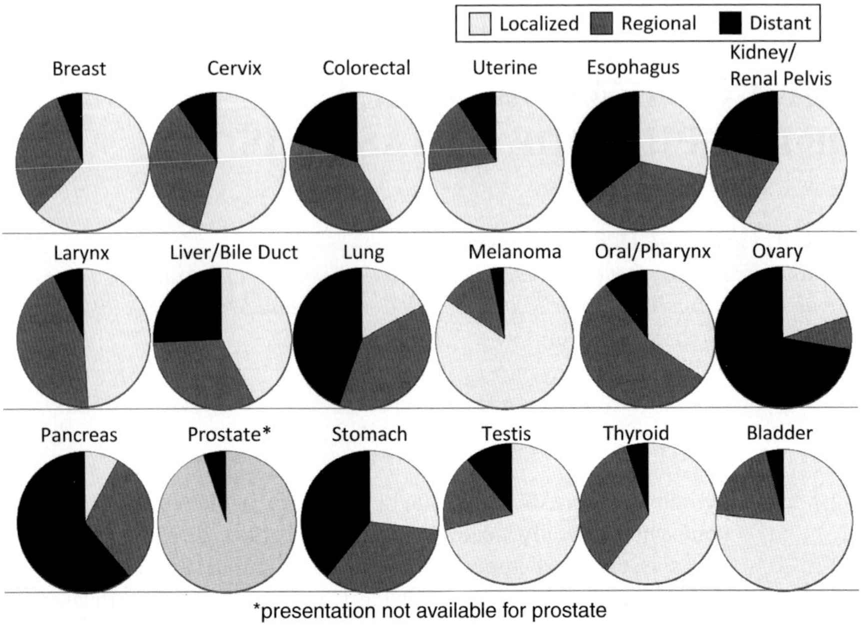

Fig. 5.1 Disease distribution at presentation: Localized, Regional, Distant. Data taken from SEER Summary Statistics. http://seer.cancer.gov/csr/1975_2004/sections.html Accessed September 8, 2007

5.1 Clinical Evaluation of Regional Nodes: Palpation and Imaging Techniques

Lymph nodes can be examined by palpation, ultrasonography (with or without fine needle aspiration biopsy), computed tomography (CT), magnetic resonance imaging (MRI), or positron emission tomography (PET). Physical examination is simplest and least invasive; results depend on location of the nodes, size of their metastatic foci and habitus of the patient. Although most nodal metastases detected by palpation are relatively large, smaller metastases may be palpated in thinner patients and in superficial locations such as the neck and groin.

Ultrasonography may improve staging accuracy, depending on the skill and experience of the examiner, the specifications of the equipment (high frequency probes and Doppler capability) and the criteria used to identify suspicious lymph nodes [1]. Suspicious characteristics include a rounded (as opposed to elongated) nodal shape with a length to width ratio of less than 2, loss of the normal fatty hilus, asymmetric cortex, increased vascular signature, and/or a hypoechoic parenchymal focus [2–4]. The absolute size of a node appears to be relatively unimportant [5]. The feasibility of ultrasonography depends on anatomic location, although endoscopic

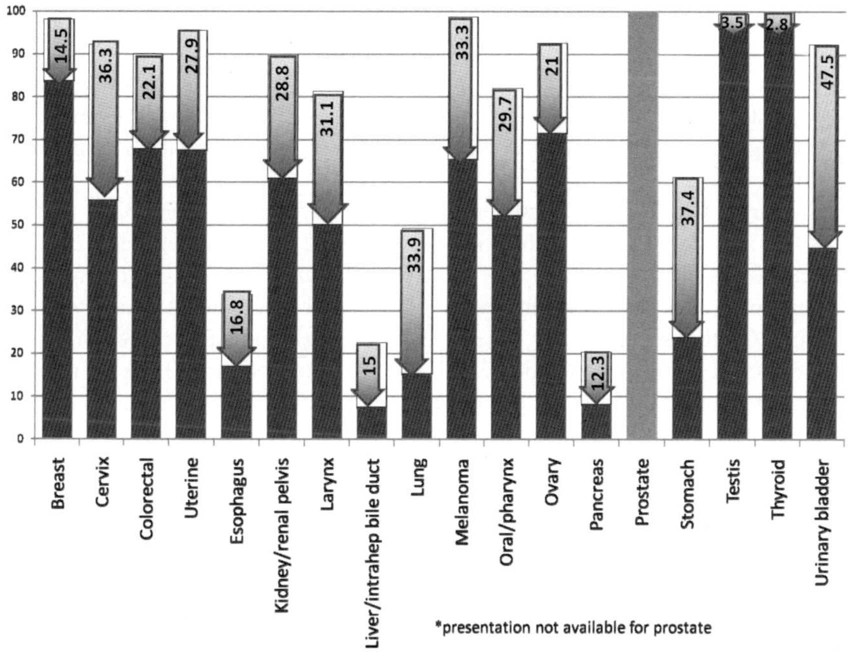

Fig. 5.2 Impact on five-year survival of lymph node metastasis (Localized → Regional). Data taken from SEER Summary Statistics. http://seer.cancer.gov/csr/1975_2004/sections.html Accessed September 8, 2007

probes can be used for many deep locations such as perirectal nodes. A relatively low cost and the absence of ionizing radiation are additional advantages to the test.

CT scans often provide useful information regarding the size of lymph nodes in the basins of interest. However, as noted above, size is often poorly correlated with the presence of metastasis. Though benign characteristics such as a normal fatty hilus are sometimes seen by CT, functional evaluation is not optimal. Contrast enhancement is present in only about one third of involved nodes [6].

MRI can demonstrate the presence of lymph nodes, and in some clinical situations such as rectal cancer, it is standard for initial nodal staging. PET scanning appears useful only for detection of lesions at least 1 cm in diameter and only for lesions with high rates of glucose metabolism.

In general, the accuracy of all imaging techniques depends in part on the skill of the operator and the quality of the equipment. Ultrasound is the least expensive, but is most operator dependent. CT and MRI scans are significantly more expensive, and quality varies with the generation of scanner used. However, they are less operator dependent and can examine more anatomic sites simultaneously. In general, palpation is a component of all initial physical examinations and imaging studies are tailored to each clinical scenario.

5.1.1 Breast Cancer

Palpation of axillary and supraclavicular nodal basins is essential for evaluation of patients with breast cancer. However, because of its relatively low sensitivity and specificity for metastasis [7], palpation generally should not be the sole basis for treatment decisions.

The sensitivity of preoperative ultrasonography ranges from 32% to 82% [2, 8–11]. A recent study of 726 patients found that 21% of all metastases could be detected using ultrasonography (8% of the entire population), potentially expediting this group's treatment by avoiding nodal biopsy prior to a complete axillary dissection [9]. Another report of 209 patients showed ultrasonographic detection of 56% of metastases (27% of population) [12]. Fine needle aspiration (FNA) is required before complete axillary dissection. Ultrasound-guided core biopsies have also been studied; in a group of 179 patients, metastases were found by ultrasound-guided core biopsy in 55 patients (31%) [13]. Ultrasonography may be more efficient in patients with relatively large primary tumors, probably due to the increased likelihood of nodal involvement [14, 15].

PET has a reported sensitivity of 79–83% and specificity of 95–100% for locally advanced tumors [16, 17]. However its sensitivity and specificity drop to 61% and 80%, respectively, in patients with smaller primary tumors [18] PET is therefore not recommended for routine preoperative staging of breast cancer. CT scans have only moderate sensitivity (73%) for preoperative staging of breast cancer [19].

5.1.2 Melanoma

Palpation of regional nodes is important for initial staging of melanoma and for postoperative monitoring. Because postoperative clinical examination is often paired with ultrasonography, the sensitivity of palpation varies widely [3, 20–25]. In this comparative setting, even a modest improvement through imaging results in a marked decrease in the measured sensitivity of palpation.

Ultrasonography is also used preoperatively, with reported sensitivities of 39% to 94% [26–30]. This variability, which may reflect differences in the size of nodal metastases and the definition of a positive test, means that ultrasound-guided needle biopsy eliminates only about 10% of sentinel node biopsies. A study from the Sydney Melanoma Unit used lymphoscintigraphy for focused ultrasonographic evaluation of nodal basins. Metastases as small as .45 mm were seen, but disease was frequently not visible when tumor foci were less than 4–4.5 mm in diameter [4]. The second Multicenter Selective Lymphadenectomy Trial (MSLT II) will determine the ability of ultrasonography to detect metastases prior to sentinel node biopsy.

As indicated above, ultrasonography appears to be significantly more sensitive than physical examination for postoperative monitoring. Although high sensitivities (> 90%) have been reported [31, 32, 52, 124], these reports only consider patients at the time of recurrence. Thus the high sensitivity indicates ultrasound is as good or better than other clinical staging techniques, but does not necessarily indicate a low risk for nodal recurrence in the future. MSLT II will prospectively evaluate nodal ultrasonography for follow-up of patients who have positive sentinel nodes but do not undergo completion lymphadenectomy.

5.1.3 Head and Neck Cancer

With a reported sensitivity of 48–76%, palpation should not be used to avoid neck dissection or limit the extent of dissection [32–35]. Similarly, ultrasonography cannot reliably rule out the presence of nodal disease [36]; its sensitivity for initial nodal staging is 48–90% in head and neck cancer [32,35,37,38] and only 20–54% in thyroid and oral cancer [39]. However, ultrasonography is superior to palpation for metastases smaller than 1.5 cm [40], and its sensitivity during postoperative follow-up is 93%, as compared with 77% for palpation [41].

Either CT or MRI scanning may be part of the standard preoperative work up for tumors of the salivary gland, oropharynx, hypopharynx, nasopharynx, and larynx. Sensitivity ranges from 38% to 84%, with a meta-analysis finding of 83% [32–34, 37,38,42–44]. MRI does not appear to be superior to contrast-enhanced CT [45].

PET scanning has high sensitivity (96–100%) and has upstaged as many as 20.8% of cancers initially assessed by standard imaging techniques [42, 44, 46]. PET scanning may also be useful during follow-up [47].

5.1.4 Lung Cancer

As might be expected, palpation of lymph nodes has been less commonly studied in lung cancer. One study reported that the mean diameter of supraclavicular nodes detected by palpation was 25.2 mm, and palpation achieved 50% sensitivity only for nodes at least 22.3 mm in size [48]. CT scanning is standard for initial staging [49]. Sensitivity is moderate (57–77%), and mediastinal nodes suspicious for metastasis should be sampled by mediastinoscopy or other technique to confirm pathologic abnormality [50–52]. In the presence of suspicious hilar lymph nodes, mediastinal nodal sampling is also recommended prior to surgical resection. PET scanning has demonstrated superior staging performance for mediastinal lymph nodes over CT scanning alone. The utility of PET appears to be relatively high in patients with larger tumors and remains an area of investigation in the earliest lung cancers [53,211].

Both endoscopic ultrasonography (EUS) and endobronchial ultrasonography, with or without FNA, have been used to evaluate mediastinal lymph nodes, usually after CT scanning. Sensitivity in this setting is 84–94% [51, 54–56]. Approximately 25% of patients with a normal CT scan reportedly will have ultrasound evidence of disease [57]. However, ultrasound cannot evaluate either the primary tumor or non-nodal sites of metastasis and therefore cannot replace CT scanning.

5.1.5 *Esophageal Cancer*

The initial staging evaluation for patients with esophageal cancer includes CT scanning, which has a relatively low sensitivity for nodal metastasis (8%) but helps define the primary tumor and indicate the possible presence of distant metastases [58]. Endoscopic ultrasound appears to be better than either CT or PET scanning for the staging of the primary tumor and regional nodes. FNA is an important component of the ultrasound evaluation. EUS is also operator dependent and access may be limited in the setting of esophageal stricture [59]. PET scanning appears to add a modest incremental improvement to CT for nodal evaluation, and is superior for systemic staging [212]. Some centers also utilize laparoscopic staging which has demonstrated similar accuracy to EUS [60].

5.1.6 *Rectal Cancer*

In addition to CT scanning, EUS or MRI with endorectal coil is standard for nodal staging in rectal cancer, and provides additional information regarding the tumor stage [61,62]. Accuracy of nodal staging is considerably higher with EUS than with CT scan alone (85% vs. 45%) [63]. The ability to detect nodal metastases seems to increase with the primary tumor's stage. The overall staging utility of MRI is similar to that of EUS. There may be some advantage in primary tumor staging for EUS and nodal staging for MRI, but these technologies are both site- and operator-dependent [62,213].

5.1.7 *Genitourinary Tumors*

In vulvar cancer, clinical examination of inguinal lymph nodes has a sensitivity of only 35% preoperatively and 72% intraoperatively. However, the sensitivity of CT scanning also is not impressive (59%). Ultrasonography has better sensitivity (87%) but lower specificity [64]. Because MRI has a reported sensitivity of 85.7% and a negative predictive value of 93.9%, it may be useful to select patients who could avoid surgical nodal evaluation [65].

In prostate cancer, nodal imaging with EUS or endorectal MRI has not been successful; sensitivity is 60% or less [66]. Operative evaluation remains an essential part of nodal evaluation for these patients.

5.2 Surgical Evaluation of Lymph Nodes

When results of noninvasive techniques are not definitive, nodal sampling may be necessary to determine nodal status. Needle biopsy or excisional surgical biopsy is practical for clinically evident nodal disease. If the regional lymph node basins are clinically normal, nodal assessment technique depends on the type of malignancy. Lymph nodes proximal to primary colon cancer are usually removed en bloc with the primary tumor. By contrast, regional nodes that drain breast cancer and melanoma must be removed during a separate procedure. In this case, if the risk of regional metastasis is very low, clinical nodal observation is reasonable. Most sarcomas and certain other tumors rarely metastasize to regional lymph nodes.

If surgical evaluation of the nodes is necessary, the traditional approach was complete dissection of the basin that drains the primary tumor. However, elective complete lymphadenectomy of a clinically normal lymph node basin introduces the risk of considerable morbidity and often proves to be unnecessary because most patients will not have nodal metastasis. This dilemma led to development of a less invasive but highly selective nodal sampling technique called sentinel node biopsy (SNB). The sentinel node (SN) is the first lymph node on the direct drainage pathway from a primary tumor. Because it is the first node to receive drainage from the primary tumor, its tumor status should reflect that of the entire basin; if the SN is tumor-free, lymphatic metastasis is highly unlikely.

5.2.1 Sentinel Node Biopsy

SNB is preceded by injection of lymphatic mapping agents at the primary tumor site. These agents, which may be dyes or radiopharmaceuticals, drain through lymphatic channels to the SN. The SN is identified intraoperatively by visualization of the dye in a node and/or by measurement of radioactivity levels over the node. Each dye-stained and/or radioactive SN is excised for pathologic evaluation. If all SNs are free of tumor, the nodal basin should be tumor-free and the risk of distant metastasis is low. Conversely, the presence of SN metastasis introduces the possibility of metastasis to other nodes in the same basin and indicates an increased likelihood of distant spread.

When compared to complete lymphadenectomy, SNB is significantly less morbid [67, 68] and significantly more accurate. This is not surprising because removal of fewer nodes decreases morbidity, while histopathologic scrutiny of a smaller but higher-risk specimen improves detection.

The tumor status of the SN has been found to be the most important prognostic variable in clinically localized melanoma [69] and breast cancer [70–72], although the prognostic significance of very small breast cancer metastases is unclear (see below). Even when en bloc resection of lymph nodes with the primary tumor eliminates the potential for additional procedural morbidity, as is the case for pulmonary and gastrointestinal malignancies, intraoperative (in vivo) or even postoperative (ex vivo) mapping of the SN may improve staging by allowing the pathologist to focus on a smaller number of lymph nodes.

Accurate results with SNB require experience. In the earliest reports of the technique in melanoma, a dye-stained SN was identified in 82% of basins [73]. With experience the success of dye-directed SN mapping increased to 95%; addition of radioactive tracers further improved identification rates. The current identification rate at experienced centers is over 99% [74]. In the Multicenter Selective Lymphadenectomy Trial I (MSLT-I) in melanoma the rate of false-negative SNs decreased from 10.3% among the first 25 cases to 5.3% thereafter [67]. This suggests that a 50-case learning curve may be necessary for optimal lymphatic mapping of the SNs that drain primary cutaneous melanoma. In breast cancer there is more debate about the length and even the existence of a learning curve; a minimum of 25 cases has been suggested to achieve competency [75, 76].

In summary, SNB was validated initially for melanoma and subsequently for breast cancer; it is now standard for those diseases. SNB may prove to have a role for staging colon, thyroid, head and neck squamous cell, genitourinary and non-melanoma skin cancers. SNB has also been investigated for gastric and esophageal cancers, as a staging tool and as a method of tailoring nodal dissection based upon lymphatic drainage patterns [77, 78]. Finally, SNB may improve staging of lung cancer [79–81]. As mentioned above, it generally does not change the operation performed, although it might reduce the extent of nodal sampling [82].

5.3 Pathologic Evaluation of Lymph Nodes

The accuracy of nodal staging reflects not only the nature and number of nodes removed but also the techniques for pathologic evaluation of these nodes. The processing of complete nodal dissection specimens [83, 84] begins with palpation to identify any enlarged nodes. After 6–8 hours of fixation, the nodes are dissected free of surrounding tissue. Care should be taken to identify small nodes (< 5 mm), which are often the only site of disease. Each node is bisected and stained with hematoxylin and eosin (H&E). More extensive sectioning or staining techniques are not practical for the large number of nodes removed during complete lymphadenectomy.

Because the SNB specimen is so much smaller than a complete lymphadenectomy specimen, each SN can be bivalved and then cut into further sections that are processed for standard H&E or for immunohistochemical (IHC) staining. IHC stains are chosen according to tumor type. Anti-cytokeratin antibodies are used to stain adenocarcinomas; concomitant staining for E-cadherin may reduce false-positive results [85]. Anti-S100 antibodies are the most sensitive stain for melanoma but they also stain nevi and dendritic cells and therefore should be used with antibodies to Mart-1, HMB45 and/or tyrosinase [83].

Focused pathologic examination of the nodes has increased identification of nodal micrometastases (0.2 and 2 mm) and isolated tumor cells (ITC; < 0.2 mm), but the prognostic significance of ITC is unknown in breast cancer [86–88] and not certain in melanoma – although retrospective data indicate that ITC may adversely affect melanoma-specific survival [89]. The prognostic significance of tumor foci detected by molecular techniques is also unclear, probably reflecting variations in technique. Reverse transcriptase polymerase chain reaction (RT-PCR) assays for nodal metastasis of breast cancer, colon cancer, melanoma, and oral/oropharyngeal cancer have utilized different methods of tissue sampling, different tissue preparations (fresh frozen vs. fixed) and different molecular markers. In general, RT-PCR assays based on multiple markers are more sensitive than those based on a single marker [90], and use of formalin-fixed, paraffin-embedded specimens is preferable to fresh frozen tissue.

As with histopathologic techniques, molecular techniques appear to have more prognostic relevance in melanoma [91] than in breast cancer [92, 93]. In MSLT II, a multimarker RT-PCR assay is being used to assess paraffin-embedded nodal tissue in patients with melanoma. Results of RT-PCR assays for colon cancer are mixed; a recent meta-analysis reported prognostic significance for RT-PCR assays but not for IHC staining [94]. At present, the most likely advantage for RT-PCR assessment of SNs associated with several solid tumors is its negative predictive value; patients whose nodes are tumor-negative by both histopathologic and molecular techniques are extremely unlikely to develop recurrent disease [95, 96, 214]. However, additional prospective studies will be required to fully establish its utility in clinical practice.

5.4 Prognostic Impact of Nodal Disease

The presence of tumor in regional nodes demonstrates a tumor's ability to exit the primary site, traffic to another location, extravasate and survive. From a clinical standpoint, nodal status impacts the management of many cancers. The prognostic impact of regional metastasis is not directly related its incidence, so that the presence of such disease may have prognostic importance even if it occurs rarely.

5.4.1 Breast Cancer

Approximately one third of breast cancers are initially diagnosed with lymph node metastasis. The risk of nodal disease is related to the primary tumor's size (Fig. 5.3), grade, histology, ploidy, hormone receptor expression, lymphovascular invasion and location in the breast; and to the patient's age and ethnicity (Tables 5.1 and 5.2) [97–101]. Other factors such as MIB-1 index and family history may prove to be relevant. Unfortunately, no factor or combination of factors can reliably exclude patients from nodal staging.

The tumor status of regional nodes is the most powerful predictor of outcome for primary breast cancer (Table 5.3) [102]. Data from the Surveillance Epidemiology and End Results (SEER) registries show that 5-year survival drops from 98% to 84% if nodes are involved. Because the number of tumor-involved nodes is directly related to mortality, AJCC staging guidelines use prognostic categories of 0, 1–3, 4–9 and 10 or more tumor-involved nodes (Fig. 5.4) [103].

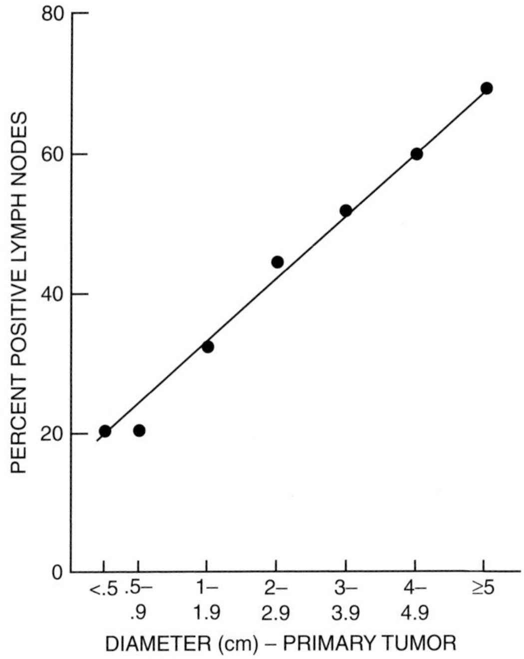

Fig. 5.3 Rate of axillary lymph node metastasis as a function of primary tumor diameter. From Carter CL, Allen C, Henson D. "Relation of tumor size, lymph node status, and survival in 24,740 breast cancer cases." *Cancer*. 63(1): 181–7, 1989

Table 5.1 Association between Incidence of Axillary Node Metastases and 11 Clinical/Pathologic Factors by Univariate and Multivariate Analysis. From Silverstein, *Cancer* 1997

Variable	Category	No.	% Lymph nodes positive	Univariate *P* value	Multivariate *P* value
Lymph/vascular invasion	Yes	116	46%	<0.0001	0.0000001
	No	754	19%		
Tumor palpable	Yes	656	28%	<0.0001	0.00004
	No	262	10%		
Nuclear grade	1	148	9%	<0.0001	0.0004
	2	510	21%		
	3	237	33%		
Tumor size	T1a	92	4%	<0.0001	0.01
	T1b	245	17%		
	T1c	581	28%		

5.4.2 Cervical Cancer

SEER data show a 36% rate of regional metastases at initial diagnosis of cervical cancer. Nodal metastasis decreases 5-year survival rate from 92% to 56%. The impact on prognosis is independent of the primary tumor's size, grade, and histologic type and independent of the patient's age [104, 105]. Lymphatic or vascular invasion by the primary tumor is associated with a 5-fold increase in nodal metastasis [106].

Table 5.2 Variables Significantly Related to Nodal Involvement in Stepwise Logistic Regression. From Gajdos, *Ann Surg*, 1999

	Nodal Involvement (%)		
	No	Yes	*P* Value
Lymphatic Invasion			
No	545 (81)	124 (19)	< 0.001
Yes	89 (49)	92 (51)	
Tumor Size			
T1a	97 (92)	8 (8)	< 0.001
T1b	233 (85)	42 (15)	
T1c	304 (65)	166 (35)	
Age			
< 40	40 (64)	23 (37)	0.009
40–49	146 (74)	52 (26)	
50–59	167 (72)	66 (28)	
60–69	157 (82)	35 (18)	
70+	124 (76)	40 (24)	

Table 5.3 Ranking of Prognostic Variables in Selected Multivariate Analyses.* From Donegan, *CA Cancer J Clin*

		Ranking of Prognostic Variables				
References	No. of Cases	1	2	3	4	5
Contesso et al[216]	612	AX nodes	Grade†	Tumor size		
Fisher et al[37]	620	AX nodes	Tumor size	Nipple invol.		
Noguchi et al[217]	128	AX nodes	IM nodes			
Axelsson et al[218]	220	AX nodes	ER	Grade†		
Fisher et al[219]	1.531	AX nodes	Grade†	Age	ER	PR
Meyer and Province[220]	414	AX nodes	Tumor size	Log TLI	Log ER	
Shek and Godolphin[221]	859	AX nodes	Stage‡	ER	Necrosis	
Meyer and Province[222]	341	AX nodes	Nuclear size	Tumor size		
Wenger et al[223]	15,877	AX nodes	Tumor size	SPF	ER	PR
Pujol et al[224]	125	AX nodes	Cathepsin-D	PR		
NIH Consensus[225]	165	AX nodes	Ploidy	EGFR	HER-2	
Duffy et al[226]	230	AX nodes	HER-2	p53		
Mansour et al[227]	156	AX nodes	p53	ER	Grade	

*Based on survival, disease-free survival, or recurrence rates. Only selected variables were included in the studies.

†Grade = histologic or nuclear grade.

‡Stage = TNM stage.

AX = axillary; TLI = thymidine labeling index; ER = estrogen receptor; EGFR = epidermal growth factor receptor; PR = progesterone receptor

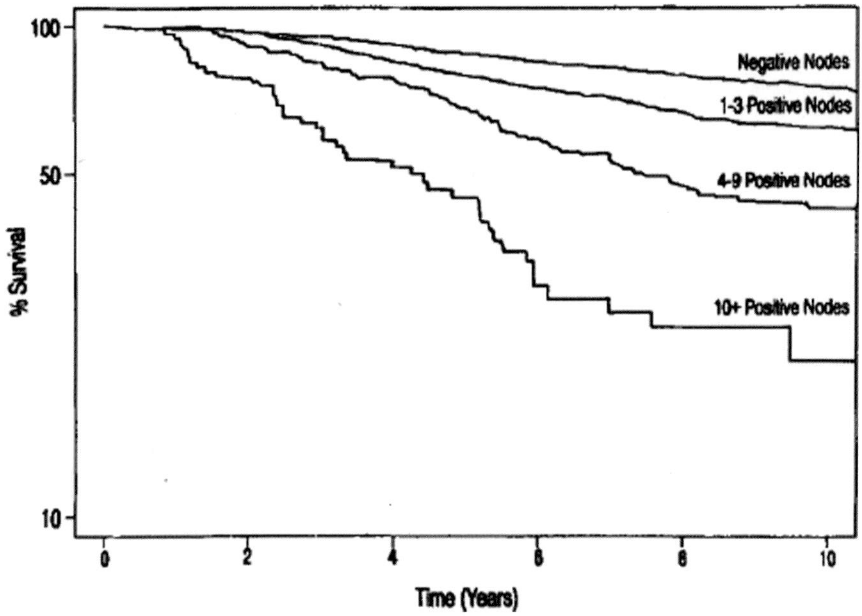

Fig. 5.4 Life table plots of survival according to nodal status (NSABP Protocol B–06). From Fisher ER, Constantino J, Fisher B, Redmond C. "Pathologic findings from the National Surgical Adjuvant Breast Project (Protocol 4). Discriminants for 15-year survival." *Cancer*.71(6 Suppl): 2141–50, 1993

5.4.3 Colorectal Cancer

Depth of bowel wall invasion and lymph node status are the most powerful prognostic variables in colorectal cancer. Approximately 38% of patients presenting with colorectal cancer in the United States will have nodal involvement, which decreases 5-year survival from 90% to 78%. The current AJCC staging system distinguishes three prognostic nodal categories: N0 (no involvement), N1 (1–3 nodes positive) and N2 (4 or more nodes positive) (Fig. 5.5) [107]. The risk of nodal disease increases not only with the depth of bowel wall invasion but also with decreased thymidylate synthase and a micropapillary histologic pattern [108, 109]. Small, regular deposits of tumor within the mesentery are considered lymph node metastases; irregular deposits are considered vascular invasion. Nodal micrometastasis of colorectal cancer may be linked to increased risk of recurrence [94, 96] and may influence a decision regarding adjuvant therapy.

5.4.4 Uterine

Approximately one in five patients presenting with uterine cancer will have nodal metastases. The risk of nodal metastasis is higher for carcinomas than mesenchymal

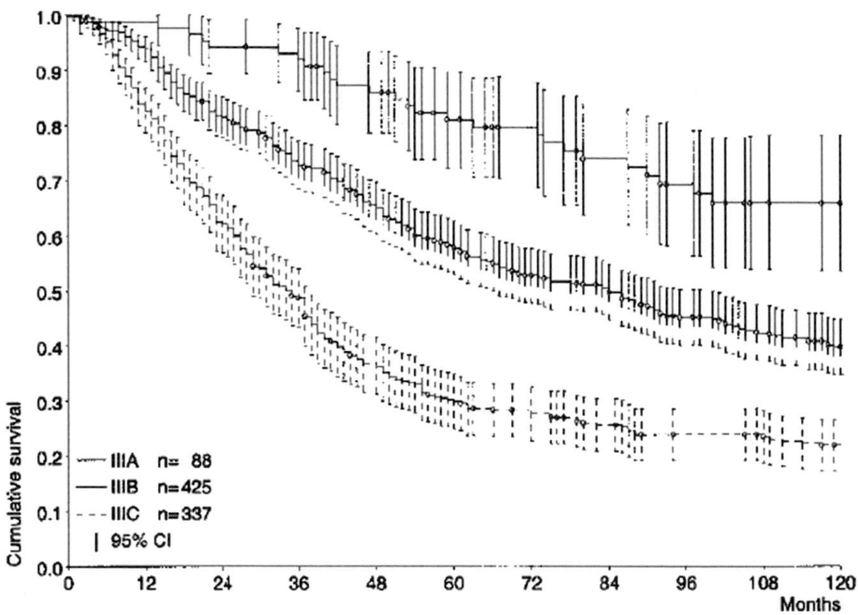

Fig. 5.5 Survival curves (observed with 95% CI) for substages based on pTpN. Data of the ER-CRC. Substage IIIA: pT1,2 pN1 M0, n = 88. Substage IIIB: pt3,4 pN1 M0 and pT1,2 pN2 M0, n = 425. Substage IIIC: pT3,4 pN2 M0, n = 337. Merkel S, Mansmann U, Papadopuoulos T, Wittekind C, Hohenberger W, Hermanek P. "The prognostic inhomogeneity of colorectal carcinomas Stage III: a proposal for subdivision of Stage III." *Cancer*. 92(11): 2754–9, 2001

tumors, and higher for non-endometrioid tumors than endometrioid tumors [110]. Other risk factors include lymphovascular space invasion, myometrial invasion, cervical invasion, and tumor diameter [111–113]. Involvement of the pelvic nodes increases the risk of para-aortic metastasis, which decreases survival [114]. Five-year survival rate is 60% with subclinical nodal metastases but only 20% with clinically positive nodes [115].

5.4.5 Esophagus

Two-thirds of patients with esophageal cancer will have nodal or distant metastases at diagnosis. Approximately half of those have regional node metastasis without distant disease. Regional nodes are defined as cervical, intrathoracic esophageal, or gastroesophageal junction according to the site of the primary tumor; metastasis to nodes outside the corresponding field is distant disease. Nodal metastasis is associated with reduced overall 5-year survival from approximately 34% to 17%.

The primary tumor's level of invasion and possibly its histology influence the risk of nodal metastasis [116]. Squamous cell carcinoma has a greater risk of nodal disease, even at very early levels of invasion, and often these metastases are farther from the primary tumor [215].

5.4.6 Kidney

Approximately 20% of patients presenting with kidney or renal pelvis cancers will have nodal involvement, which decreases 5-year survival from approximately 90% to 60%. Regional metastasis to renal hilar, paracaval and aortic nodes is categorized as N1 (a single node) or N2 (more than one node). Retroperitoneal lymphadenopathy is associated with a worse outcome and is less likely to respond to immunotherapy with interleukin-2 [117, 118]. Although primary tumor characteristics, patient age and symptom type have been linked to nodal metastasis [119, 120], nodal staging is often not performed [119]. In a review of nodal staging, nearly 40% of patients were found to be pNx after no pathologic nodal staging [121].

5.4.7 Laryngeal, Oral, and Pharyngeal

Although laryngeal and oral/pharyngeal cancers are separated in the SEER database, survival rates are similar. Over 40% of Americans with laryngeal cancer present with regional metastasis, which decreases 5-year survival rate from 81% to 50%.

Predictors of nodal disease include tumor stage, degree of differentiation, host inflammatory response, epidermal growth factor receptor expression [122], DNA content [123], proliferating cell nuclear antigen, MIB-1, and E-cadherin [124, 125]. EGFR and cyclin D1 may also be predictive in oral cancer [126]. SNB is still under evaluation for head and neck cancer and may become standard in certain tumors [127].

Patients without clinically evident nodal disease should undergo neck dissection for definitive nodal staging, although some have proposed close clinical observation [128]. The extent of nodal dissection remains controversial; selective neck dissections (e.g. supraomohyoid dissection in oral cancer) are a potential alternative to comprehensive dissections. More than one involved lymph node is an indication for adjuvant radiotherapy; three or more involved nodes is predictive of distant metastases [129].

5.4.8 Liver/Intrahepatic Bile Duct

The 5-year survival rate for hepatocellular carcinoma (HCC), gallbladder cancer, and cholangiocarcinoma is 22% when disease is localized and 7% when it involves regional sites [130]. The majority of regionally advanced cases are related to direct local extension, but approximately 7% of patients undergoing liver transplantation for HCC will have hilar nodal disease. These patients have a higher risk of recurrence and death [131, 132].

In gallbladder cancer, approximately 45% of patients will have regional nodal involvement at presentation [133]. The approximately 13% rate of nodal metastasis for T1 lesions [134] increases with tumor stage.

Many patients diagnosed with intrahepatic cholangiocarcinoma will be determined to be unresectable due to the local extent of tumor. Among those who are deemed able to undergo resection, nodal metastasis is the most important prognostic factor [135, 136]. There are few survivors in the setting of nodal spread, which is associated with histologic type [137]. Nodal status is also an independent prognostic variable in distal bile duct carcinoma [138].

5.4.9 Lung

Thirty-five percent of patients are diagnosed with regional metastases. These patients have a 15% rate of 5-year survival, as compared with 49% for those with localized disease. Predictors of nodal disease include histology (adenocarcinoma or large cell), tumor location (central or right upper lobe), large tumor size, high tumor grade, and young age [139, 140].

The prognostic significance of nodal disease is determined by the location of involved nodes. N1 nodes are ipsilateral bronchopulmonary or hilar, N2 nodes are ipsilateral or subcarinal mediastinal, and N3 nodes are contralateral. These N stages determine initial treatment: surgery for N0 or N1 disease, neoadjuvant or definitive chemoradiation for N2 disease, and chemoradiation for N3 disease. Five-year survival rate is reportedly about 43% with 1–3 involved nodes, 30% with 4–14 nodes, and 12% with >14 nodes [141].

5.4.10 Melanoma

Nodal status is the most important prognostic factor in melanoma, and nodal metastasis is identified at initial diagnosis in approximately 12% of patients in the United States. Five-year survival is 98% without metastasis and 65% with nodal disease. Risk factors include thick primary tumors, younger age, male sex, ulceration and a high mitotic rate [142]. In addition, there may be an immune component to nodal metastasis; increased infiltration of lymphocytes into primary tumors appears to decrease risk of metastasis, whereas tumor-induced regional immunosuppression may favor metastasis [143, 144]. Nodal metastasis of melanomas < 1 mm thick is less common but nodal staging may be indicated in selected patients [145].

The prognostic implications of nodal disease (AJCC stage III melanoma) are based on the size and number of metastases (Fig. 5.6) [146]. A single positive node is staged as N1, two or three positive nodes are N2, and four or more nodes are N3. The presence of nodal metastasis diminishes the importance of other prognostic factors.

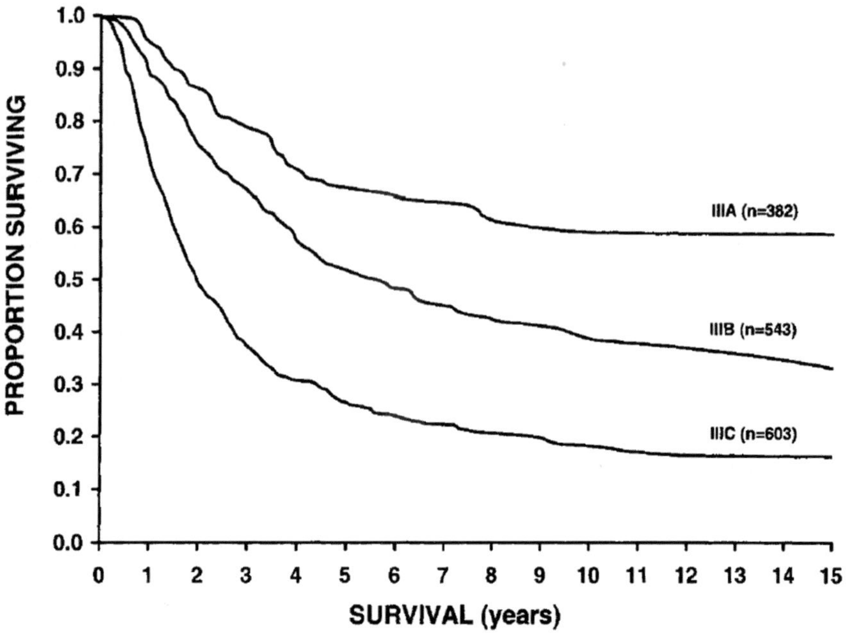

Fig. 5.6 Fifteen-year survival curves for the stage groupings of patients with regional metastatic melanoma (Stage III). Numbers of patients from the AJCC Melanoma Staging Database are shown in parentheses. The differences between the survival curves are highly significant ($P < 0.0001$). From Balch C, et al. "An evidence-based staging system for cutaneous melanoma." *CA Cancer J Clin.* 54: 131–49, 2004

5.4.11 Ovarian Cancer

Most (72%) ovarian cancers present with distant metastases; only about 7% are regionally limited. Regional metastasis of localized disease decreases survival by about 21% (to 71% from 92%) at 5 years.

Staging of ovarian cancer by the FIGO (Federation Internationale de Gynecologie et d'Obstetrique) system is determined by the extent of peritoneal involvement, the tumor status of the regional lymph nodes, and the presence of tumor at distant sites such as the liver parenchyma or pleural cavity. Stage IIIA or IIIB indicates peritoneal disease up to 2 cm in size beyond the pelvis; stage IIIC is peritoneal disease > 2 cm beyond the pelvis and/or involvement of regional lymph nodes, although the latter is somewhat controversial [147]. However, among patients whose disease would otherwise be stage I or II, nodal involvement has a significant impact on survival.

Predictors of regional nodal involvement include the primary tumor's extent and histology [148,149]. Serous tumors are more likely to metastasize to more proximal, para-aortic nodes, and non-serous tumors to pelvic nodes [150]. Approximately 10–20% of patients with disease clinically limited to the pelvis will have nodal involvement [151,152]. Among patients with larger-volume peritoneal disease, two-thirds have nodal involvement [153, 154].

5.4.12 Pancreatic Cancer

Similar to ovarian cancer, most (61%) patients with pancreatic cancer have distant metastatic disease at presentation. Of the remaining patients, three-fourths will have regional involvement. Nodal status is an important prognostic variable [155] that decreases 5-year survival to 8%, as compared with 20% for localized disease. Median survival after resection is approximately 24–33 months with negative nodes and 12–16 months with positive nodes [156–158].

5.4.13 Prostate Cancer

Many patients with prostate cancer do not undergo pathologic evaluation of regional nodes. Among those who undergo nodal staging during surgical resection of clinically localized disease, the rate of nodal involvement is about 4% to 25% [160,161]. The risk of nodal disease has been linked to PSA level > 10.5 ng/ml, Gleason score < 7 and clinical T stage [161–163]. Population factors may influence the risk of nodal disease. For example, European men reportedly have almost a ninefold higher risk of nodal disease than North American men [164].

The prognostic impact of nodal metastasis is directly related to the volume of disease in the node [165] and number of involved nodes [160, 166]. Nodal disease was previously felt to be equivalent to distant spread. However, approximately 20% of patients with positive nodes will not recur at 10 years, and 80% will not die of prostate cancer by that time [167]. By comparison, men with distant metastases have a 32% five-year survival.

5.4.14 Gastric Cancer

In the United States, approximately a third of gastric cancers are initially diagnosed with metastases limited to regional lymph nodes. Their 5-year survival is only 24%, as compared with 61% in the absence of metastasis. Both the site and the number of involved nodes have prognostic significance [168, 169].

Due to the high incidence of gastric cancer in Japan and Korea, more screening is performed and many patients are diagnosed with early lesions limited to the mucosa and/or submucosa. Is nodal dissection indicated for these early gastric cancers? Most studies concur that submucosal invasion, depth of invasion, size of primary tumor, and tumor differentiation are associated with nodal metastasis; lymphovascular invasion is possibly the strongest independent indicator [170–175]. Minimally invasive mucosal resection without formal gastrectomy has been proposed for small tumors (< 2–3 cm) limited to the mucosa.

5.4.15 Thyroid

SEER-based analyses suggest that more than a third of thyroid cancers are diagnosed with regional disease, but more than 50% of patients with well differentiated thyroid cancer do not undergo nodal evaluation [176]. In addition, it is not clear that all nodal metastases identified pathologically are clinically important. For example, Wada et al. demonstrated nodal disease in approximately 60% of patients with papillary microcarcinoma (< 1 cm primary tumor), but the rate of nodal recurrence among patients who did not undergo nodal evaluation was less than 1% at a mean follow-up of 53 months [177]. Moreover, SEER data indicate that nodal metastasis of localized disease decreases 5-year survival by less than 3%. Among patients with well differentiated thyroid cancer, nodal metastasis plus older age tends to decrease disease-free survival but not necessarily overall survival. The AJCC/UICC staging system does not use nodal status for patients with well differentiated cancers if the patient is less than 45 years old. Nodal staging is also not considered for anaplastic cancers. However lymph node involvement corresponds to stage III disease in older (≥ 45 years) patients with papillary and follicular cancers and in patients with medullary cancers,

5.4.16 Sarcoma

Nodal metastasis is relatively uncommon in most sarcomas. Large series report rates of less than 4% [178, 179]. The frequency of nodal metastasis is higher for angiosarcoma (11–13%), rhabdomyosarcoma (13–19%), epithelioid sarcoma (16–20%), clear cell sarcoma (melanoma of soft parts) (10–20%) [178–180], and possibly synovial sarcoma [181]. Rates of nodal metastasis for osteosarcoma are 6–10% [182, 183]. By comparison, nodal metastasis is seen in less than 3% of liposarcoma, synovial cell sarcoma, chondrosarcoma, and malignant fibrous histiocytoma.

Lymph node metastasis of sarcoma is classified as stage IV disease, although postoperative rates of survival reportedly reach 34% at 5 years or even 71% at 4 years for isolated disease in the nodes [178, 179]. Resection should be accompanied by SNB when tumor histology is associated with relatively high rates of nodal disease [184].

5.5 Therapeutic Impact of Lymphadenectomy

This is the most controversial aspect of lymph node metastasis. The therapeutic impact of lymphadenectomy theoretically depends on whether lymph nodes represent an early site of protected tumor growth prior to systemic spread (the so-called incubator hypothesis) or whether nodal metastasis is merely an indicator of more aggressive tumor biology (marker hypothesis). If the incubator hypothesis is correct, early excision of involved regional lymph nodes will improve survival; if the marker hypothesis is correct, early excision/evaluation merely provides prognostic information.

5.5.1 Number of Evaluated Lymph Nodes

The number of resected/evaluated lymph nodes has been linked with the clinical outcome of patients with colorectal cancer [185, 186], breast cancer [187, 188], cervical cancer [189], invasive bladder cancer [190], pancreatic cancer [157, 191, 192], esophageal cancer [193,194], and gastric cancer [195]. In colorectal cancer, a SEER-based study [185] suggested 15 lymph nodes as a reasonable prognostic cutoff. The number of resected nodes also impacts recommendations for adjuvant therapy in colon cancer; guidelines of the American Society of Clinical Oncology recommend consideration of adjuvant chemotherapy for patients with stage II colon cancer if fewer than 12 lymph nodes are pathologically evaluated [196]. In breast cancer, data from early randomized multicenter trials showed a direct relationship between number of evaluated nodes and outcome [187]. These findings have been reproduced in numerous subsequent studies [188, 197].

Recent trends suggest an increase in the number of nodes evaluated, possibly reflecting an impact of recent treatment guidelines and the use of node counts as surrogate quality measures by third party payers. However, nodal count reflects not only the extent of surgical resection and the thoroughness of pathologic examination, but also intrinsic differences in lymphatic function (Fig. 5.10) and physiology. Certain patients may have fewer lymph nodes or at least fewer identifiable lymph nodes.

5.5.2 Randomized Trials

Among the earliest studies, Fisher et al. randomized women with breast cancer to modified radical mastectomy, total mastectomy, or segmental mastectomy with adjuvant radiation [198]. Although there was an increase in locoregional recurrence among patients who did not undergo radical mastectomy, there was no statistically significant difference in overall survival. However, more recent data in breast cancer suggest that locoregional control is an important determinant of overall survival [199].

In esophageal cancer, survival has been examined after two- vs. three-field nodal dissection, and after transhiatal resection vs. transthoracic resection that includes en bloc nodal dissection. Three-field dissection has relatively high morbidity and is not widely used, although it might yield better survival than two-field dissection [200, 201]. There is no definitive evidence for a survival difference between transhiatal and transthoracic esophagectomy (Fig. 5.7) [202], although the latter technique reportedly improves survival among patients with 1 to 8 positive lymph

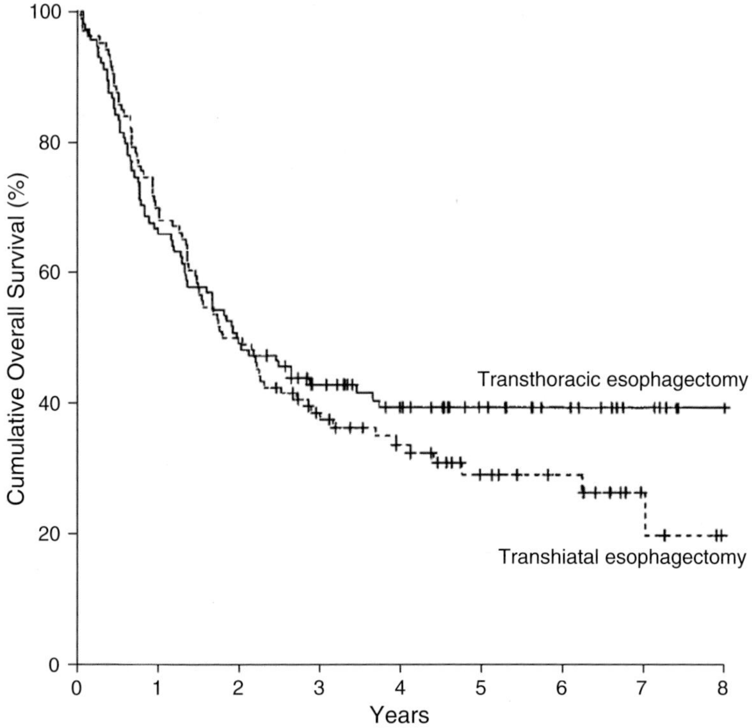

No. at Risk								
Transhiatal esophagectomy	106	74	53	35	25	16	11	4
Transthoracic esophagectomy	114	76	57	42	31	20	14	7

Fig. 5.7 Kaplan-Meier Curves Showing Overall Survival among Patients Randomly Assigned to Transhiatal Esophagectomy or Transthoracic Esophagectomy with Extended en Bloc Lymphadenectomy. Hulscher JB, et al. "Extended transthoracic resection compared with limited transhiatal resection for adenocarcinoma of the esophagus." *N Engl J Med.* 347(21): 1662–9, 2002

nodes (23% vs. 64% at 5 years.) [203]. However, the transthoracic approach has higher operative morbidity.

Japan remains at the forefront of surgical treatment for gastric cancer; nodal dissection tends to be more extensive. In the West, the debate has centered on D1 versus D2 nodal dissection. D1 dissection routinely removes only the perigastric lymph nodes, whereas D2 dissection also removes nodes along the left gastric, celiac, hepatic and splenic arteries. The Dutch Gastric Cancer Group reported that the operative mortality of D2 dissection diminished its potential therapeutic benefit (Fig. 5.8) [204]. However, extended dissections may be of benefit if morbidity and mortality can be avoided.

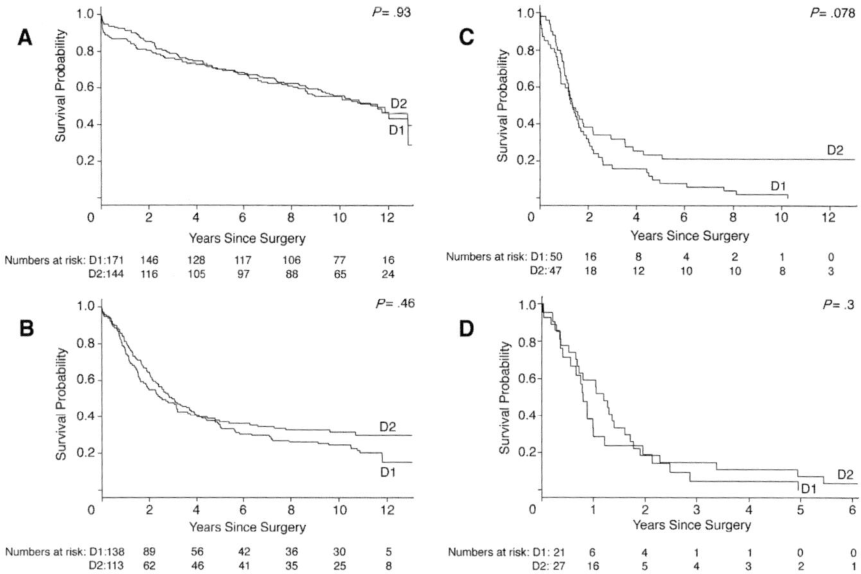

Fig. 5.8 Survival of patients treated with curative intent according to N stage. (A), N0; (B), N1; (C), N2; (D), N3. D1, limited lymph node dissection group; D2, extended lymph node dissection group. From Hartgrink HH, et al. "Extended lymph node dissection for gastric cancer: who may benefit? Final results of the randomized Dutch gastric cancer group trial." *J Clin Oncol.* 22(11): 2069–77, 2004

More extensive nodal dissection might actually impair the outcome of patients with pancreatic cancer due to increased perioperative morbidity [205, 206]. Randomized trial data suggest that although a pylorus-preserving procedure removes fewer lymph nodes it does not compromise survival [207]. The number of lymph nodes evaluated may impact staging accuracy.

Total mesorectal excision (TME) has dramatically improved locoregional control and overall survival of patients with rectal cancer. TME, which uses the natural anatomic boundaries of the mesorectum to ensure complete extirpation of regional nodes, is associated with local recurrence rates of only 4–9% and 5-year survival rates of 62–75% [208]. It the setting of such dramatically improved local recurrence rates, it would be difficult to ethically justify conducting a randomized trial to prove the technique. In addition, due to the decreased blood loss and increased ease of dissection with TME, a randomized trial seems unnecessary [209].

By contrast, the results of randomized trials are anxiously awaited to settle the controversy regarding nodal management of clinically localized melanoma. Routine complete lymphadenectomy is not satisfactory because approximately 80% of these patients will not have histopathologic evidence of nodal involvement. To a certain extent, Breslow thickness of the primary melanoma can be used to identify candidates for lymphadenectomy: patients with intermediate-thickness melanoma are most likely to benefit; those with thin melanoma have a very low risk of nodal metastasis, whereas those with thick lesions have a relatively high risk of concurrent systemic metastasis. The Intergroup Melanoma Surgical Trial randomized patients with melanoma 1–4 mm in thickness to elective lymphadenectomy or nodal observation. The trial showed a trend to overall survival benefit for the entire study (p = 0.12) and demonstrated significant benefit in several subgroups: patients with extremity melanoma, those with non-ulcerated melanoma, those 60 years old or younger, and those with tumors 1–2 mm in thickness [210].

The advent of SNB, as described earlier in this chapter, changed the approach to the regional lymph nodes by allowing accurate staging with minimal morbidity. Results of MSLT-I demonstrated a significant increase in disease-free survival, in part by reducing regional nodal recurrences, but, as of the third interim analysis, have not shown an overall survival benefit for the entire population of patients whose nodes are managed by SNB instead of observation. However, among the subgroup of patients with tumor-positive SNB specimens versus clinical evidence of nodal recurrence during observation, the risk of death from melanoma was reduced almost 50% (Fig. 5.9). The trial also confirmed the prognostic impact of sentinel node status, and SNB is now standard care for patients with intermediate-

B

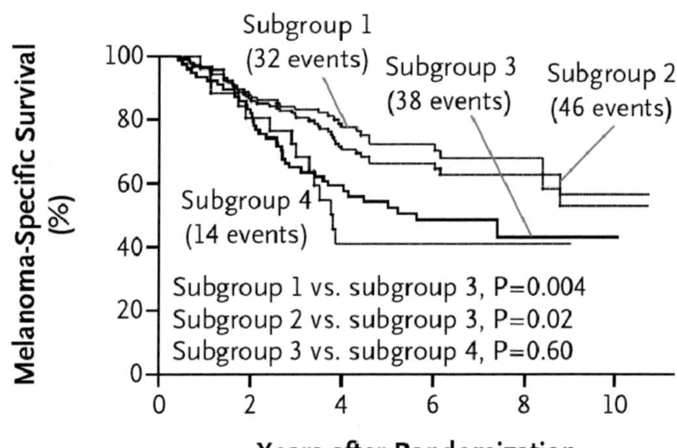

No. at Risk

Subgroup 1	122	100	65	38	15	2
Subgroup 2	148	120	73	43	18	2
Subgroup 3	78	63	37	23	5	1
Subgroup 4	26	20	8	5	3	0

Fig. 5.9 Melanoma-specific survival among patients with nodal metastases. Subgroup 1 comprised patients with a tumor-positive sentinel node; subgroup 2, the patients in subgroup 1 plus those in subgroup 4 with a nodal recurrence after a negative result on biopsy; subgroup 3, those with nodal recurrence during observation; and subgroup 4, those with nodal recurrence after a negative result on biopsy. From Morton DL, et al. "Sentinel-node biopsy or nodal observation in melanoma." *N Engl J Med*. 355(13): 1307–17, 2006 copyright permission of the Massachusetts Medical Society

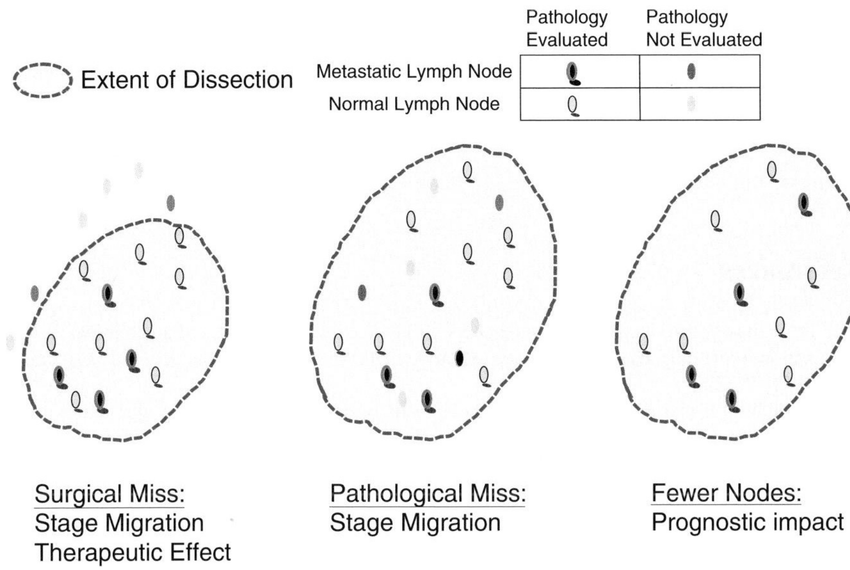

Fig. 5.10 Schematic representing reasons for association between node count and prognosis: (1) surgical miss: Inadequate surgery leaves potentially diseased nodes behind. In this setting patients may also be understaged. So improved outcomes related to increased node counts result may result from a therapeutic effect of improved surgery or through stage migration. (2) Pathological miss: surgery is adequate, but not all nodes are found pathologically. All of the improved outcome is related to stage migration. (3) Host factor: Some patients may have fewer lymph nodes. This may be a prognostic factor per se

thickness melanoma. Because most patients with a positive SNB specimen do not have further (non-sentinel) nodal involvement, MSLT II is accruing patients to determine if there is a therapeutic effect of completion lymph node dissection in patients whose sentinel node contains histopathologic/molecular evidence of tumor.

5.6 Conclusion

It is impossible to summarize the current diagnostic, prognostic and therapeutic implications of lymph node metastasis in a single chapter, especially since the field is changing so rapidly. Each improvement in technology increases the sensitivity of detection, decreases the morbidity of assessment, and/or expands the prognostic armamentarium. The only constant in this field is the unquestioned importance of lymph node status as a predictor of outcome for most solid tumors. Removal of metastases limited to regional lymph nodes will remain standard care in most malignancies; the timing and extent of dissection continue to be a topic of controversy and active research.

Acknowledgments Supported by grant CA29605 from the National Cancer Institute and by funding from the Amyx Foundation, Inc. (Boise, ID), the Wayne and Gladys Valley Foundation (Oakland, CA), Mr. and Mrs. Louis Johnson, (Stanfield, AZ), Berton M. Kirshner (Los Angeles, CA), Todd Kirshner (Los Angeles, CA), Mr. and Heather and Jim Murren (Las Vegas, NV), Mrs. Marianne Reis (Lake Forest, CA), and the Wallis Foundation (Los Angeles, CA).

References

1. de Freitas R, Jr., Costa MV, Schneider SV, et al. Accuracy of ultrasound and clinical examination in the diagnosis of axillary lymph node metastases in breast cancer. Eur J Surg Oncol 1991, 17: 240–4.
2. Kebudi A, Caliskan C, Yetkin G, et al. The role of pre-operative B mode ultrasound in the evaluation of the axillary lymph node metastases in the initial staging of breast carcinoma. Acta Chir Belg 2005, 105: 511–4.
3. Saiag P, Bernard M, Beauchet A, et al. Ultrasonography using simple diagnostic criteria vs palpation for the detection of regional lymph node metastases of melanoma. Arch Dermatol 2005, 141: 183–9.
4. Starritt EC, Uren RF, Scolyer RA, et al. Ultrasound examination of sentinel nodes in the initial assessment of patients with primary cutaneous melanoma. Ann Surg Oncol 2005, 12: 18–23.
5. Atula TS, Varpula MJ, Kurki TJ, et al. Assessment of cervical lymph node status in head and neck cancer patients: palpation, computed tomography and low field magnetic resonance imaging compared with ultrasound-guided fine-needle aspiration cytology. Eur J Radiol 1997, 25: 152–61.
6. van den Brekel MW, Pameijer FA, Koops W, et al. Computed tomography for the detection of neck node metastases in melanoma patients. Eur J Surg Oncol 1998, 24: 51–4.
7. Lanng C, Hoffmann J, Galatius H, Engel U. Assessment of clinical palpation of the axilla as a criterion for performing the sentinel node procedure in breast cancer. Eur J Surg Oncol 2007, 33: 281–4.
8. Mathijssen IM, Strijdhorst H, Kiestra SK, Wereldsma JC. Added value of ultrasound in screening the clinically negative axilla in breast cancer. J Surg Oncol 2006, 94: 364–7.
9. van Rijk MC, Deurloo EE, Nieweg OE, et al. Ultrasonography and fine-needle aspiration cytology can spare breast cancer patients unnecessary sentinel lymph node biopsy. Ann Surg Oncol 2006, 13: 31–5.
10. Brancato B, Zappa M, Bricolo D, et al. Role of ultrasound-guided fine needle cytology of axillary lymph nodes in breast carcinoma staging. Radiol Med (Torino) 2004, 108: 345–55.
11. Kuenen-Boumeester V, Menke-Pluymers M, de Kanter AY, et al. Ultrasound-guided fine needle aspiration cytology of axillary lymph nodes in breast cancer patients. A preoperative staging procedure. Eur J Cancer 2003, 39: 170–4.
12. Gilissen F, Oostenbroek R, Storm R, et al. Prevention of futile sentinel node procedures in breast cancer: ultrasonography of the axilla and fine-needle aspiration cytology are obligatory. Eur J Surg Oncol 2008, 34: 497–500.
13. Nathanson SD, Burke M, Slater R, Kapke A. Preoperative identification of the sentinel lymph node in breast cancer. Ann Surg Oncol 2007, 14: 3102–10.
14. de Kanter AY, van Eijck CH, van Geel AN, et al. Multicentre study of ultrasonographically guided axillary node biopsy in patients with breast cancer. Br J Surg 1999, 86: 1459–62.
15. Sato K, Tamaki K, Tsuda H, et al. Utility of axillary ultrasound examination to select breast cancer patients suited for optimal sentinel node biopsy. Am J Surg 2004, 187: 679–83.
16. Avril N, Adler LP. F-18 fluorodeoxyglucose-positron emission tomography imaging for primary breast cancer and loco-regional staging. Radiol Clin North Am 2007, 45: 645–57, vi.

17. Mustafa S, Marshall C, Griffiths PA, et al. 18F-FDG dual-headed gamma camera PET in detection of axillary nodal disease in patients with large or locally advanced breast cancer: Possible alternative staging of axilla. Oncol Rep 2007, 18: 1545–9.
18. Wahl RL, Siegel BA, Coleman RE, Gatsonis CG. Prospective multicenter study of axillary nodal staging by positron emission tomography in breast cancer: a report of the staging breast cancer with PET Study Group. J Clin Oncol 2004, 22: 277–85.
19. Pamilo M, Soiva M, Lavast EM. Real-time ultrasound, axillary mammography, and clinical examination in the detection of axillary lymph node metastases in breast cancer patients. J Ultrasound Med 1989, 8: 115–20.
20. Bafounta ML, Beauchet A, Chagnon S, Saiag P. Ultrasonography or palpation for detection of melanoma nodal invasion: a meta-analysis. Lancet Oncol 2004, 5: 673–80.
21. Brountzos EN, Panagiotou IE, Bafaloukos DI, Kelekis DA. Ultrasonographic detection of regional lymph node metastases in patients with intermediate or thick malignant melanoma. Oncol Rep 2003, 10: 505–10.
22. Ulrich J, Voit C. Ultrasound in dermatology. Part II. Ultrasound of regional lymph node basins and subcutaneous tumours. Eur J Dermatol 2001, 11: 73–9.
23. Voit C, Kron M, Schafer G, et al. Ultrasound-guided fine needle aspiration cytology prior to sentinel lymph node biopsy in melanoma patients. Ann Surg Oncol 2006, 13: 1682–9.
24. Voit C, Mayer T, Kron M, et al. Efficacy of ultrasound B-scan compared with physical examination in follow-up of melanoma patients. Cancer 2001, 91: 2409–16.
25. Machet L, Nemeth-Normand F, Giraudeau B, et al. Is ultrasound lymph node examination superior to clinical examination in melanoma follow-up? A monocentre cohort study of 373 patients. Br J Dermatol 2005, 152: 66–70.
26. van Rijk MC, Teertstra HJ, Peterse JL, et al. Ultrasonography and fine-needle aspiration cytology in the preoperative evaluation of melanoma patients eligible for sentinel node biopsy. Ann Surg Oncol 2006, 13: 1511–6.
27. Sibon C, Chagnon S, Tchakerian A, et al. The contribution of high-resolution ultrasonography in preoperatively detecting sentinel-node metastases in melanoma patients. Melanoma Res 2007, 17: 233–7.
28. Hocevar M, Bracko M, Pogacnik A, et al. The role of preoperative ultrasonography in reducing the number of sentinel lymph node procedures in melanoma. Melanoma Res 2004, 14: 533–6.
29. Rossi CR, Mocellin S, Scagnet B, et al. The role of preoperative ultrasound scan in detecting lymph node metastasis before sentinel node biopsy in melanoma patients. J Surg Oncol 2003, 83: 80–4.
30. Testori A, Lazzaro G, Baldini F, et al. The role of ultrasound of sentinel nodes in the pre- and post-operative evaluation of stage I melanoma patients. Melanoma Res 2005, 15: 191–8.
31. Dalle S, Paulin C, Lapras V, et al. Fine-needle aspiration biopsy with ultrasound guidance in patients with malignant melanoma and palpable lymph nodes. Br J Dermatol 2006, 155: 552–6.
32. Haberal I, Celik H, Gocmen H, et al. Which is important in the evaluation of metastatic lymph nodes in head and neck cancer: palpation, ultrasonography, or computed tomography? Otolaryngol Head Neck Surg 2004, 130: 197–201.
33. Hao SP, Ng SH. Magnetic resonance imaging versus clinical palpation in evaluating cervical metastasis from head and neck cancer. Otolaryngol Head Neck Surg 2000, 123: 324–7.
34. Kau RJ, Alexiou C, Stimmer H, Arnold W. Diagnostic procedures for detection of lymph node metastases in cancer of the larynx. ORL J Otorhinolaryngol Relat Spec 2000, 62: 199–203.
35. Knappe M, Louw M, Gregor RT. Ultrasonography-guided fine-needle aspiration for the assessment of cervical metastases. Arch Otolaryngol Head Neck Surg 2000, 126: 1091–6.
36. Hoft S, Muhle C, Brenner W, et al. Fine-needle aspiration cytology of the sentinel lymph node in head and neck cancer. J Nucl Med 2002, 43: 1585–90.

37. Rottey S, Petrovic M, Bauters W, et al. Evaluation of metastatic lymph nodes in head and neck cancer: a comparative study between palpation, ultrasonography, ultrasound-guided fine needle aspiration cytology and computed tomography. Acta Clin Belg 2006, 61: 236–41.
38. Akoglu E, Dutipek M, Bekis R, et al. Assessment of cervical lymph node metastasis with different imaging methods in patients with head and neck squamous cell carcinoma. J Otolaryngol 2005, 34: 384–94.
39. Kouvaraki MA, Shapiro SE, Fornage BD, et al. Role of preoperative ultrasonography in the surgical management of patients with thyroid cancer. Surgery 2003, 134: 946–54; discussion 954–5.
40. Eichhorn T, Schroeder HG, Glanz H, Schwerk WB. [Histologically controlled comparison of palpation and sonography in the diagnosis of cervical lymph node metastases]. Laryngol Rhinol Otol (Stuttg) 1987, 66: 266–74.
41. Steinkamp HJ, Knobber D, Schedel H, et al. [Palpation and sonography in after-care of head-neck tumor patients: comparison of ultrasound tumor entity parameters]. Laryngorhinootologie 1993, 72: 431–8.
42. Myers LL, Wax MK. Positron emission tomography in the evaluation of the negative neck in patients with oral cavity cancer. J Otolaryngol 1998, 27: 342–7.
43. Takes RP, Righi P, Meeuwis CA, et al. The value of ultrasound with ultrasound-guided fine-needle aspiration biopsy compared to computed tomography in the detection of regional metastases in the clinically negative neck. Int J Radiat Oncol Biol Phys 1998, 40: 1027–32.
44. Stokkel MP, ten Broek FW, Hordijk GJ, et al. Preoperative evaluation of patients with primary head and neck cancer using dual-head 18fluorodeoxyglucose positron emission tomography. Ann Surg 2000, 231: 229–34.
45. Anzai Y, Brunberg JA, Lufkin RB. Imaging of nodal metastases in the head and neck. J Magn Reson Imaging 1997, 7: 774–83.
46. Kresnik E, Mikosch P, Gallowitsch HJ, et al. Evaluation of head and neck cancer with 18F-FDG PET: a comparison with conventional methods. Eur J Nucl Med 2001, 28: 816–21.
47. Salaun P, Abgral R, Querellou S, et al. Does (18) fluoro-fluorodeoxyglucose positron emission tomography improve recurrence detection in patients treated for head and neck squamous cell carcinoma with negative clinical follow-up? Head Neck 2007, 29: 1115–20.
48. van Overhagen H, Brakel K, Heijenbrok MW, et al. Metastases in supraclavicular lymph nodes in lung cancer: assessment with palpation, US, and CT. Radiology 2004, 232: 75–80.
49. Ettinger D, Bepler G, Cox J, et al. Non-small cell lung cancer. NCCN: Practice Guidelines Oncol 2002, 1.
50. Gress FG, Savides TJ, Sandler A, et al. Endoscopic ultrasonography, fine-needle aspiration biopsy guided by endoscopic ultrasonography, and computed tomography in the preoperative staging of non-small-cell lung cancer: a comparison study. Ann Intern Med 1997, 127: 604–12.
51. Fritscher-Ravens A, Bohuslavizki KH, Brandt L, et al. Mediastinal lymph node involvement in potentially resectable lung cancer: comparison of CT, positron emission tomography, and endoscopic ultrasonography with and without fine-needle aspiration. Chest 2003, 123: 442–51.
52. Yasufuku K, Nakajima T, Motoori K, et al. Comparison of endobronchial ultrasound, positron emission tomography, and CT for lymph node staging of lung cancer. Chest 2006, 130: 710–8.
53. Lardinois D, Weder W, Hany TF, et al. Staging of non-small-cell lung cancer with integrated positron-emission tomography and computed tomography. N Engl J Med 2003, 348: 2500–7.
54. Eloubeidi MA, Cerfolio RJ, Chen VK, et al. Endoscopic ultrasound-guided fine needle aspiration of mediastinal lymph node in patients with suspected lung cancer after positron emission tomography and computed tomography scans. Ann Thorac Surg 2005, 79: 263–8.
55. Wallace MB, Silvestri GA, Sahai AV, et al. Endoscopic ultrasound-guided fine needle aspiration for staging patients with carcinoma of the lung. Ann Thorac Surg 2001, 72: 1861–7.
56. Rintoul RC, Skwarski KM, Murchison JT, et al. Endobronchial and endoscopic ultrasound-guided real-time fine-needle aspiration for mediastinal staging. Eur Respir J 2005, 25: 416–21.

57. Wallace MB, Ravenel J, Block MI, et al. Endoscopic ultrasound in lung cancer patients with a normal mediastinum on computed tomography. Ann Thorac Surg 2004, 77: 1763–8.
58. Reed CE, Mishra G, Sahai AV, et al. Esophageal cancer staging: improved accuracy by endoscopic ultrasound of celiac lymph nodes. Ann Thorac Surg 1999, 67: 319–21; discussion 322.
59. Vazquez-Sequeiros E, Wiersema MJ, Clain JE, et al. Impact of lymph node staging on therapy of esophageal carcinoma. Gastroenterology 2003, 125: 1626–35.
60. Kaushik N, Khalid A, Brody D, et al. Endoscopic ultrasound compared with laparoscopy for staging esophageal cancer. Ann Thorac Surg 2007, 83: 2000–2.
61. Landmann R, Wong W, Hoepfl J, et al. Limitations of early rectal cancer nodal staging may explain failure after local excision. Dis Colon Rectum 2007, 50: 1520–5.
62. Bianchi P, Ceriani C, Palmisano A, et al. A prospective comparison of endorectal ultrasound and pelvic magnetic resonance in the preoperative staging of rectal cancer. Ann Ital Chir 2006, 77: 41–6.
63. Shami VM, Parmar KS, Waxman I. Clinical impact of endoscopic ultrasound and endoscopic ultrasound-guided fine-needle aspiration in the management of rectal carcinoma. Dis Colon Rectum 2004, 47: 59–65.
64. Land R, Herod J, Moskovic E, et al. Routine computerized tomography scanning, groin ultrasound with or without fine needle aspiration cytology in the surgical management of primary squamous cell carcinoma of the vulva. Int J Gynecol Cancer 2006, 16: 312–7.
65. Singh K, Orakwue CO, Honest H, et al. Accuracy of magnetic resonance imaging of inguinofemoral lymph nodes in vulval cancer. Int J Gynecol Cancer 2006, 16: 1179–83.
66. Artifon EL, Sakai P, Ishioka S, et al. EUS for locoregional staging of prostate cancer – a pilot study. Gastrointest Endosc 2007, 65: 440–7.
67. Morton DL, Cochran AJ, Thompson JF, et al. Sentinel node biopsy for early-stage melanoma: accuracy and morbidity in MSLT-I, an international multicenter trial. Ann Surg 2005, 242: 302–11; discussion 311–3.
68. Lucci A, McCall LM, Beitsch PD, et al. Surgical complications associated with sentinel lymph node dissection (SLND) plus axillary lymph node dissection compared with SLND alone in the American College of Surgeons Oncology Group Trial Z0011. J Clin Oncol 2007, 25: 3657–63.
69. Gershenwald JE, Thompson W, Mansfield PF, et al. Multi-institutional melanoma lymphatic mapping experience: the prognostic value of sentinel lymph node status in 612 stage I or II melanoma patients. J Clin Oncol 1999, 17: 976–83.
70. Vanderveen KA, Schneider PD, Khatri VP, et al. Upstaging and improved survival of early breast cancer patients after implementation of sentinel node biopsy for axillary staging. Ann Surg Oncol 2006, 13: 1450–6.
71. Marrazzo A, Taormina P, Gebbiab V, et al. Is sentinel lymph node biopsy more accurate than axillary dissection for staging nodal involvement in breast cancer patients? Chir Ital 2007, 59: 693–9.
72. Wong SL, Chao C, Edwards MJ, et al. The use of cytokeratin staining in sentinel lymph node biopsy for breast cancer. Am J Surg 2001, 182: 330–4.
73. Morton DL, Wen DR, Wong JH, et al. Technical details of intraoperative lymphatic mapping for early stage melanoma. Arch Surg 1992, 127: 392–9.
74. Morton DL, Thompson JF, Essner R, et al. Validation of the accuracy of intraoperative lymphatic mapping and sentinel lymphadenectomy for early-stage melanoma: a multicenter trial. Multicenter Selective Lymphadenectomy Trial Group. Ann Surg 1999, 230: 453–63; discussion 463–5.
75. Cox CE, Bass SS, Boulware D, et al. Implementation of new surgical technology: outcome measures for lymphatic mapping of breast carcinoma. Ann Surg Oncol 1999, 6: 553–61.
76. Clarke D, Newcombe RG, Mansel RE. The learning curve in sentinel node biopsy: the ALMANAC experience. Ann Surg Oncol 2004, 11: 211S–5S.

77. Kitagawa Y, Kitano S, Kubota T, et al. Minimally invasive surgery for gastric cancer – toward a confluence of two major streams: a review. Gastric Cancer 2005, 8: 103–10.
78. Kitajima M, Kitagawa Y. Surgical treatment of esophageal cancer – the advent of the era of individualization. N Engl J Med 2002, 347: 1705–9.
79. Faries MB, Bleicher RJ, Ye X, et al. Lymphatic mapping and sentinel lymphadenectomy for primary and metastatic pulmonary malignant neoplasms. Arch Surg 2004, 139: 870–6; discussion 876–7.
80. Liptay MJ. Sentinel node mapping in lung cancer. Ann Surg Oncol 2004, 11: 271S–4S.
81. Rzyman W, Hagen OM, Dziadziuszko R, et al. Intraoperative, radio-guided sentinel lymph node mapping in 110 nonsmall cell lung cancer patients. Ann Thorac Surg 2006, 82: 237–42.
82. Muraoka M, Akamine S, Oka T, et al. Sentinel node sampling limits lymphadenectomy in stage I non-small cell lung cancer. Eur J Cardiothorac Surg 2007, 32: 356–61.
83. Cochran AJ, Starz H, Ohsie SJ, et al. Pathologic reporting and special diagnostic techniques for melanoma. Surg Oncol Clin N Am 2006, 15: 231–51.
84. Cserni G. Pathological evaluation of sentinel lymph nodes. Surg Oncol Clin N Am 2007, 16: 17–34.
85. Turner RR, Nora DT, Trocha SD, Bilchik AJ. Colorectal carcinoma nodal staging. Frequency and nature of cytokeratin-positive cells in sentinel and nonsentinel lymph nodes. Arch Pathol Lab Med 2003, 127: 673–9.
86. Chen SL, Hoehne FM, Giuliano AE. The prognostic significance of micrometastases in breast cancer: a SEER population-based analysis. Ann Surg Oncol 2007, 14: 3378–84.
87. Grabau D, Jensen MB, Rank F, Blichert-Toft M. Axillary lymph node micrometastases in invasive breast cancer: national figures on incidence and overall survival. Apmis 2007, 115: 828–37.
88. Kuijt GP, Voogd AC, van de Poll-Franse LV, et al. The prognostic significance of axillary lymph-node micrometastases in breast cancer patients. Eur J Surg Oncol 2005, 31: 500–5.
89. Scheri RP, Essner R, Turner RR, et al. Isolated tumor cells in the sentinel node affect long-term prognosis of patients with melanoma. Ann Surg Oncol 2007, 14: 2861–6.
90. Martinez SR, Mori T, Hoon DS. Molecular upstaging of sentinel lymph nodes in melanoma: where are we now? Surg Oncol Clin N Am 2006, 15: 331–40.
91. Takeuchi H, Morton DL, Kuo C, et al. Prognostic significance of molecular upstaging of paraffin-embedded sentinel lymph nodes in melanoma patients. J Clin Oncol 2004, 22: 2671–80.
92. Gillanders WE, Mikhitarian K, Hebert R, et al. Molecular detection of micrometastatic breast cancer in histopathology-negative axillary lymph nodes correlates with traditional predictors of prognosis: an interim analysis of a prospective multi-institutional cohort study. Ann Surg 2004, 239: 828–37, discussion 837–40.
93. Sakaguchi M, Virmani A, Dudak MW, et al. Clinical relevance of reverse transcriptase-polymerase chain reaction for the detection of axillary lymph node metastases in breast cancer. Ann Surg Oncol 2003, 10: 117–25.
94. Iddings D, Ahmad A, Elashoff D, Bilchik A. The prognostic effect of micrometastases in previously staged lymph node negative (N0) colorectal carcinoma: a meta-analysis. Ann Surg Oncol 2006, 13: 1386–92.
95. Stojadinovic A, Nissan A, Protic M, et al. Prospective randomized study comparing sentinel lymph node evaluation with standard pathologic evaluation for the staging of colon carcinoma: results from the United States Military Cancer Institute Clinical Trials Group Study GI-01. Ann Surg 2007, 245: 846–57.
96. Bilchik AJ, Hoon DS, Saha S, et al. Prognostic impact of micrometastases in colon cancer: interim results of a prospective multicenter trial. Ann Surg 2007, 246: 568–75; discussion 575–7.
97. Gajdos C, Tartter PI, Bleiweiss IJ. Lymphatic invasion, tumor size, and age are independent predictors of axillary lymph node metastases in women with T1 breast cancers. Ann Surg 1999, 230: 692–6.

98. Silverstein MJ, Gierson ED, Waisman JR, et al. Predicting axillary node positivity in patients with invasive carcinoma of the breast by using a combination of T category and palpability. J Am Coll Surg 1995, 180: 700–4.
99. Barth A, Craig PH, Silverstein MJ. Predictors of axillary lymph node metastases in patients with T1 breast carcinoma. Cancer 1997, 79: 1918–22.
100. Rivadeneira DE, Simmons RM, Christos PJ, et al. Predictive factors associated with axillary lymph node metastases in T1a and T1b breast carcinomas: analysis in more than 900 patients. J Am Coll Surg 2000, 191: 1–6; discussion 6–8.
101. Orucevic A, Reddy VB, Bloom KJ, et al. Predictors of lymph node metastasis in T1 breast carcinoma, stratified by patient age. Breast J 2002, 8: 349–55.
102. Donegan WL. Tumor-related prognostic factors for breast cancer. CA Cancer J Clin 1997, 47: 28–51.
103. Fisher ER, Costantino J, Fisher B, Redmond C. Pathologic findings from the National Surgical Adjuvant Breast Project (Protocol 4). Discriminants for 15-year survival. National Surgical Adjuvant Breast and Bowel Project Investigators. Cancer 1993, 71: 2141–50.
104. Tsang RW, Juvet S, Pintilie M, et al. Pretreatment proliferation parameters do not add predictive power to clinical factors in cervical cancer treated with definitive radiation therapy. Clin Cancer Res 2003, 9: 4387–95.
105. Kosary CL. FIGO stage, histology, histologic grade, age and race as prognostic factors in determining survival for cancers of the female gynecological system: an analysis of 1973–87 SEER cases of cancers of the endometrium, cervix, ovary, vulva, and vagina. Semin Surg Oncol 1994, 10: 31–46.
106. Sakuragi N, Takeda N, Hareyama H, et al. A multivariate analysis of blood vessel and lymph vessel invasion as predictors of ovarian and lymph node metastases in patients with cervical carcinoma. Cancer 2000, 88: 2578–83.
107. Merkel S, Mansmann U, Papadopoulos T, et al. The prognostic inhomogeneity of colorectal carcinomas Stage III: a proposal for subdivision of Stage III. Cancer 2001, 92: 2754–9.
108. Haupt B, Ro JY, Schwartz MR, Shen SS. Colorectal adenocarcinoma with micropapillary pattern and its association with lymph node metastasis. Mod Pathol 2007, 20: 729–33.
109. Artinyan A, Essani R, Lake J, et al. Molecular predictors of lymph node metastasis in colon cancer: increased risk with decreased thymidylate synthase expression. J Gastrointest Surg 2005, 9: 1216–21; discussion 1221.
110. Kalogiannidis I, Lambrechts S, Amant F, et al. Role of lymphadenectomy and pelvic radiotherapy in patients with clinical FIGO stage I endometrial adenocarcinoma: an analysis of 208 patients. Int J Gynecol Cancer 2006, 16: 1885–93.
111. Kamura T, Yahata H, Shigematsu T, et al. Predicting pelvic lymph node metastasis in endometrial carcinoma. Gynecol Oncol 1999, 72: 387–91.
112. Cheewakriangkrai C, Panggid K, Siriaungkul S, et al. Lymphovascular space invasion as a prognostic determinant in uterine cancer. Asian Pac J Cancer Prev 2007, 8: 363–6.
113. Tang X, Tanemura K, Ye W, et al. Clinicopathological factors predicting retroperitoneal lymph node metastasis and survival in endometrial cancer. Jpn J Clin Oncol 1998, 28: 673–8.
114. Nomura H, Aoki D, Suzuki N, et al. Analysis of clinicopathologic factors predicting para-aortic lymph node metastasis in endometrial cancer. Int J Gynecol Cancer 2006, 16: 799–804.
115. Creasman WT, Odicino F, Maisonneuve P, et al. Carcinoma of the corpus uteri. J Epidemiol Biostat 2001, 6: 47–86.
116. Stein HJ, Feith M, Bruecher BL, et al. Early esophageal cancer: pattern of lymphatic spread and prognostic factors for long-term survival after surgical resection. Ann Surg 2005, 242: 566–73; discussion 573–5.
117. Pantuck AJ, Zisman A, Dorey F, et al. Renal cell carcinoma with retroperitoneal lymph nodes. Impact on survival and benefits of immunotherapy. Cancer 2003, 97: 2995–3002.
118. Vasselli JR, Yang JC, Linehan WM, et al. Lack of retroperitoneal lymphadenopathy predicts survival of patients with metastatic renal cell carcinoma. J Urol 2001, 166: 68–72.

119. Blute ML, Leibovich BC, Cheville JC, et al. A protocol for performing extended lymph node dissection using primary tumor pathological features for patients treated with radical nephrectomy for clear cell renal cell carcinoma. J Urol 2004, 172: 465–9.
120. Hutterer GC, Patard JJ, Jeldres C, et al. Patients with distant metastases from renal cell carcinoma can be accurately identified: external validation of a new nomogram. BJU Int 2008, 101: 39–43.
121. Ward JF, Blute ML, Cheville JC, et al. The influence of pNx/pN0 grouping in a multivariate setting for outcome modeling in patients with clear cell renal cell carcinoma. J Urol 2002, 168: 56–60.
122. Almadori G, Bussu F, Paludettii G. Predictive factors of neck metastases in laryngeal squamous cell carcinoma. Towards an integrated clinico-molecular classification. Acta Otorhinolaryngol Ital 2006, 26: 326–34.
123. Wolf GT, Fisher SG, Truelson JM, Beals TF. DNA content and regional metastases in patients with advanced laryngeal squamous carcinoma. Department of Veterans Affairs Laryngeal Study Group. Laryngoscope 1994, 104: 479–83.
124. Franchi A, Gallo O, Boddi V, Santucci M. Prediction of occult neck metastases in laryngeal carcinoma: role of proliferating cell nuclear antigen, MIB-1, and E-cadherin immunohistochemical determination. Clin Cancer Res 1996, 2: 1801–8.
125. Schipper JH, Unger A, Jahnke K. E-cadherin as a functional marker of the differentiation and invasiveness of squamous cell carcinoma of the head and neck. Clin Otolaryngol Allied Sci 1994, 19: 381–4.
126. Myo K, Uzawa N, Miyamoto R, et al. Cyclin D1 gene numerical aberration is a predictive marker for occult cervical lymph node metastasis in TNM Stage I and II squamous cell carcinoma of the oral cavity. Cancer 2005, 104: 2709–16.
127. Civantos F, Zitsch R, Bared A. Sentinel node biopsy in oral squamous cell carcinoma. J Surg Oncol 2007, 96: 330–6.
128. Kowalski LP, Bagietto R, Lara JR, et al. Prognostic significance of the distribution of neck node metastasis from oral carcinoma. Head Neck 2000, 22: 207–14.
129. Moe K, Wolf GT, Fisher SG, Hong WK. Regional metastases in patients with advanced laryngeal cancer. Department of Veterans Affairs Laryngeal Cancer Study Group. Arch Otolaryngol Head Neck Surg 1996, 122: 644–8.
130. Sun HC, Zhuang PY, Qin LX, et al. Incidence and prognostic values of lymph node metastasis in operable hepatocellular carcinoma and evaluation of routine complete lymphadenectomy. J Surg Oncol 2007, 96: 37–45.
131. Sotiropoulos GC, Malago M, Molmenti EP, et al. Hilar lymph nodes sampling at the time of liver transplantation for hepatocellular carcinoma: to do or not to do? Meta-analysis to determine the impact of hilar lymph nodes metastases on tumor recurrence and survival in patients with hepatocellular carcinoma undergoing liver transplantation. Transpl Int 2007, 20: 141–6.
132. Klintmalm GB. Liver transplantation for hepatocellular carcinoma: a registry report of the impact of tumor characteristics on outcome. Ann Surg 1998, 228: 479–90.
133. Boerma EJ. Towards an oncological resection of gall bladder cancer. Eur J Surg Oncol 1994, 20: 537–44.
134. Pawlik TM, Gleisner AL, Vigano L, et al. Incidence of finding residual disease for incidental gallbladder carcinoma: implications for re-resection. J Gastrointest Surg 2007, 11: 1478–86; discussion 1486–7.
135. Shirabe K, Shimada M, Harimoto N, et al. Intrahepatic cholangiocarcinoma: its mode of spreading and therapeutic modalities. Surgery 2002, 131: S159–64.
136. Kokudo N, Makuuchi M. Extent of resection and outcome after curative resection for intrahepatic cholangiocarcinoma. Surg Oncol Clin N Am 2002, 11: 969–83.
137. Yamamoto M, Takasaki K, Yoshikawa T. Lymph node metastasis in intrahepatic cholangiocarcinoma. Jpn J Clin Oncol 1999, 29: 147–50.
138. Murakami Y, Uemura K, Hayashidani Y, et al. Prognostic significance of lymph node metastasis and surgical margin status for distal cholangiocarcinoma. J Surg Oncol 2007, 95: 207–12.

139. Shafazand S, Gould MK. A clinical prediction rule to estimate the probability of mediastinal metastasis in patients with non-small cell lung cancer. J Thorac Oncol 2006, 1: 953–9.
140. Sawyer TE, Bonner JA, Gould PM, et al. Predictors of subclinical nodal involvement in clinical stages I and II non-small cell lung cancer: implications in the inoperable and three-dimensional dose-escalation settings. Int J Radiat Oncol Biol Phys 1999, 43: 965–70.
141. Lee JG, Lee CY, Park IK, et al. Number of metastatic lymph nodes in resected non-small cell lung cancer predicts patient survival. Ann Thorac Surg 2008, 85: 211–5.
142. Paek SC, Griffith KA, Johnson TM, et al. The impact of factors beyond Breslow depth on predicting sentinel lymph node positivity in melanoma. Cancer 2007, 109: 100–8.
143. Taylor RC, Patel A, Panageas KS, et al. Tumor-infiltrating lymphocytes predict sentinel lymph node positivity in patients with cutaneous melanoma. J Clin Oncol 2007, 25: 869–75.
144. Cochran AJ, Huang RR, Lee J, et al. Tumour-induced immune modulation of sentinel lymph nodes. Nat Rev Immunol 2006, 6: 659–70.
145. Karakousis GC, Gimotty PA, Czerniecki BJ, et al. Regional nodal metastatic disease is the strongest predictor of survival in patients with thin vertical growth phase melanomas: a case for SLN Staging biopsy in these patients. Ann Surg Oncol 2007, 14: 1596–603.
146. Balch C, Soong S, Atkins M, et al. An evidence-based staging system for cutaneous melanoma. CA Cancer J Clin 2004, 54: 131–49.
147. Cliby WA, Aletti GD, Wilson TO, Podratz KC. Is it justified to classify patients to Stage IIIC epithelial ovarian cancer based on nodal involvement only? Gynecol Oncol 2006, 103: 797–801.
148. Ushijima K. Management of retroperitoneal lymph nodes in the treatment of ovarian cancer. Int J Clin Oncol 2007, 12: 181–6.
149. Morice P, Joulie F, Camatte S, et al. Lymph node involvement in epithelial ovarian cancer: analysis of 276 pelvic and paraaortic lymphadenectomies and surgical implications. J Am Coll Surg 2003, 197: 198–205.
150. Takeshima N, Hirai Y, Umayahara K, et al. Lymph node metastasis in ovarian cancer: difference between serous and non-serous primary tumors. Gynecol Oncol 2005, 99: 427–31.
151. Onda T, Yoshikawa H, Yasugi T, et al. Patients with ovarian carcinoma upstaged to stage III after systematic lymphadenctomy have similar survival to Stage I/II patients and superior survival to other Stage III patients. Cancer 1998, 83: 1555–60.
152. Suzuki M, Ohwada M, Yamada T, et al. Lymph node metastasis in stage I epithelial ovarian cancer. Gynecol Oncol 2000, 79: 305–8.
153. Eisenkop SM, Spirtos NM. The clinical significance of occult macroscopically positive retroperitoneal nodes in patients with epithelial ovarian cancer. Gynecol Oncol 2001, 82: 143–9.
154. Spirtos NM, Gross GM, Freddo JL, Ballon SC. Cytoreductive surgery in advanced epithelial cancer of the ovary: the impact of aortic and pelvic lymphadenectomy. Gynecol Oncol 1995, 56: 345–52.
155. Nakagohri T, Kinoshita T, Konishi M, et al. Nodal involvement is strongest predictor of poor survival in patients with invasive adenocarcinoma of the head of the pancreas. Hepatogastroenterology 2006, 53: 447–51.
156. House MG, Gonen M, Jarnagin WR, et al. Prognostic significance of pathologic nodal status in patients with resected pancreatic cancer. J Gastrointest Surg 2007, 11: 1549–55.
157. Pawlik TM, Gleisner AL, Cameron JL, et al. Prognostic relevance of lymph node ratio following pancreaticoduodenectomy for pancreatic cancer. Surgery 2007, 141: 610–8.
158. Sohn TA, Yeo CJ, Cameron JL, et al. Resected adenocarcinoma of the pancreas-616 patients: results, outcomes, and prognostic indicators. J Gastrointest Surg 2000, 4: 567–79.
159. Tomlinson JS, Jain S, Bentrem DJ, et al. Accuracy of staging node-negative pancreas cancer: a potential quality measure. Arch Surg 2007, 142: 767–723; discussion 773–4.
160. Bader P, Burkhard FC, Markwalder R, Studer UE. Disease progression and survival of patients with positive lymph nodes after radical prostatectomy. Is there a chance of cure? J Urol 2003, 169: 849–54.

161. Cagiannos I, Karakiewicz P, Eastham JA, et al. A preoperative nomogram identifying decreased risk of positive pelvic lymph nodes in patients with prostate cancer. J Urol 2003, 170: 1798–803.
162. Partin AW, Kattan MW, Subong EN, et al. Combination of prostate-specific antigen, clinical stage, and Gleason score to predict pathological stage of localized prostate cancer. A multi-institutional update. JAMA 1997, 277: 1445–51.
163. Briganti A, Chun FK, Salonia A, et al. Validation of a nomogram predicting the probability of lymph node invasion among patients undergoing radical prostatectomy and an extended pelvic lymphadenectomy. Eur Urol 2006, 49: 1019–26; discussion 1026–7.
164. Briganti A, Shariat SF, Chun FK, et al. Differences in the rate of lymph node invasion in men with clinically localized prostate cancer might be related to the continent of origin. BJU Int 2007, 100: 528–32.
165. Cheng L, Bergstralh EJ, Cheville JC, et al. Cancer volume of lymph node metastasis predicts progression in prostate cancer. Am J Surg Pathol 1998, 22: 1491–500.
166. Daneshmand S, Quek ML, Stein JP, et al. Prognosis of patients with lymph node positive prostate cancer following radical prostatectomy: long-term results. J Urol 2004, 172: 2252–5.
167. Scher H, Leibel S, Fuks Z, et al. Cancer of the postate. In: Cancer: Principles and Practice of Oncology (DeVita V, Hellman S, Rosenberg S, Eds.), pp. 1192–1259. Lippincott Williams & Wilkins, Philadelphia, 2005.
168. Yokota T, Ishiyama S, Saito T, et al. Lymph node metastasis as a significant prognostic factor in gastric cancer: a multiple logistic regression analysis. Scand J Gastroenterol 2004, 39: 380–4.
169. Folli S, Morgagni P, Roviello F, et al. Risk factors for lymph node metastases and their prognostic significance in early gastric cancer (EGC) for the Italian Research Group for Gastric Cancer (IRGGC). Jpn J Clin Oncol 2001, 31: 495–9.
170. Nasu J, Nishina T, Hirasaki S, et al. Predictive factors of lymph node metastasis in patients with undifferentiated early gastric cancers. J Clin Gastroenterol 2006, 40: 412–5.
171. An JY, Baik YH, Choi MG, et al. Predictive factors for lymph node metastasis in early gastric cancer with submucosal invasion: analysis of a single institutional experience. Ann Surg 2007, 246: 749–53.
172. Abe N, Watanabe T, Suzuki K, et al. Risk factors predictive of lymph node metastasis in depressed early gastric cancer. Am J Surg 2002, 183: 168–72.
173. Kunisaki C, Shimada H, Nomura M, Akiyama H. Appropriate lymph node dissection for early gastric cancer based on lymph node metastases. Surgery 2001, 129: 153–7.
174. Abe N, Sugiyama M, Masaki T, et al. Predictive factors for lymph node metastasis of differentiated submucosally invasive gastric cancer. Gastrointest Endosc 2004, 60: 242–5.
175. Borie F, Millat B, Fingerhut A, et al. Lymphatic involvement in early gastric cancer: prevalence and prognosis in France. Arch Surg 2000, 135: 1218–23.
176. Hundahl SA, Cady B, Cunningham MP, et al. Initial results from a prospective cohort study of 5583 cases of thyroid carcinoma treated in the United States during 1996. U.S. and German Thyroid Cancer Study Group. An American College of Surgeons Commission on Cancer Patient Care Evaluation Study. Cancer 2000, 89: 202–17.
177. Wada N, Duh QY, Sugino K, et al. Lymph node metastasis from 259 papillary thyroid microcarcinomas: frequency, pattern of occurrence and recurrence, and optimal strategy for neck dissection. Ann Surg 2003, 237: 399–407.
178. Fong Y, Coit DG, Woodruff JM, Brennan MF. Lymph node metastasis from soft tissue sarcoma in adults. Analysis of data from a prospective database of 1772 sarcoma patients. Ann Surg 1993, 217: 72–7.
179. Riad S, Griffin AM, Liberman B, et al. Lymph node metastasis in soft tissue sarcoma in an extremity. Clin Orthop Relat Res 2004, 426: 129–34.
180. Ariel IM. Incidence of metastases to lymph nodes from soft-tissue sarcomas. Semin Surg Oncol 1988, 4: 27–9.
181. Gaakeer HA, Albus-Lutter CE, Gortzak E, Zoetmulder FA. Regional lymph node metastases in patients with soft tissue sarcomas of the extremities, what are the therapeutic consequences? Eur J Surg Oncol 1988, 14: 151–6.

182. Mazeron JJ, Suit HD. Lymph nodes as sites of metastases from sarcomas of soft tissue. Cancer 1987, 60: 1800–8.
183. Caceres E, Zaharia M, Calderon R. Incidence of regional lymph node metastasis in operable osteogenic sarcoma. Semin Surg Oncol 1990, 6: 231–3.
184. Blazer DG, 3rd, Sabel MS, Sondak VK. Is there a role for sentinel lymph node biopsy in the management of sarcoma? Surg Oncol 2003, 12: 201–6.
185. Chen SL, Bilchik AJ. More extensive nodal dissection improves survival for stages I to III of colon cancer: a population-based study. Ann Surg 2006, 244: 602–10.
186. Baxter NN, Virnig DJ, Rothenberger DA, et al. Lymph node evaluation in colorectal cancer patients: a population-based study. J Natl Cancer Inst 2005, 97: 219–25.
187. Fisher B, Slack NH. Number of lymph nodes examined and the prognosis of breast carcinoma. Surg Gynecol Obstet 1970, 131: 79–88.
188. Blancas I, Garcia-Puche JL, Bermejo B, et al. Low number of examined lymph nodes in node-negative breast cancer patients is an adverse prognostic factor. Ann Oncol 2006, 17: 1644–9.
189. Pieterse QD, Kenter GG, Gaarenstroom KN, et al. The number of pelvic lymph nodes in the quality control and prognosis of radical hysterectomy for the treatment of cervical cancer. Eur J Surg Oncol 2007, 33: 216–21.
190. Herr HW, Bochner BH, Dalbagni G, et al. Impact of the number of lymph nodes retrieved on outcome in patients with muscle invasive bladder cancer. J Urol 2002, 167: 1295–8.
191. Slidell MB, Chang DC, Cameron JL, et al. Impact of total lymph node count and lymph node ratio on staging and survival after pancreatectomy for pancreatic adenocarcinoma: a large, population-based analysis. Ann Surg Oncol 2008, 15: 165–74.
192. Schwarz RE, Smith DD. Extent of lymph node retrieval and pancreatic cancer survival: information from a large US population database. Ann Surg Oncol 2006, 13: 1189–200.
193. Greenstein AJ, Litle VR, Swanson SJ, et al. Effect of the number of lymph nodes sampled on postoperative survival of lymph node-negative esophageal cancer. Cancer 2008, Mar 15, 112: 1239–46.
194. Mariette C, Piessen G, Briez N, Triboulet JP. The number of metastatic lymph nodes and the ratio between metastatic and examined lymph nodes are independent prognostic factors in esophageal cancer regardless of neoadjuvant chemoradiation or lymphadenectomy extent. Ann Surg 2008, 247: 365–71.
195. Shen JY, Kim S, Cheong JH, et al. The impact of total retrieved lymph nodes on staging and survival of patients with pT3 gastric cancer. Cancer 2007, 110: 745–51.
196. Benson AB, 3rd, Schrag D, Somerfield MR, et al. American Society of Clinical Oncology recommendations on adjuvant chemotherapy for stage II colon cancer. J Clin Oncol 2004, 22: 3408–19.
197. Iyer RV, Hanlon A, Fowble B, et al. Accuracy of the extent of axillary nodal positivity related to primary tumor size, number of involved nodes, and number of nodes examined. Int J Radiat Oncol Biol Phys 2000, 47: 1177–83.
198. Fisher B, Jeong JH, Anderson S, et al. Twenty-five-year follow-up of a randomized trial comparing radical mastectomy, total mastectomy, and total mastectomy followed by irradiation. N Engl J Med 2002, 347: 567–75.
199. Clarke M, Collins R, Darby S, et al. Effects of radiotherapy and of differences in the extent of surgery for early breast cancer on local recurrence and 15-year survival: an overview of the randomised trials. Lancet 2005, 366: 2087–106.
200. Kato H, Watanabe H, Tachimori Y, Iizuka T. Evaluation of neck lymph node dissection for thoracic esophageal carcinoma. Ann Thorac Surg 1991, 51: 931–5.
201. Nishihira T, Hirayama K, Mori S. A prospective randomized trial of extended cervical and superior mediastinal lymphadenectomy for carcinoma of the thoracic esophagus. Am J Surg 1998, 175: 47–51.
202. Hulscher JB, van Sandick JW, de Boer AG, et al. Extended transthoracic resection compared with limited transhiatal resection for adenocarcinoma of the esophagus. N Engl J Med 2002, 347: 1662–9.

203. Omloo JM, Lagarde SM, Hulscher JB, et al. Extended transthoracic resection compared with limited transhiatal resection for adenocarcinoma of the mid/distal esophagus: five-year survival of a randomized clinical trial. Ann Surg 2007, 246: 992–1000; discussion 1000–1.
204. Hartgrink HH, van de Velde CJ, Putter H, et al. Extended lymph node dissection for gastric cancer: who may benefit? Final results of the randomized Dutch gastric cancer group trial. J Clin Oncol 2004, 22: 2069–77.
205. Farnell MB, Pearson RK, Sarr MG, et al. A prospective randomized trial comparing standard pancreatoduodenectomy with pancreatoduodenectomy with extended lymphadenectomy in resectable pancreatic head adenocarcinoma. Surgery 2005, 138: 618–28; discussion 628–30.
206. Yeo CJ, Cameron JL, Sohn TA, et al. Pancreaticoduodenectomy with or without extended retroperitoneal lymphadenectomy for periampullary adenocarcinoma: comparison of morbidity and mortality and short-term outcome. Ann Surg 1999, 229: 613–22; discussion 622–4.
207. Seiler CA, Wagner M, Bachmann T, et al. Randomized clinical trial of pylorus-preserving duodenopancreatectomy versus classical Whipple resection-long term results. Br J Surg 2005, 92: 547–56.
208. Havenga K, Enker WE, Norstein J, et al. Improved survival and local control after total mesorectal excision or D3 lymphadenectomy in the treatment of primary rectal cancer: an international analysis of 1411 patients. Eur J Surg Oncol 1999, 25: 368–74.
209. Mortenson MM, Khatri VP, Bennett JJ, Petrelli NJ. Total mesorectal excision and pelvic node dissection for rectal cancer: an appraisal. Surg Oncol Clin N Am 2007, 16: 177–97.
210. Balch C, Soong S, Ross M, et al. Long-term results of a multi-institutional randomized trial comparing prognostic factors and surgical results for intermediate thickness melanomas (1.0–4.0 mm). Ann Surg Oncol 2000, 7: 87–97.
211. Ung YC, Maziak DE, Vanderveen JA, et al. 18 Fluorodeoxyglucose positron emission tomography in the diagnosis and staging of lung cancer: a systematic review. J Natl Cancer Inst 2007, 99: 1753–67.
212. Dam HQ, Manzone TM, Sagar VV. Evolving role of 18F-Fluorodeoxyglucose in the management of esophageal carcinoma. Surg Oncol Clin N Am 2006, 15: 733–49.
213. Myenberger C, Huch Boni RI, Bertschinger P, Zala GF, Klotz HP, Krestin GP. Endoscopic ultrasound and endorectal magnetic resonance imaging: a prospective, comparative study for preoperative staging and follow-up of rectal cancer. Endoscopy 1995, 27: 469–79.
214. Takeuchi H, Morton DL, Kuo C, et al. Prognostic significance of molecular upstaging of paraffin-embedded sentinel nodes in melanoma patients. J Clin Oncol 2004, 22: 2671–80.
215. Stein HJ, Feith M, Bruecher BL, Naehrig J, Sarbia M, Siewert JR. Early esophageal cancer: pattern of lymphatic spread and prognostic factors for long-term survival after surgical resection. Ann Surg 2005, 242: 566–73.
216. Contesso G, Mouriesse H, Friedman S, Genin J, Sarrazin D, Rouesse J. The importance of histologic grade in long-term prognosis of breast cancer: a study of 1,010 patients, uniformly treated at the Institut Gustave-Roussy. J Clin Oncol 1987; 5: 1378.
217. Noguchi M, Minami M, Earashi M, Taniya T, Miyazaki I, Mizukami Y, Nonomura A. Intraoperative histologic assessment of surgical margins and lymph node metastasis in breast-conserving surgery. Arch Surg 1993, 128: 242.
218. Axelsson K, Ljung BME, Moore DH II, Thor AD, Chew KL, Edgerton SM, Smith HS, Mayall BH. Tumor angiogenesis as a prognostic assay for invasive ductal breast carcinoma. J Natl Cancer Inst 1995, 87: 997–1008.
219. Meyer JS, Province M. Proliferative index of breast carcinoma by thymidine labeling: prognostic power independent of stage, estrogen and progesterone receptors. Breast Cancer Res Treat 1988 Oct, 12: 191–204.
220. Fisher B, Fisher ER, Redmond C, et al. Tumor nuclear grade, estrogen receptor, and progesterone receptor: their value alone or in combination as indicators of outcome following adjuvant therapy for breast cancer. Breast Cancer Res Treat 1986, 7: 147.

221. Shek LL, Godolphin W: Survival with breast cancer: the importance of estrogen receptor quantity. Eur J Cancer Clin Oncol 1989, 25: 243–50.
222. Meyer JS, Province MA. S-phase fraction and nuclear size in long term prognosis of patients with breast cancer. Cancer. 1994 Oct 15, 74: 2287–2299.
223. Wenger CR, Beardslee S, Owens MA, Pounds S, Oldaker T, Vendely P, et al. DNA ploidy, S-phase, and steroid receptors in more than 127,000 breast cancer patients. Breast Cancer Res Treat 1993, 28: 9–20.
224. Pujol P, Maudelonde T, Daures JP, Rouanet P, Brouillet JP, Pujol H, Rochefort H. A prospective study of the prognostic value of cathepsin D levels in breast cancer cytosol. Cancer 1993, 71: 2006–12.
225. National Institutes of Health Consensus, Development Conference Statement. June 18–21, 1990, 8(6).
226. Duffy MJ, Reilly D, McDermott E, O'Higgins N, Fennally JJ, Andreasen PA. Urokinase plasminogen activator as a prognostic marker in different subgroups of patients with breast cancer. Cancer 1994, 74: 2276–80.
227. Mansour EG, Ravdin PM, Dressler L. Prognostic factors in early breast carcinoma. Cancer. 1994 Jul 1, 74: 381–400.

Chapter 6

CLINICOPATHOLOGICAL ANALYSIS OF LYMPHATIC VESSELS AND OF LYMPHANGIOGENESIS IN HUMAN CANCER

Ilse Van der Auwera, Peter B. Vermeulen and Luc Y. Dirix
Translational Cancer Research Group, Laboratory of Pathology, University of Antwerp/University Hospital Antwerp, Antwerp, Belgium; Oncology Centre, General Hospital Sint-Augustinus, 2610 Wilrijk, Belgium,
e-mail: ilse.van.der.auwera@telenet.be

Abstract: Novel prognostic factors are needed to enhance identification of those patients with early-stage node-negative cancer that are at increased risk for relapse and should be considered for adjuvant treatment. Lymphatic invasion, lymphatic vessel density and lymphatic growth factor expression have been proposed as reliable prognostic indicators for several human malignancies. However, controversy concerning the precise role of lymphangiogenesis-associated parameters in predicting patients' outcome still exists and this is mainly due to differences in patient selection and applied methodology and to the lack of standardization. In this chapter, we provide an overview of the current applied techniques for evaluating tumor lymphangiogenesis in solid human tumors and discuss the biological relevance of lymphangiogenesis for progression of different human malignancies.

Key words: Lymphangiogenesis · Lymphangiogenic growth factor · Lymphatic vessel density · Lymph node metastasis · Prognosis

6.1 Introduction

With a few exceptions, all cancers can metastasize. The metastatic spread of tumor cells is the major cause of cancer mortality. Tumor dissemination represents a series of complex processes, including: (i) local invasion into surrounding stromal tissue, (ii) direct seeding of body cavities, (iii) systemic metastasis via tumor-associated blood vessels to distant organs, and (iv) lymphatic metastasis via tumor-associated lymphatic vessels to regional lymph nodes. Clinical and pathological observations suggest that for many carcinomas, the most common route of tumor cell metastasis

S.A. Stacker, M.G. Achen (eds.), *Lymphangiogenesis in Cancer Metastasis*,
Cancer Metastasis – Biology and Treatment 13, DOI 10.1007/978-90-481-2247-9_6,

is the lymphatic route, which occurs early and frequently with patterns of spread via afferent vessels following routes of natural drainage [180].

The extent of lymph node involvement is a major determinant for the staging and the prognosis of many human cancers and often guides therapeutic decisions. Accurate detection of metastasis to lymph nodes is therefore critical. Regional nodal basins have traditionally been assessed by full nodal dissections, either in the axilla, the groin, the pelvis or the neck, depending on the location of the primary tumor. However, the problem with this approach is that radical lymphadenectomy is mainly performed as a staging procedure and carries a risk of additional morbidity for patients without lymph node metastasis. In 1992, Morton et al. developed a novel technique involving intra-operative lymphatic mapping for cutaneous melanoma [135]. This technique revolutionized the assessment of regional lymph nodes and is now known as sentinel lymph node (SLN) mapping, performed in combination with a SLN biopsy. Morton et al. were the first to demonstrate that lymphatic drainage from a melanoma can be mapped by injecting the skin around the tumor with blue dye. Injected blue dye was shown to travel through lymphatic channels to the first or 'sentinel' lymph node that drains the tumor. After intense histological analysis of the SLN, these investigators were able to show that the pathologic status of the SLN accurately reflects the pathologic status of the entire regional nodal basin. Theoretically, a negative SLN in malignant tumors thus demonstrates the negative status of other lymph nodes as well. Complete lymphadenectomy is performed only for those patients with SLN metastases, and patients with negative SLN are consequently spared the morbidity of further lymphablation surgery, which could lead to undesirable side effects such as lymphedema. The SLN concept has become a standard technique in the care of patients with melanoma and breast cancer and has also been proven potentially valuable in other cancers [208].

Although lymph node evaluation is an integral part in the assessment of tumor spread for many carcinomas, a significant proportion of lymph node-negative cancer patients will subsequently develop recurrent disease. Thus, additional methods are needed to enhance identification of those patients with early-stage node-negative cancer that are at increased risk for relapse and should be included in a more intensive follow-up schedule or be considered for adjuvant treatment. Lymphangiogenesis-related parameters such as lymphatic invasion, lymphatic vessel density or lymphatic growth factor expression are promising prognostic indicators that could increase the efficiency of staging for early-stage cancer.

Lymphatic invasion by cancer cells may be the first stage of lymph node metastasis. The presence of carcinomatous lymphatic invasion is a highly significant risk factor for tumor recurrence and a predictor of shorter disease-free or overall survival in several node-negative carcinomas, including melanoma, breast cancer, prostate cancer, gastric cancer, colon cancer, non-small cell lung cancer, bladder cancer, esophageal cancer and cervical squamous cell carcinoma. These findings support the routine evaluation of lymphatic invasion in cancer specimens and provide the option for its incorporation into nomograms predictive of patients' outcome. Indeed, peritumoral lymphatic invasion has been included as an adverse prognostic factor in a series of guidelines and recommendations for postoperative adjuvant systemic therapies of early-stage breast cancer, such as these developed by an International

Consensus Panel during the St Gallen Conference in 2005 [55]. The standard method for assessing lymphatic invasion has been light microscopic examination of haematoxylin and eosin (H&E) stained sections, after which lymphatic invasion is identified as the presence of tumor emboli within vascular channels distinctly lined by a single layer of endothelial cells. Pitfalls in this technique include the inability to discern tumor emboli that obliterate the lumen of lymphatics and to distinguish retraction artifacts that isolate tumor aggregates due to tissue shrinkage during fixation from true tumor emboli in lymphatic spaces. Many studies have shown that the use of an immunohistochemical marker of lymphatic endothelium or dual immunostaining for epithelial and endothelial cell markers increases the accuracy of the immunohistochemical detection of lymphatic invasion compared to conventional H&E staining [8, 18, 84, 147, 197, 209, 219]. However, no standard method for immunohistochemical detection of lymphatic invasion has so far been proposed and current variability in the assessment of lymphatic invasion might limit the value of this measurement for clinical decision-making and impairs comparison of results obtained at different institutes.

Recent developments in lymphatic biology and research, especially the discovery of unique molecular markers with some specificity for lymphatic endothelial cells, such as LYVE-1, podoplanin and Prox-1 have provided exciting new insights into the mechanisms by which tumors exploit the lymphatics for metastasis. In the past few years, it has become apparent that lymphangiogenesis, controlled by a complex network of growth factors, cytokines and chemokines, can contribute actively to tumor metastasis. The dimension of tumor-endothelial interface is mainly reflected by lymphatic vessel density (the number of intratumoral or peritumoral lymphatics) and theoretically, increased tumor-related lymphatic vessel density facilitates the access of tumor cells to the lymphatics. An increasing number of clinicopathological studies have shown a direct relationship between tumor expression of the vascular endothelial growth factors (VEGF-C/VEGF-D) and metastatic tumor spread in human tumors. However, the association between lymphatic vessel density and the presence of nodal metastases and aggressive behavior in human cancer is still inconsistent. The controversial issues about the role of lymphatic vessel density in tumor progression are mainly due to differences in patient selection and applied methodology and to the lack of standardization. Recently, an international consensus report regarding the quantification of lymphangiogenesis in solid human tumors was published [209]. This report aimed at improving standardization of lymphangiogenesis assessment in order to allow for meta-analyses. In the first part of this chapter, we will provide an overview of the current techniques and describe the findings of this consensus report in detail. Since the biological relevance of peritumoral and intratumoral lymphangiogenesis for tumor progression can differ significantly in tumors of different types or anatomical locations, we will discuss the association between lymphatic vessel density or lymphangiogenic growth factor expression and clinicopathological parameters separately for each tumor type, including breast cancer, cervical cancer, ovarian cancer, prostate cancer, bladder cancer, renal cell cancer, melanoma, head and neck cancer, thyroid cancer, esophageal cancer, colorectal cancer, gastric cancer, pancreatic cancer, hepatocellular cancer and lung cancer.

6.2 Methodology of Lymphangiogenesis Quantification in Solid Human Tumors

Studies on the usefulness of lymphangiogenesis as a prognostic tumor marker have yielded inconsistent conclusions. Discrepancies between various reports are mainly due to differences in the methods used for scoring lymphatic vessel density (random field versus hot spot counting), differences between patient numbers and selection criteria and random variation in the sites of tumor formation (lymphatic-poor versus lymphatic-rich areas) [76]. Recently, an international consensus report regarding the quantification of lymphangiogenesis in solid human tumors was published [211]. The aim of this report was to improve the standardization of the estimation of the ongoing lymphangiogenesis in histological tumor sections in order to allow tumor-associated lymphangiogenesis to be applied as a prognostic indicator. The findings of this report and currently applied technology for lymphangiogenesis assessment are described below.

6.2.1 Lymphatic Vessel Density

The number of lymphatics in a microscopic field is the net result of previous phases of tumor lymphangiogenesis and of lymphatic vessel remodeling or regression a tumor went through. Quantification of tumor lymphatics for the purpose of tumor staging has long been problematic. Although morphology can sometimes distinguish lymphatic vessels from blood vessels in histological sections by the frequent absence of a basement membrane and lack of erythrocytes in lymphatics, neither is a reliable method for routine use. However, major research efforts during the last ten years have lead to the discovery of several markers that allow the distinction between lymphatic and blood vessels at the capillary level, resulting in marked advances in the study of lymphatics (Table 6.1). VEGFR-3 was one of the first lymphatic markers to be identified in healthy tissue. However, the observation of VEGFR-3 expression on blood vessels in tumors and wound granulation tissue has meant that it is less appropriate for specific identification of lymphatics in these conditions [102, 160, 207]. Other molecules that have been proposed as markers of the lymphatic endothelium include podoplanin, a glomerular podocyte membrane mucoprotein [24], Prox-1, a homeobox gene product involved in regulating the embryonic development of the lymphatic system [228], LYVE-1, a lymphatic vascular endothelial cell receptor for hyaluronan [16] and desmoplakin, a glycoprotein that locates exclusively to the intracellular junctions between the endothelial cells of lymphatic vessels [170]. However, none of these markers fulfils the criteria of an ideal lymphatic vessel marker, which should be exclusively found on all types of lymphatic endothelial cells in all pathological conditions.

The selection of the optimal marker of the lymphatic endothelium is clearly a critical step in the assessment of lymphatic vessel count since false data arising from low specificity of the staining must be avoided. To date, the majority of the experimental studies of tumor lymphatics have employed LYVE-1 antibodies.

Table 6.1 Specific markers for the lymphatic endothelium

Markers	Function	Sites of expression
VEGFR-3	Transmembrane tyrosine kinase receptor for VEGF-C and VEGF-D	Mainly expressed on lymphatic endothelium in adult tissue, but also reactivated in blood vessels in pathological conditions
Podoplanin	Glomerular podocyte membrane glycoprotein	Expressed in lymphatic capillaries and in osteoblastic cells, lung alveolar type I cells and kidney podocytes
Prox-1	Homeobox protein required for embryologic lymphatic development	Expressed in lymphatic endothelial cells and in non-endothelial cells in lens, heart, liver, pancreas and nervous system
LYVE-1	Receptor for extracellular matrix/lymph fluid hyaluronan	Expressed on lymphatic endothelium and in blood sinusoidal endothelial cells in liver and spleen and placental syncytiotrophoblasts
Desmoplakin	Associates with desmosomal cadherins to form a cell adhesion complex	Small lymphatic endothelium, epithelial cells and cardiac muscle

However, the expression of LYVE-1 can be down-modulated in some tissues, for example, in response to inflammation [80], and is absent in some tumor-associated lymphatics [20,165,185,213]. In 2002, Kahn et al. introduced D2-40 as a new selective marker of lymphatic endothelium and reported its value in detecting lymphatic invasion in human malignancies [83,84]. Later it became apparent that this antibody recognizes a fixation-resistant epitope on podoplanin [168]. An increasing number of studies have used this antibody to visualize lymphatics in tumor sections due to its high sensitivity and specificity for the lymphatic endothelium [44]. However, the fact that podoplanin appears to be only expressed in small lymphatics and not in larger ones that have smooth-muscle cells [183] and is also expressed in other cell types, such as osteoblastic cells, kidney podocytes and lung alveolar type I cells [24, 226], underlines the importance of utilizing multiple markers to characterize lymphatic vessels in comprehensive studies of lymphangiogenesis. The best combination of markers of the lymphatic endothelium could vary on the tissue type.

Some studies have assessed absolute tumor-associated lymphatic vessel density by counting all immunostained vessels in histological tumor sections, whereas other studies have determined only the lymphatic vessel density of the 'hot spots' of lymphatics within or surrounding the tumor. A hot spot is an area giving the impression at low magnification of containing numerous microvessels as defined by Weidner et al. in 1991 [225]. The reproducibility of the assignment of these hot spots is a critical variable in the analysis of lymphatic vessel density and the success of finding the relevant hot spot depends on the training and experience of the investigator [216]. Vascular hot spots are thought to represent localized areas of biological importance since they potentially originate from tumor cell clones with the highest angiogenic potential which might preferentially enter the circulation and give rise to vascularized metastases. Localized changes in oxygen tension are indeed a strong angiogenic drive. The methodology of counting the number of microvessel entities in regions with an elevated vascular density has been adapted for the assessment of lymphatic vessel density, although this is based on the assumption that a functional increase in lymphatic vessels occurs in hot spots. Since data on the association of lymphangiogenesis with hypoxia are still contradictory, the relevance of counting lymphatic vessels in hot spots, as opposed to an overall lymphatic vessel count, has been questioned [176].

6.2.2 Computerized Image Analysis Systems

The major drawbacks of the visual lymphatic vessel counting method are its inherent subjectivity and the difficulty of standardization between different laboratories. In contrast, image cytometry is more objective and reproducible and moreover, many image cytometry software packages allow additional information on vessel luminal area and vessel luminal perimeter. However, the widespread application of image cytometry is hampered by the need for specialized equipment to perform the analyses. Another limitation of this method is the possibility of confounding signals of non-endothelial structures in the stromal compartment. Choi et al. [29] performed a direct comparison of visual and image cytometric lymphatic vessel density assessment on podoplanin-immunostained sections of invasive breast carcinoma. An automated scanning microscope and an automated image analysis application was used that identified stained ring-like structures based on color and morphometry in areas marked during direct microscopic microvessel counting. Lymphatic vessel densities determined by direct microscopy and image cytometry were significantly correlated. However, tumor stage only correlated with image cytometric lymphangiogenesis, while lymph node status and VEGF family gene expression only correlated with visual data.

6.2.3 Chalkley Count

The Chalkley point overlap morphometric technique has abolished one of the highly observer-dependent steps of measuring lymphatic vessel density, namely the frequent decision an observer has to make whether two immunostained and adjacent structures are the reflection of one single or two separate lymphatic vessels. This technique involves the use of an eyepiece graticule containing 25 randomly positioned dots, which is rotated so that the maximum number of points is on or within the vessels of the vascular hotspot. Thus, instead of counting the individual microvessel, the overlaying dots are counted. The Chalkley count is a reflection of the relative area taken by the lymphatic vasculature and offers a suitable alternative for lymphatic vessel density assessment.

6.2.4 Lymphatic Endothelial Cell Proliferation

Both the sprouting of lymphatics and the enlargement of lymphatics are accompanied by the proliferation of lymphatic endothelial cells. Lymphatic endothelial cell proliferation is generally assessed by a double immunostaining of tumor sections with antibodies directed at a lymphatic endothelial cell marker and a marker of proliferating cells (e.g. Ki-67 or PCNA). However, also a triple immunostaining method to detect proliferating lymphatics has been developed, utilizing antibodies against a marker of cell proliferation, antibodies against an epithelial cell marker (cytokeratin) and antibodies against a lymphatic-specific marker [154]. This method has the advantage of allowing the distinction between proliferating lymphatics and proliferating carcinoma cells that are trapped within the lymphatic lumen.

6.2.5 Tumor/Circulating Levels of Lymphangiogenic Growth Factors

The expression of lymphangiogenic factors, such as VEGF-C and VEGF-D, is closely related to tumor-induced lymphatic dilatation or lymphangiogenesis and thereby to lymph node metastasis [163]. For the detection of VEGF-C and VEGF-D at the protein level antibodies for immunohistochemistry or Western blot analysis are commercially available and have been widespread used. Other investigators have correlated mRNA expression levels of lymphangiogenic growth factors with tumor characteristics employing RT-PCR and Northern blot procedures. From a practical point of view, the detection of circulating levels of lymphangiogenic growth factors in preoperative blood samples of cancer patients might be a useful prognostic indicator. The quantitative measurement of serum protein levels can be performed easily and frequently because of their minimal invasiveness compared with examinations of surgically obtained tissue specimens and offer a more objective approach for lymphangiogenesis assessment. Duff et al. [39] were the first to describe the development of an indirect enzyme-linked immunosorbent (ELISA) assay for the quantification of VEGF-C in plasma [39]. Capture of VEGF-C was achieved using a goat anti-human VEGF-C antibody, followed by detection with a rabbit anti-human VEGF-C antibody. The antibody combination used in the ELISA recognized both the partially processed and the fully mature form of the protein. The sensitivity of the assay was amplified using the biotin-avidin and enhanced chemiluminescence systems. The assay was highly sensitive and reproducible with a detection range of 0.4–100 U/ml and the intra- and inter-assay variations were less than 8%. Substitution tests demonstrated that the assay was specific for VEGF-C without cross-reacting with VEGF-A or VEGF-D. Using plasma samples from patients with colorectal cancer, a threefold increase of VEGF-C levels was found when compared to normal controls. Weich et al. [224] developed a quantitative sandwich ELISA for VEGF-C that can be used to detect and measure VEGF-C in plasma and cell and tissue lysates [224]. Different antibodies were combined to detect processed and partially processed VEGF-C in a specific manner. The ELISA was able to detect human VEGF-C with a minimum detection limit of 100 pg/ml. These studies have demonstrated that ELISA is a useful tool for investigations concerning the significance of VEGF-C in predicting patient's prognosis.

Bando et al. [14] described the development of a specific indirect ELISA for the quantification of VEGFR-3 in different human cell and tissue lysates [14]. A combination of the goat anti-VEGFR-3 antibody and the mouse monoclonal anti-human VEGFR-3 antibody was used. The assay was highly sensitive and reproducible with a detection range of 0.2–25 ng/ml. The assay was specific for VEGFR-3, with no cross-reactivity to VEGFR-1 or VEGFR-2. In this report, the level of VEGFR-3 protein detected in the ELISA correlated significantly with the number of VEGFR-3-positive vessels observed in histochemical sections, suggesting that the ELISA assay may be a reliable surrogate of measuring VEGFR-3-positive vessel density.

6.3 Prognostic Value of Lymphangiogenesis in Human Malignancies

In the following part of this chapter we will describe current views on the prognostic impact of lymphangiogenesis-associated parameters, such as tumoral and circulating lymphatic growth factor levels and lymphatic counts in several human malignancies. A summary is also provided in Tables 6.2, 6.3 and 6.4.

6.3.1 Breast Cancer

Breast cancer is one of the most common malignancies among women, accounting for nearly one in three cancers diagnosed among women in the United States [181]. Tumor spread to the lymph nodes is a frequent complication in breast cancer and the degree of axillary lymph node involvement at the time of diagnosis remains the most valuable prognostic factor for breast cancer survival. It is standard practice to administer systemic therapy to all patients with lymph node-positive disease. As a result of more widespread screening, more women are diagnosed as having breast cancer at an early, node-negative stage. The majority of these patients are cured with total mastectomy or breast conservative treatment. However, 20%–30% of the patients ultimately develop disease recurrence in distant sites. In randomized clinical trials, adjuvant hormonal therapy and polychemotherapy have been shown to reduce the rate of recurrence in these patients. However, adjuvant systemic therapy has associated risks and proper selection of patients for adjuvant therapy is thus necessary to avoid exposing many patients with low risk of recurrence to treatments for whom the benefit is not justified by the toxicity and the cost.

Only recently, tumor lymphangiogenesis has gained an increasing interest as a potential prognostic indicator for patients with breast cancer. A PubMed search for the terms 'lymphangiogenesis' and 'breast cancer' identifies 64 citations, which all postdate 2000. However, so far the prognostic significance of lymphangiogenesis for breast cancer metastasis has remained largely unknown, which is mainly caused by the discrepancies between studies performed at different institutes.

Table 6.2 Association of tumor lymphangiogenesis-associated parameters with lymph node metastasis

Human tumor tissues	Tumor VEGF-C	Serum VEGF-C	Tumor VEGF-D	Serum VEGF-D	ILV D	PLV D
Breast cancer	+	x	+	X	−	+
Cervical cancer	+	−	x	X	−	+
Ovarian cancer	+	x	+	x	x	x
Prostate cancer	+	x	+	x	−	+
Bladder cancer	+	x	x	x	x	x
Renal cell cancer	−	x	−	x	−	−
Cutaneous melanoma	+	+	−	−	+	+
Head and neck cancer	+	x	−	x	+	+
Papillary thyroid cancer	+	x	+	x	+	x
Follicular thyroid cancer	−	x	−	x	−	−
Esophageal cancer	+	+	x	x	−	−
Colorectal cancer	+	+	+/−	−	+	+
Gastric cancer	+	+	+	x	+*	
Pancreatic endocrine tumors	−	x	−	x	−	−
Pancreatic ductal adenocarcinoma	+/−	x	+/−	x	−	−
Hepatocellular cancer	+	x	−	x	−	−
Lung cancer	+/−	+	−	X	−	+

(ILVD = intratumoral lymphatic vessel density, PLVD = peritumoral lymphatic vessel density, + = association, − = no association, +/− = no consensus, x = not investigated, * = no distinction made between intratumoral and peritumoral lymphatics)

Studies on lymphatic growth factor expression in human breast cancer have reported that VEGF-C is located in the cytoplasm of human intraductal and invasive breast cancer cells [167, 207] and that high VEGF-C expression levels are strongly associated with tumor cell proliferation, lymphatic invasion, lymphatic vessel density, lymph node metastasis, distant metastasis and an unfavorable prognosis [15, 60, 70, 72, 95, 104, 115, 133, 142, 145, 245]. Moreover, VEGF-C expression correlates with c-erbB2 expression in breast carcinomas, which suggests the existence of a functional relationship and may, at least in part, explain the aggressive phenotype associated with c-erbB2-positive breast tumors [68, 236]. In contrast, no prognostic value for plasma VEGF-C levels in patients with breast cancer has been reported [7]. No significant association between plasma levels of VEGF-C and age, tumor size, tumor grade, HER2 status, ER status, PR status or disease-free and overall survival has been found. Not only VEGF-C, but also VEGF-D has been detected in tumor cells and endothelium in ductal carcinoma of the breast [1]. Moreover,

Table 6.3 Association of tumor lymphangiogenesis-associated parameters with patients' survival

Human tumor tissues	Tumor VEGF-C	Serum VEGF-C	Tumor VEGF-D	Serum VEGF-D	ILVD	PLVD
Breast cancer	+	x	+	x	−	+
Cervical cancer	+	x	x	x	x	+
Ovarian cancer	+	x	+	x	−	−
Prostate cancer	x	x	x	x	x	X
Bladder cancer	+	x	x	x	x	+
Renal cell cancer	x	x	x	x	x	X
Cutaneous melanoma	x	−	x	x	+	+
Head and neck cancer	+	x	x	x	+	+
Papillary thyroid cancer	x	x	x	x	x	x
Follicular thyroid cancer	x	x	x	x	x	x
Esophageal cancer	+	x	x	x	+*	
Colorectal cancer	+	x	+	x	+*	
Gastric cancer	+	+	+	x	+*	
Pancreatic endocrine tumors	x	x	x	x	−	−
Pancreatic ductal adenocarcinoma	+/−	x	+	x	−	−
Hepatocellular cancer	+	x	x	x	−	−
Lung cancer	+	x	x	x	x	+

(ILVD = intratumoral lymphatic vessel density, PLVD = peritumoral lymphatic vessel density, + = association, − = no association, +/− = no consensus x = not investigated, *= no distinction made between intratumoral and peritumoral lymphatics

VEGF-D-positivity of breast tumor cells significantly correlates with lymph node metastasis, high c-erbB2 expression and disease-free and overall survival of patients with breast carcinoma [145,153]. These findings suggest that VEGF-D and VEGF-C may be useful in the treatment of breast cancer as decision-making biomarkers for aggressive treatment after surgery, although prospective studies in a larger population should be carried out to demonstrate their clinical significance.

To date, there are only limited data concerning the clinicopathological significance of VEGFR-3 expression by breast cancer cells. Gunnigham et al. [58] investigated the long and short isoforms of VEGFR-3 mRNA in normal and tumor tissue and observed a significant loss of the long isoform in breast tumors compared with normal breast tissue. This difference was largely accounted for by the reduction of long VEGFR-3 in node-positive tumors. The authors therefore suggest that the measurement of the VEGFR-3 isoform expression in breast tumors might identify a patient group that is likely to have node-positive disease and therefore benefit form additional treatment.

Table 6.4 Presence of proliferating lymphatic endothelial cells in human cancer

Human tumor tissues	LECP	PLECP %	ILECP %	References
Breast cancer	–			[4, 214, 229]
	+	2.2%	1.83%	[212]
Cervical cancer	x			
Ovarian cancer	x			
Prostate cancer	x			
Bladder cancer	x			
Renal cell cancer	+	6.5%	2.6%	
Cutaneous melanoma	+	no percentages reported		[32, 186]
Head and neck cancer	+	no percentages reported		
Papillary thyroid cancer	x			
Follicular thyroid cancer	x			
Esophageal cancer	x			
Colorectal cancer	+		13.3%*	
Gastric cancer	x			
Pancreatic endocrine tumors	+		0.7–3%	[178]
Pancreatic ductal adenocarcinoma	x			
Hepatocellular cancer	x			
Lung cancer	+	4.16%	1.3%	[163]

(ILECP %= fraction of intratumoral lymphatic endothelial cell proliferation, PLECP % = fraction of peritumoral lymphatic endothelial cell proliferation, + = presence, − = absence, +/− = no consensus x = not investigated, *= no distinction made between intratumoral and peritumoral lymphatics)

Several papers have reported that intratumoral lymphatic are sparsely present in only a minority of early ductal breast cancers and that the density of peritumoral lymphatics may have a greater clinical significance [4, 20, 88, 172, 214, 217]. Bono et al. found that the presence of down-regulated LYVE-1 expressing intratumoral lymphatics was not associated with axillary nodal status or patients' survival, although intratumoral lymphatics were more commonly present in poorly differentiated ductal breast cancers than in well differentiated ones [20]. The above findings differ from other studies in which podoplanin-positive intratumoral lymphatics were observed within the majority of breast tumors [29, 142, 212]. This discrepancy could be due to differences in methodology. In contrast to intratumoral lymphatic vessel density, high peritumoral lymph vessel counts have been associated with a high number of metastatic axillary lymph nodes at the time of diagnosis and with poor distant disease-free and overall survival [20, 142, 145, 172]. However, this effect on survival has been limited to the subgroup of axillary node-positive breast cancer.

Few of the comprehensive studies of lymphangiogenesis in breast cancer investigated the presence of proliferating lymphatic endothelial cells. Most authors found no evidence of proliferation [4, 214, 229], although in one study, proliferating lymphatic endothelial cells, identified using a Ki-67/D2-40 doublestain, were observed both within breast tumors and in peritumoral areas, albeit in limited numbers [212].

While the prognostic role of lymphangiogenesis in patients with node-negative breast disease is still a subject of debate, it has been suggested that a high peritumoral lymphatic count might be a potential unfavorable prognostic factor in patients with node-positive breast cancer. Interestingly, more lymphatics are also present in metastatically involved lymph nodes when compared to uninvolved lymph nodes and these show increased fractions of proliferating lymphatic endothelial cells [210]. Moreover, a higher lymphatic vessel density in the lymph node metastases is related to a shorter survival of patients with breast cancer [214].

6.3.2 Cervical Cancer

Cervical cancer is the second most common cancer among women worldwide [198]. Fortunately, due to cytological screening, cervical cancer is frequently diagnosed at early stages. Although most patients with stage I disease have a favorable outcome, approximately 20% to 35% are expected to die from their disease [150]. In addition to International Federation of Obstetrics and Gynecology (FIGO) stage and tumor volume, lymphatic invasion and nodal metastasis are known predictors of shorter disease-free and overall survival in carcinoma of the uterine cervix [30, 35, 201, 218]. The utility of lymphangiogenesis-related parameters for predicting patients' outcome has been investigated in cervical cancer, which is discussed below.

Serum VEGF-C concentrations are elevated in cervical cancer when compared to healthy controls [120, 130]. In an analysis of squamous cell carcinoma, the pretherapeutic circulating levels of VEGF-A and VEGF-C have been shown to correlate with advanced FIGO stage, large tumor size, disease recurrence or persistence after treatment, but not with lymph node metastasis [130]. In the same study, both serum VEGF-A and VEGF-C levels significantly decreased after treatment. Thus, the serum levels of VEGF-A and VEGF-C have potential usefulness as biologic markers of squamous cell carcinoma of the uterine cervix. VEGF-C expression is also detected in the majority of cervical carcinoma cells, mainly at the tumor periphery [56, 64, 120, 215]. Several studies have indicated that high VEGF-C expression by tumor cells at the invasive edge induces lymphangiogenesis and contributes to high peritumoral lymphatic vessel density, leading to increased lymphatic invasion and pelvic lymph node metastasis [56, 64, 204]. VEGF-C has been shown to independently affect patient's survival in cervical carcinomas. Overall survival rates for patients with strong VEGF-C staining tumors are lower than those for patients with weak VEGF-C staining tumors [205]. So far, the prognostic significance of VEGF-C expression in the subgroup of node-negative breast tumors remains unknown. The survival data suggest that the measurement of VEGF-C expression in cervical tumor biopsies may be useful as a tumor marker for patients' prognosis. Furthermore, in a subset of node-positive uterine cervical cancers, the VEGF-C level in involved lymph nodes is remarkably increased when compared to the primary tumor, and this phenomenon is associated with a worse survival [49].

Few studies have investigated the association of lymphatic vessel density with clinicopathological tumor characteristics in cervical cancer and these have yielded inconsistent findings. One study has reported that D2-40-positive intratumoral lymphatics are present in cervical cancers and are unevenly distributed throughout the tumor with a significantly higher density compared with normal cervical tissue [56]. These vessels appear to be small and flattened with a close lumen, contrasting the widely open lymphatics in peritumoral regions. Severe dysplasia/carcinoma in situ has also been associated with increased lymphatic vessel density and high lymphatic growth factor expression, suggesting an essential role for lymphangiogenesis in the progression to invasive behavior [56,215]. However, no association between intratumoral lymphatic vessel density and nodal metastasis or lymphatic invasion has been observed. In contrast to the above-mentioned study, Schoppmann et al. [174] have reported that podoplanin and LYVE-1 double-positive lymphatics are exclusively found within the peritumoral stroma, both in non-invasive and pT1b1-stage invasive cervical carcinomas [174, 215]. The local density of peritumoral lymphatics has been shown to be significantly increased over normal tissues, to be correlated with high tumor stage, lymphatic invasion, and nodal metastasis, and to be independently predictive of poor survival and/or shorter recurrence-free survival, suggesting that high peritumoral lymphatic vessel density may be an independent prognostic factor in early-stage cervical cancer [56, 175, 215].

The local immunological response, evident by inflammatory stromal reaction, has been shown to play a role in inducing lymphangiogenesis in early-stage cervical cancer [173, 174]. A strong correlation between lymphatic vessel density and inflammatory stromal reaction and lymphatic invasion by tumor cells has been encountered in specimens of cervical cancer. The density of activated tumor-associated macrophages, expressing large amounts of VEGF-C and VEGF-D, has been shown to correlate with peritumoral inflammatory stromal reaction, lymphatic vessel density, and indirectly with peritumoral lymphatic invasion and frequency of lymph node metastasis, suggesting an important role of these cells in peritumoral lymphangiogenesis and cancer dissemination [174].

6.3.3 Ovarian Cancer

Epithelial ovarian carcinomas are the most lethal gynecologic malignancies. Owing to the paucity of symptoms and their insidious onset and the absence of any convincing screening method, about two thirds of patients with ovarian cancer present with advanced stage disease, involving sites such as the upper abdomen, pleural space, and para-aortic lymph nodes [25]. Despite the highly lethal nature of epithelial ovarian cancer, the clinical course of advanced disease can be difficult to predict in an individual patient. A small fraction of patients will be cured with surgery followed by chemotherapy, another group will experience relapse after a relatively long time interval, others will relapse and succumb to this disease within months of completing first-line therapy, and some will exhibit primary resistance to first-line chemotherapy. Currently available clinical and molecular prognostic factors provide an imperfect assessment of prognosis for patients with epithelial ovarian cancer.

The prognostic impact of lymphangiogenesis has remained largely uninvestigated in ovarian cancer. VEGF-C, VEGF-D and VEGFR-3 expression levels in ovarian cancer correlate significantly with lymph node metastasis and peritoneal spread in the upper abdomen [36,204,238]. Moreover, VEGF-C expression correlates well with c-erbB2 expression, matrix metalloproteinase-2 gene expression, angiogenesis and a low apoptotic index [69,204]. VEGF-C and VEGF-D have emerged as strong predictors for poorer survival [204, 238], which suggests that the measurement of lymphatic growth factors may improve prospective identification of ovarian cancer patients with a poor prognosis.

Only one study has examined whether lymphatic vessel density has any value or relevance with respect to predicting the disease course in ovarian carcinomas. Immunostaining of ovarian tumors with antibodies to LYVE-1 has identified irregularly shaped, thin walled lymphatics in the capsular and intratumoral regions of the tumor with capsular lymphatics being more frequent than intratumoral lymphatics [188]. Lymphatic count has not been associated with age, residual disease, histological subtype, FIGO stage or vascular count, nor has it predicted any difference in survival curves for patients with ovarian cancer.

6.3.4 Prostate Cancer

Prostate carcinoma is the most commonly diagnosed malignancy and the second leading cause of cancer-related mortality among men in the United States [19, 126]. Despite these high death rates, prostate cancer is often an indolent disease, and patients can remain asymptomatic for years. The introduction of serum prostate specific antigen (PSA) testing in the late 80's has led to a major shift towards the diagnosis of this malignancy at much earlier stages than in previous decades with the possibility of cure. Currently, prognostication and treatment stratification at the time of diagnosis are based on clinical stage, biopsy Gleason grade (a measure of tumor differentiation), and serum PSA levels. In cases treated by radical prostatectomy, prognosis can be refined by using pathological stage and grade. However, these prognostic indicators do not accurately predict clinical outcome for individual patients. Improved markers are needed to determine which patients might benefit from a more aggressive treatment, and which patients might be spared unnecessary and potentially harmful interventions.

Several studies have pointed to an important role of lymphatic growth factor expression in the metastatic process of human prostate cancer. VEGF-C protein expression is lower in benign prostate epithelium than in adjacent carcinoma [77,235,243]. Patients with lymph node metastasis have a significantly higher expression of VEGF-C than patients without lymph node metastasis [37, 77, 200, 203]. The correlation of VEGF-C with lymph node status suggests a role for the development of lymph node metastasis. VEGF-D is highly expressed and localized to both cancer epithelial cells and stromal cells in prostate carcinoma [89, 243]. Its expression is associated with advanced stages of prostate cancer disease [89]. VEGF-D expression levels are significantly elevated in primary tumors with lymph node involvement when compared to those without lymph node involvement [184]. Moreover, circulating levels of VEGF-D protein are highly significantly different in plasma from patients with advanced-stage and early-stage prostate cancer and seem to correlate with clinical observations of widespread metastasis to bone and lymph nodes [89]. VEGFR-3 expression is upregulated in prostate carcinomas when compared to prostate hyperplasia and normal prostate [112, 235]. Furthermore, its expression is correlated with pre-operative PSA, Gleason score and lymph node metastasis [77, 112]. One study has found augmented tyrosine phosphorylation of VEGFR-3 in advanced versus early stage (node negative) prostate cancer [89], whereas another found upregulation of a truncated form of VEGFR-3, but not the full-length receptor [184].

Although lymphatics are detected in prostate cancer [164, 200, 203, 243, 244], the role of intratumoral lymphatics in mediating lymph node metastasis has been controversial. The lymphatic vessel density in prostate adenocarcinoma regions is significantly decreased compared to that in peritumoral and normal prostate regions [164, 199, 200]. Increased peritumoral lymphatic vessel density, but not intratumoral lymphatic vessel density, has been correlated with lymph node metastasis and high Gleason score, a marker of more aggressive tumors [243, 244], suggesting that peritumoral lymphatics might be functionally more important than intratumoral lymphatics. In addition, tumor emboli are observed in peritumoral lymphatics and lymphatic invasion involving either the peritumoral or the intratumoral compartment is strongly associated with regional lymph node metastasis [164, 243, 244]. Lymphatic invasion has been strongly correlated with biochemical failure after radical prostatectomy in patients with node-negative prostate cancer, supporting the routine evaluation of lymphatic invasion status in radical prostatectomy specimens and providing the option for its incorporation into nomograms predictive of clinical outcome [28, 116, 125, 164].

6.3.5 Bladder Cancer

Bladder cancer is the second most common cancer of the urogenital region. The majority of newly diagnosed bladder cancers, 70% to 80%, are classified as superficial disease [128]. The remaining tumors initially present as muscle-invasive or metastatic disease. The first-line treatment of patients with bladder cancer, clinically diagnosed as superficial disease, is transurethral resection, due to its relative indolent nature and low malignant potential. Patients with superficial bladder cancer who have a high risk of progression receive adjunctive intravesical therapy, which has been shown to be more effective than transurethral resection alone in preventing tumor recurrence [42]. Radical cystectomy has evolved in the most common therapeutic modality for the 20% to 30% of all patients with bladder cancer that are diagnosed with muscle-invasive tumors [108], which is an aggressive malignancy that is widely believed to have a high propensity for distant metastasis [43]. Owing to the variability in the clinical behavior of bladder cancer after radical cystectomy, prognostic indicators are crucial for identifying patients who are at a high risk for disease progression and recurrence and therefore should receive adjuvant therapies. The clinical and pathological significance of tumor lymphangiogenesis in human bladder cancer tissues remains to be fully investigated.

Both VEGF-D and VEGF-C expressions are associated with lymphatic vessel density in human bladder cancer [131]. Multivariate analysis has shown that VEGF-C expression in transitional cell cancer of the bladder is an exclusive independent factor influencing pelvic lymph node metastasis and that patients with high VEGF-C expression have a markedly poorer prognosis than those with no or low expression [189,247]. Therefore, examination of VEGF-C expression in biopsy specimens might be beneficial in predicting pelvic lymph node metastasis.

There is paucity of lymphangiogenesis studies in relation to bladder cancer. In one study, the number of D2-40-positive lymphatics in stromal tissue of human bladder cancer was higher than in normal bladder tissue [131]. However, only in a minority of this type of tumors intratumoral lymphatics have been detected and almost all of these vessels had collapsed. Only peritumoral lymphatic vessel density has been associated with tumor grade in transitional cell carcinomas of the bladder. The 5-year metastasis-free survival rate in patients with low peritumoral lymphatic vessel density appears to be significantly higher than that of patients with high lymphatic vessel density. However, analysis of the independent predictive value of lymphatic vessel density for metastasis-free survival using a multivariate analysis model including pathological tumor stage, tumor grade, and adjuvant therapy, identified pathological tumor stage as the only independent and significant predictive factor, whereas lymphatic vessel density was not.

Although these studies suggest that tumor lymphangiogenesis might be a useful tool for the selection of postoperative management and treatment strategies in patients with bladder cancer, the independent prognostic significance of lymphangiogenesis parameters should be confirmed in future large-scale and prospective studies.

6.3.6 Renal Cell Cancer

Renal cell carcinoma preferentially extends beyond the kidney by invasion of the renal sinus [21,23]. This is usually associated with invasion of renal sinus veins and is likely responsible for subsequent development of hematogenous metastases to lung, liver, bone and other sites. However, lymph node dissections have demonstrated that 7%–17% of patients have hilar or locoregional lymph node metastases, indicating that lymphatic spread also occurs in these tumors [123,129].

VEGF-C and its receptors VEGFR-2 and VEGFR-3 are detected in normal and neoplastic kidney tissues [59]. In two studies, no upregulation of VEGF-C has been found in tumor samples, nor has its expression been associated with tumor grade, patient sex, patient age or tumor size [59, 202]. However, this has recently been contradicted in another study, where VEGF-C mRNA expression, but not VEGF-D or VEGFR-3 expression, was higher in clear cell renal carcinoma when compared to normal renal tissue [13]. However, increased VEGF-C expression did not correlate with lymphatic vessel density, lymphatic endothelial cell proliferation or lymphatic metastasis.

Studies on lymphangiogenesis in renal cell tumors are sparse and conflicting. In two small sample sized studies, few or no lymphatics were detected in the central area of the tumor [22, 75]. Clusters of small to medium-sized lymphatics were mainly identified within inflamed cortex and inflamed renal sinus both outside a pseudocapusle and intermingled with tumor cells at an invasive front [22]. On the other hand, in another study about one third of renal tumors had D2-40-positive intratumoral lymphatics, frequently located in inflammatory response areas [13]. However, peritumoral lymphatic vessel density was higher than intratumoral lymphatic vessel density. Although proliferating lymphatic endothelial cells were observed in renal cell tumors, this fraction was significantly lower than in normal renal tissue. These results do not suggest an important role for lymphangiogenesis in this type of tumors.

6.3.7 Melanoma

Malignant melanoma of the skin is an aggressive, therapy-resistant malignancy of melanocytes. Despite early detection, both the incidence and mortality of cutaneous melanoma are still increasing worldwide [62], resulting in an increasing public health problem. Initially, the primary tumor grows horizontally through the epidermis, but later, for reasons poorly understood, it will begin to invade vertically, with a direct correlation between the thickness of this vertical growth phase component of the tumor and the likelihood of metastasis [65]. Because about half of all cutaneous melanoma patients with tumor progression first develop regional lymph node metastases [127], it is thought that the tumor preferentially spreads through the lymphatic system. The early identification of metastatic disease is important, as it determines the requirement for adjuvant therapy and further management. Tumor thickness is currently the most sensitive parameter for predicting the metastatic risk of cutaneous melanoma [12]. However, the prognostic significance of tumor thickness is limited because a considerable proportion (15%) of patients with thin tumors (< 1 mm) also go on to develop metastatic disease, whereas other patients with thick melanomas have long-term survival period [87]. There is currently no consensus on the frequency of follow-up or recommendations for surveillance testing for all patients with melanoma, since there is no effective method to identify the small subgroup of patients with thin but aggressive melanoma. It would be helpful, therefore, to find a prognostic indicator to detect the high-risk patients in the group of thin melanoma.

VEGF-D expression is upregulated in human melanomas compared with melanocytes [2]. No association of VEGF-D expression with lymph node metastasis has been observed [31]. Two studies have found only low-level expression of

VEGF-C by tumor cells that is not associated with metastatic potential in human melanomas [31, 119]. However, in other studies VEGF-C mRNA and protein levels have been correlated with the extent of peritumoral lymphangiogenesis and lymph node metastasis [31, 57, 169]. Moreover, tumor cell expression of bFGF has been associated with the presence of peritumoral lymphatics and with lymphatic vessel dilation in human melanomas [186].

At present, the relevance of tumor lymphangiogenesis in predicting prognosis of patients with cutaneous melanoma remains unclear. The presence of high peritumoral and intratumoral lymphatic vessel densities has been associated with lymph node metastasis and shorter survival, also in thin melanomas [32, 119, 176]. Hot spots of proliferating intratumoral and peritumoral lymphatics are detected in a large number of melanomas, suggestive for active lymphangiogenesis [32, 186]. Multivariate risk analysis has revealed that the lymphatic vascular area of primary melanomas is the most sensitive prognostic marker for sentinel lymph node metastasis, and is able to more accurately predict which tumors are metastatic to sentinel lymph nodes than the currently used method of measuring tumor thickness [32]. A prognostic index, calculated using lymphatic vessel density, lymphatic invasion and tumor thickness, has been shown to clearly discriminate between those tumors that have metastasized and those that have not done so after at least 6 years [176]. However, further work is required to refine this prognostic index and to evaluate the incidence of false-negative and false-positive prediction. Counterintuitively and contradictory to the above-mentioned studies, however, high numbers of LYVE-1-positive lymphatics in peritumoral and intratumoral areas of cutaneous melanoma have been associated with improved survival on multivariate analysis in another study [186]. An explanation for this finding might be that, for an immunogenic tumor like melanomas, the presence of a large and functional lymphatic network might provide an increased T-cell mediated immune response to tumor cells. Indeed, increased lymphocytic infiltration has been associated with high lymphatic vessel densities. Moreover, a reduction of LYVE-1-positive lymphatics has been found in thicker tumors with a high proliferative rate [186].

6.3.8 Head and Neck Cancer

Head and neck squamous cell carcinoma (HNSCC) is a heterogeneous group of squamous cell cancers arising from different anatomical locations in the oral cavity, pharynx and larynx. Although potentially curable by local radiotherapy and surgical resection, the overall 5-year survival rate is only around 50% [50], largely because of the propensity of some HNSCC tumors to disseminate via the lymphatics [162]. Tumor metastasis to cervical nodes is the single most important prognostic factor in patients with HNSCC. An accurate clinical assessment of lymph node status is crucial in treatment planning. There is general agreement that neck dissection is indicated when there are clinically detectable lymph node metastases. However, controversies remain about the management of the clinically N0 neck. The N0 neck is at risk of harboring occult metastases and is often treated electively if this risk is considered to be above 20%. However, elective neck dissections cause overtreatment for the majority of these patients. The development of reliable prognostic markers could enable more individualized treatment planning and effective therapy for patients with HNSCC. Hence, there is urgent need to identify characteristics of the primary tumor that might predict tumor dissemination via the lymphatics. The significances of both lymphatic vessel density and the expressions of lymphangiogenic growth factors in tumor cells as predictors of outcome in HNSCC have been determined.

HNSCC cell lines and tumors produce increased levels of lymphangiogenic factors such as VEGF-C and VEGF-A when compared to normal cells [17, 26, 111]. High mRNA and protein expressions of VEGF-C are correlated with lymph node metastasis and poorer survival in patients with HNSCC [11,26,67,107]. Multivariate analysis has demonstrated that tumor thickness (distance from the surface of the epithelium to the deepest invading tumour island or cell), nuclear pleomorphism, pattern of invasion and immunohistochemical expression of VEGFR-3 and VEGF-C are associated with delayed neck metastasis in early stage tumors with a clinically N0 neck [223]. Analysis of lymphatic growth factor expression may therefore help to identify patients who would benefit from a neck dissection or irradiation by predicting the likelihood of lymph node metastasis.

In HNSCC, the intratumoral localization of lymphatics is clearly strongly associated with nodal metastasis and a higher risk for local relapse as well as with poor disease-specific prognosis [10, 47, 67, 106, 124, 140]. In oropharyngeal carcinoma, discrete hotspots of intratumoral small proliferating lymphatics have been observed, and a high intratumoral lymphatic vessel density is found to be associated with neck node metastases and an infiltrating margin of tumor invasion [17]. Double immunostaining with LYVE-1 or podoplanin and a proliferation marker has shown that actively proliferating intratumoral lymphatics are present in HNSCC [17, 106]. A small proportion of the intratumoral proliferating lymphatics contain cancer emboli [106]. In contrast, no dividing nuclei appear in lymphatics either in the normal or VEGF-C-expressing peritumoral tissues, suggesting that the intratumoral lymphatics are proliferating new vessels rather than preexisting lymphatics that have merely been surrounded and entrapped by aggressive tumor mass [17]. The intratumoral lymphatic vessel density might be used as a criterion to separate patients at higher risk of an adverse clinical outcome or as a discriminator in predicting the outcome of patients with no nodal metastases.

Results concerning the role of peritumoral lymphatics in HNSCC dissemination remain controversial. In one study, high peritumoral lymphatic vessel density has offered a markedly better survival capacity for patients with HNSCC [124]. The presence of LYVE-1-positive lymphatics in the peritumoral region was even more favorable for the patient than a total absence of LYVE-1-positive lymphatics. The rationale behind this would be that peritumoral lymphatics facilitate the recruitment of antigen-presenting cells, such as dendritic cells, which then cross-prime cytotoxic T cells in draining lymph nodes. However, other investigators have suggested a significant association between high peritumoral lymphatic vessel density or relative vascular area and lymph node metastasis [47, 106].

In HNSCC, the presence of intratumoral lymphatic vessels was found to be associated with a significantly higher risk of local disease recurrence and a poorer prognosis, but only two studies found these correlations on multivariate analysis [67, 106]. However, results of the above-mentioned studies support the possibility of using the determination of tumor lymphangiogenesis to identify patients with HNSCC who are at risk of developing cervical lymph node metastasis. If a positive correlation between lymphangiogenesis and cervical lymph node metastasis is confirmed in further studies, this parameter could be useful for selecting HNSCC patients who are more susceptible to metastatic spread via lymphatic route to undergo elective cervical lymph node dissection.

6.3.9 Thyroid Cancer

Papillary and follicular carcinomas account for 90% of all thyroid carcinomas [41]. Most thyroid cancers are slow-growing, easily treatable tumors with an excellent prognosis after surgical resection and targeted medical therapy. Unfortunately, a considerable number of patients, approximately 30% have recurrent disease [187]. Distant metastases are present in about 20% of patients with recurrent cancer. It is thus of utmost importance for clinicians to identify tumors with more aggressive biology and treat them accordingly with more aggressive regimens. Strong independent prognostic factors for patients with thyroid cancer include age, gender, histological type, vascular invasion and lymph node and distant metastases [41].

Papillary and follicular carcinomas have a different propensity for lymph node metastasis: papillary thyroid cancer tends to metastasize to regional lymph nodes, whereas follicular thyroid cancer usually metastasizes by a hematogenous rather than by a lymphatic route. Accordingly, papillary thyroid cancers have increased VEGF-C expression when compared to follicular and other thyroid malignancies [33, 45, 73, 79]. High VEGF-C expression is associated with lymph node metastasis and lymphatic invasion in this type of thyroid tumors [79, 114, 195, 241]. Also VEGF-D mRNA transcript levels and VEGF-D immunoreactivity correlate with lymphatic vessel density and lymph node metastasis in papillary thyroid carcinoma [106, 142, 144].

Numerous morphologically abnormal intratumoral lymphatics in papillary thyroid carcinomas have been observed [61, 237]. The development of intratumoral lymphatics appears to be associated with multifocal disease and presence of lymph node metastases at presentation, but is not a significant predictor of tumor recurrence [61, 114]. In follicular thyroid carcinomas, intratumoral lymphatics are almost non-existent [54]. It thus seems that lymphangiogenesis and lymphatic growth factor expression are particularly high in papillary thyroid cancers, tumors prone to lymphatic metastases. The prognostic significance of tumor lymphangiogenesis in this type of cancer remains to be determined.

6.3.10 Esophageal Cancer

Esophageal adenocarcinoma is currently the most rapidly increasing cancer in the United States and Western Europe [206]. This type of tumor, as well as squamous cell carcinoma of the esophagus, are usually detected at an advanced stage and patients' survival is low, despite improvements in surgical resection and (neo-)adjuvant therapy. In fact, less than 20% of patients with advanced esophageal cancer benefit from a neoadjuvant therapy. Moreover, a significant proportion of patients with early-stage disease develop locally recurrent tumors or distant metastases within a short period after curative surgery. Therefore prognostic markers are needed that allow a proper selection of patients for chemotherapy who are at high risk for tumor recurrence after a successful resection.

Serum VEGF-C levels are elevated in patients with esophageal squamous cell carcinoma and are associated with lymph node involvement, especially in advanced cancers [101]. Serum VEGF-C as a marker of lymph node involvement has been characterized by a sensitivity of 76% and a specificity of 58%. Furthermore, VEGF-C expression is detected in the cytoplasm of esophageal carcinoma cells and stromal cells but not in normal mucosa [37, 96, 98, 132, 149, 158]. VEGF-C expression is associated with neoplastic progression in the esophageal mucosa since increased expression in Barrett's epithelium as it progresses through dysplasia to adenocarcinoma has been observed [11]. In parallel, expression of VEGFR-3 was also upregulated, particularly in the dysplasia and adenocarcinoma stages. Furthermore, in esophageal squamous cell carcinoma, VEGF-C-positivity is significantly correlated with tumor stage, histological grade, lymphatic and venous invasion, depth of tumor invasion and lymph node metastasis [37, 94, 96, 98, 117, 122, 132, 149, 158]. Moreover, lesions with VEGF-C mRNA expression have a higher microvessel density, immunohistochemically determined with an anti-CD31/-CD34 antibody, than those without VEGF-C expression [37, 96, 117]. The clinical impact of the association between VEGF-C expression and prognosis is not fully understood. Nevertheless, the prognosis for patients with VEGF-C-positive tumors is poorer than that for patients with VEGF-C-negative tumors on univariate analysis, suggesting that VEGF-C is an important predictor of biological behavior in esophageal squamous cell carcinoma [94]. However, in patients with adenocarcinoma of the esophagus, VEGF-C expression fails to give prognostic information, indicating that VEGF-C might not play an important role in progression in this type of tumors [132]. The role of VEGF-D in esophageal cancer remains to be explored.

Reports concerning the association of lymphatic vessel density with pathological variables in esophageal cancer are conflicting. Compared with normal lymphatic vessel density, peritumoral and intratumoral lymphatic densities are significantly increased in esophageal cancer [134]. In addition, peritumoral lymphatic density seems to be higher than intratumoral lymphatic vessel density of the same tumor. In one study, no correlations between lymphatic vessel density and any pathological variables were found [134]. Only lymphatic invasion in the peritumoral compartment, restricted at the mucosa and submucosa, was associated with lymph node metastasis. However, in another study, lymphatic vessel density, determined by immunohistochemistry for podoplanin, was higher in advanced tumors with lymphatic invasion [146]. Moreover, the survival rate of patients with a low lymphatic vessel density tended to be higher than that of patients with a high lymphatic vessel density. These last results suggest that the evaluation of lymphatic vessel density might be useful in predicting the prognosis in patients with esophageal carcinomas, although this needs to be further investigated.

6.3.11 Colorectal Cancer

Colorectal cancer is the second leading cause of cancer death in the United States. Approximately 75% of patients with colorectal cancer present with localized disease. Patients with colorectal cancer initially presenting with resectable tumors and tumor-free lymph nodes are generally considered as patients at low risk for recurrence. Therefore, adjuvant therapy is not recommended in these cases. Despite the low tumor stage, about 30%–40% of these patients subsequently develop recurrent disease [34]. It is in this patient group that prognostic markers may identify a patient subgroup at high risk for disease relapse who may also benefit from adjuvant therapy.

Circulating VEGF-C levels, but not VEGF-D levels, are higher in patients with colorectal carcinoma than in healthy controls and are associated with the presence of lymph node metastasis [39, 40, 53, 231]. Serum VEGF-C levels have reached a sensitivity of 81% and a specificity of 76%, suggesting that circulating VEGF-C levels might provide additional information for distinguishing the absence from the presence of lymph node involvement in patients with colorectal carcinoma [231]. Also VEGF-C expression in the cytoplasm of colorectal cancer cells is significantly increased compared to normal mucosa [5, 53, 63, 221, 230]. Both VEGF-C mRNA and protein expressions have been reported to be associated with lymphatic and venous invasion, lymph node status, Dukes' stage, liver metastasis, depth of invasion and poorer histological grade in colorectal carcinoma [5, 52, 71, 78, 90, 91, 93, 99, 118, 157, 182, 221, 232]. The survival time of patients with VEGF-C-positive tumors is significantly shorter than of patients with VEGF-C-negative tumors [52, 71, 85, 157, 182]. However, other studies have not demonstrated a relationship between VEGF-C levels of expression and lymph node metastasis and colorectal cancer [53, 159]. Also VEGFR-3 expression in colorectal cancer cells has been associated with poorer overall survival [230]. The impact of VEGF-D on tumor progression and on prognosis of colorectal cancer remains elusive. Whereas some studies have reported only low levels of VEGF-D expression in colorectal cancer compared to normal mucosa and no or an inverse correlation with lymph node metastasis [53, 63, 90, 91], others did find a significant relationship between the presence of high VEGF-D expression and lymph node metastasis and disease-free and overall survival of patients with colorectal cancer [51, 71, 157, 227].

Intratumoral lymphatics are observed in the majority of colon carcinomas [46, 105] and the proliferating activity of lymphatics is significantly increased in colorectal carcinoma tissues compared with their normal counterparts [154]. Proliferating lymphatics have been identified as doubly labeled with anti-podoplanin and Ki-67 antibodies and negative for the cytokeratin-22 antibody. No significant positive correlations between the proliferation index of lymphatics and clinical outcome have been observed. Whether a relationship exists between lymphatic vessel proliferation and expression of lymphangiogenic factors such as VEGFs remains to be investigated. Intratumoral lymphatic vessel density and lymphatic vessel density at the tumor border are higher in cases with lymph node metastasis than in cases without metastasis [105,113,121,136,166,221]. Survival rates are also significantly lower in patients with high lymphatic vessel density [121, 136].

6.3.12 Gastric Cancer

Gastric cancer is one of the leading causes of cancer deaths worldwide. Recent advances have enabled the early detection of gastric carcinoma by endoscopy. Endoscopic mucosal resection is accepted as a treatment option for cases of early gastric cancer, confined to the mucosa or submucosa, where the probability of lymph node metastasis is low [155]. However, even when the carcinoma is completely resected, additional surgery is necessary when lymph node metastasis appears likely or when cancer cells have invaded the submucosa. Preoperative staging, including endoscopic ultrasonography, is not specific enough to identify the presence of lymph node metastases [6]. Therefore, reliable markers for lymph node metastasis that could be applied to endoscopic mucosal resection specimens would be very useful. It has been shown that lymphatic vessel invasion, histological ulceration of the tumor, and tumor diameter (> 30 mm) are independent risk factors for regional lymph node metastasis [234].

The measurement of lymphatic growth factor expression in gastric cancer could be of relevance with respect to predicting disease outcome. Increased expression of VEGF-C in the primary tumor correlates with increased dissemination of tumor cells to regional lymph nodes in gastric cancer [81, 82, 97, 156, 220, 239, 242] and with lymphatic invasion [74,82,156,175,239]. Furthermore, a relationship has been found between the expression of VEGF-C in tumor tissues and poor prognosis as well as reduced survival in gastric cancers [74, 190, 220, 239]. The serum VEGF-C levels in patients with gastric cancer are higher than in controls [220]. Serum VEGF-C is particularly associated with poorly differentiated gastric adenocarcinomas, T3 and T4, lymph node metastasis, distant metastasis, and pathological TNM groups III and IV. Given that serum VEGF-C is an independent factor predicting poor prognosis and is able to predict lymph node metastasis with high sensitivity and specificity, serum VEGF-C might be a useful biomarker for lymph node metastasis in gastric cancer. Also VEGF-D expression has been detected in gastric cancer [81, 175]. The expression of VEGF-D is significantly correlated with tumor size, lymphatic and venous system invasion, lymphatic vessel counts and lymph node metastasis [81, 175]. Multivariate analysis has indicated that VEGF-D expression is an independent prognostic factor for both relapse-free survival and overall survival [81, 175].

Lymphatics in gastric cancer are not restricted to the peritumoral region, but are also present in intratumoral lesions [143]. Increased density of lymphatics of primary tumors closely correlates with lymph node metastasis and lymphatic invasion in human samples of gastric cancer and predicts poor prognosis [97, 143, 177, 220, 240, 242]. A significant correlation between lymphatic vessel density and lymphatic vessel invasion, detected by podoplanin immunohistochemistry, and lymph node metastasis has also been noted in T1 early gastric cancer [143]. These result indicate that the assessment of lymphatic vessel density may be a useful predictor of lymph node metastasis, or may become a decision making factor for additional surgery in T1 early gastric cancer.

6.3.13 Pancreatic Cancer

Pancreatic ductal adenocarcinoma, whose nomenclature derives from its histological resemblance to ductal cells, is the most common pancreatic neoplasm and accounts for $>85\%$ of pancreatic tumor cases [66]. It is the fourth leading cause of cancer death in the United States with a dismal 5-years survival rate of 3%–5%. Pancreatic adenocarcinoma is characterized by its aggressive local invasion of adjacent structures, perineural invasion and early lymph node and liver metastasis [141]. Its aggressive biology and resistance to conventional and targeted therapeutic agents leads to a typical clinical presentation of incurable disease at the time of diagnosis. Endocrine tumors of the pancreas are uncommon tumors, representing 1% to 2% of all pancreatic neoplasms [48]. The tumors tend to have an indolent behavior and long-term survival is common. The classification of these tumors remains controversial, and prognosis is difficult to predict using classical histopathological malignancy criteria (size, cellular atypia, necrosis, mitotic activity and vascular invasion), but important features include metastasis and invasion of adjacent structures.

In pancreatic adenocarcinomas, results on lymphatic growth factor expression are conflicting. In one study, no overexpression of lymphangiogenic cytokines, such as VEGF-C and VEGF-D, has been found [179]. However, other studies have shown that mRNA expression levels of VEGF-C and VEGFR-3 are indeed upregulated when compared to normal pancreatic tissue [171,196]. Immunohistochemical analysis revealed VEGF-C protein expression in the cytoplasm of ductal-like cancer cells and VEGFR-3 protein expression in pancreatic cancer stromal cells and microvessels [171]. The presence of VEGF-C in these tumors was associated with increased lymph node metastasis [103, 196] and with patients' survival [103], although this could not be confirmed in other studies [171, 196]. In pancreatic endocrine tumors, mainly tumor cells and some peritumoral pancreatic islets show VEGF-C expression [178]. Tumoral VEGF-C expression correlates with glucagon expression, pancreatic polypeptide expression and malignant phenotype. No association has been found between VEGF-C expression and functional status, lymph vessel invasion, or lymph node metastasis. VEGF-D mRNA expression is not observed in most pancreatic endocrine tumors.

Intratumoral and peritumoral lymphatic vessel densities are slightly increased in pancreatic adenocarcinoma compared with the normal pancreas [179]. However, no correlation between lymphatic vessel density and any of the biological features of pancreatic adenocarcinomas, including lymph node metastasis and patient survival, have been observed in these tumors. Also pancreatic endocrine tumors exhibit higher intratumoral lymphatic vessel density values than the normal pancreas, although the majority of the lymphatics are collapsed [165, 178]. In tumors with high lymphatic vessel density, proliferating lymphatic endothelial cells, identified via a double immunostaining for podoplanin and PCNA, have been observed [178]. The observations that high intratumoral lymphatic vessel density is associated with lymphatic invasion and with angioinvasive/metastatic tumor characteristics have indicated that intratumoral lymphangiogenesis promotes the malignant progression [165, 178]. In contrast, peritumoral lymphatic vessel density does not seem to be of any clinicopathological significance in pancreatic endocrine tumors [178].

6.3.14 Hepatocellular Cancer

Hepatocellular carcinoma is one of the most common cancers worldwide, with the highest incidence in regions with high prevalence of chronic viral hepatitis infection, especially hepatitis B infection. Hepatocellular carcinoma commonly metastasizes to lungs, lymph nodes, adrenal gland and bones, including the skull. Despite many available treatment options, the prognosis remains poor. Surgical resection or liver transplantation still represents the only potentially curative treatments for HCC.

VEGF-C expression is significantly stronger in poorly differentiated hepatocellular carcinomas than in well or moderately differentiated hepatocellular carcinomas [233]. The frequency of intrahepatic recurrence tends to be higher and extrahepatic metastasis is significantly higher in cases which have VEGF-C expression in the tumor casts of the intrahepatic portal/hepatic vein branches than other cases without the expression. Disease-free survival time tends to be shorter in cases with VEGF-C expression in tumor casts of the portal/hepatic vein than in those without VEGF-C expression. Thus, VEGF-C expression is related to the progression of hepatocellular carcinoma and VEGF-C expression in tumor casts of the intrahepatic portal/hepatic vein is considered to be a factor indicating recurrence/metastasis sites.

In human hepatocellular carcinoma and liver metastases, Prox-1-LYVE-1-double-positive lymphatics are not present in the tumor parenchyma or in the intratumoral septa of connective tissue but are restricted to the tumor margin and surrounding liver [139]. No correlation or even a trend between lymphatic vessel density and any tumor parameter, including lymphatic invasion, lymph node status and patient survival has been indicated in this type of tumors.

6.3.15 Lung Cancer

Primary lung cancer is the leading cause of cancer-related death in most industrialized countries. Non-small cell lung cancer (NSCLC) accounts for 70%–80% of primary lung cancers and is well known for its ability to involve regional lymph nodes even at the early stages of tumor growth. The tumor-node-metastasis system is generally used in the evaluation of tumor progression and nodal involvement as well as distant metastasis is the most critical factor to determine the prognosis and to guide therapeutic decisions [137]. Patients with early-stage (I and II) NSCLC tumors are treated by complete surgical resection with or without adjuvant chemotherapy while stage III patients require combined modality approaches that may include chemotherapy, radiation and surgery. Nevertheless, the overall 5-year survival rates of patients remain relatively poor, ranging from 70% for stage IA patients to 25% for stage IIIA patients whose tumors are surgically resectable [138].

The evaluation of VEGF-C in circulation has proven to give additional information for discriminating between the absence and the presence of lymph node metastasis [192–194]. NSCLC patients with lymph node metastasis have higher serum VEGF-C concentrations than patients without lymph node metastasis and circulating VEGF-C levels have reached a sensitivity of 70% to 85% and a specificity of 68% to 77% for prediction of lymph node status [192–194]. Moreover, a combination assay of circulating VEGF-C, MMP-9 and VEGF-A has improved accuracy in detecting lymph node metastasis [192]. This type of protein assays can be easily and frequently be performed because of their minimal invasiveness in comparison with examinations using surgical obtained tissue. However, its diagnostic value is not inferior to that of PET or mediastinoscopy.

Although recent clinical studies have examined the abundance of lymphatic growth factors in lung cancer, the results have been difficult to interpret. Some studies have reported a significant positive correlation between tumor-VEGF-C and lymphatic invasion and lymph node (micro-) metastasis [86, 110, 152, 246] whereas others failed to detect a significant relationship between VEGF-C expression and lymph node status [9, 151, 191]. Nevertheless, the status of VEGF-C in tumor cells appears to be a significant prognostic factor in NSCLC [9, 38, 99, 110, 151, 246]. In one study, a low ratio of VEGF-D:VEGF-C in lung adenocarcinoma has been shown to be associated with lymphatic invasion and lymph node metastasis [148]. VEGFR-3 is expressed both on tumor cells and on endothelial cells of microvessels in NSCLC, and there is a significant correlation between VEGFR-3-positive endothelial cell density and VEGFR-3 expression on tumor cells [27]. The expression of VEGFR-3 is closely related with lymph node metastasis and TNM stage [109]. Both higher VEGFR-3-positive endothelial cell density and higher tumor cell-VEGFR-3 are significant and independent prognostic factors [27, 99]. In one study, the balance of VEGF-C and VEGFR-3 expression levels in the tumor was shown to affect lymph node metastasis [191]. The VEGF-C/VEGFR-3 ratio of the node-positive group was significantly higher than that of the node-negative group. Since serum VEGF-C levels of patients with lymph node metastasis are significantly increased in NSCLC [192–194], one might speculate that when the serum VEGF-C level is excessive, and excess VEGF-C cannot bind VEGFR-3 in the cancer cells, it binds to VEGFR-3 in lymphatics, promoting lymphangiogenesis and lymph node metastasis.

In human samples of NSCLC a decrease in lymphatic vessel density from peritumoral lung tissue towards the tumor center can be observed [163]. In fact, in the study by Koukourakis et al., using LYVE-1 as a marker, lymphatics were only noted at the tumor periphery and not within the main tumor mass [100]. However, this could be explained by the observation that LYVE-1 seems to stain only a subset of lymphatics when compared with podoplanin [163]. The presence of lymphangiogenesis in NSCLC has been associated with the growth pattern [163]. In destructively growing angiogenic tumors, LYVE-1+ lymphatics are detected exclusively at the tumor periphery and in the peritumoral host tissue. In contrast, a significant proportion of nonangiogenic NSCLCs contain lymphatics positive for LYVE-1 (and also for podoplanin) both in the tumor center and at the tumor periphery. Moreover, angiogenic tumors have actively sprouting lymphatics (D2-40 and Ki-67 double-positive) in all of the investigated tumor areas, whereas nonangiogenic tumors show no Ki67 staining intratumorally. This suggests that nonangiogenic NSCLCs mainly co-opt host tissue lymphatics during their growth, in contrast to angiogenic tumors, which expand with concomitant lymphangiogenesis. Other studies have shown that the number of lymphatics (assessed using podoplanin as a marker) in lymph node-positive NSCLC tumors is increased when compared to lymph node-negative NSCLC tumors and is a significant and independent prognostic factor on multivariate analysis [3, 222].

6.4 Conclusions

The lymphatic invasion of tumor cells to regional lymph nodes is an important indicator of poor prognosis in many types of malignant tumors. The discovery of specific markers of the lymphatic endothelium in the last decade has enabled the study of lymphatic biology and the clinical implications of lymphangiogenesis. Many studies have revealed that the growth of lymphatic vessels in the vicinity of solid tumors correlates well with lymph node metastasis. VEGF-C and VEGF-D have been identified as the main drivers of the lymphangiogenic process via binding to VEGFR-3. There is now evidence to suggest that a significant correlation between the expression of these molecules and clinicopathological variables exists in several human cancers. These observations suggests that the assessment of the ongoing lymphangiogenesis might be of particular importance in improving the prognostic stratification of patients with cancer, so that patients can receive treatment at an earlier stage of the diagnosis and that unnecessary risk can be avoided to those patients who do not need additional treatment. However, results from different centers concerning the role of lymphangiogenesis-associated parameters having potential influence on tumor behavior and patients' prognosis are often contradictory. A variety of problems can explain for these discrepancies, such as general methodological differences, differences in patients' selection and the use of assays that are not standardized or reproducible. An international consensus report concerning the methodology of lymphangiogenesis quantification in solid tumors has recently been published with the intention to initiate discussion amongst researchers on the standardization of lymphangiogenesis quantification methods. This is imperative to improve the generalization of study results and for lymphangiogenesis assessment to be adopted into clinical practice.

Abbreviations

SLN	sentinel lymph node
H&E	haematoxylin and eosin
VEGF	vascular endothelial growth factor
RT-PCR	reverse-transcriptase polymerase chain reaction
ELISA	enzyme-linked immunosorbent assay
FIGO	International Federation of Obstetrics and Gynecology
PSA	prostate specific antigen
HNSCC	head and neck squamous cell carcinoma
NSCLC	non-small cell lung cancer

References

1. Achen MG, Williams RA, Baldwin ME, et al. The angiogenic and lymphangiogenic factor vascular endothelial growth factor-D exhibits a paracrine mode of action in cancer. Growth Factors 2002, 20: 99–107.
2. Achen MG, Williams RA, Minekus MP, et al. Localization of vascular endothelial growth factor-D in malignant melanoma suggests a role in tumour angiogenesis. J Pathol 2001, 193: 147–54.
3. Adachi Y, Nakamura H, Kitamura Y, et al. Lymphatic vessel density in pulmonary adenocarcinoma immunohistochemically evaluated with anti-podoplanin or anti-D2-40 antibody is correlated with lymphatic invasion or lymph node metastases. Pathol Int 2007, 57: 171–7.
4. Agarwal B, Saxena R, Morimiya A, Mehrotra S, Badve S. Lymphangiogenesis does not occur in breast cancer. Am J Surg Pathol 2005, 29: 1449–55.
5. Akagi K, Ikeda Y, Miyazaki M, et al. Vascular endothelial growth factor-C (VEGF-C) expression in human colorectal cancer tissues. Br J Cancer 2000, 83: 887–91.
6. Akahoshi K, Chijiiwa Y, Hamada S, et al. Endoscopic ultrasonography: a promising method for assessing the prospects of endoscopic mucosal resection in early gastric cancer. Endoscopy 1997, 29: 614–19.
7. Al Mowallad A, Kirwan C, Byrne G, et al. Vascular endothelial growth factor-C in patients with breast cancer. In Vivo 2007, 21: 549–51.
8. Alexander-Sefre F, Singh N, Ayhan A, et al. Detection of tumour lymphovascular space invasion using dual cytokeratin and CD31 immunohistochemistry. J Clin Pathol 2003, 56: 786–8.
9. Arinaga M, Noguchi T, Takeno S, et al. Clinical significance of vascular endothelial growth factor C and vascular endothelial growth factor receptor 3 in patients with nonsmall cell lung carcinoma. Cancer 2003, 97: 457–64.
10. Audet N, Beasley NJ, MacMillan C, et al. Lymphatic vessel density, nodal metastases, and prognosis in patients with head and neck cancer. Arch Otolaryngol Head Neck Surg 2005, 131: 1065–70.
11. Auvinen MI, Sihvo EI, Ruohtula T, et al. Incipient angiogenesis in Barrett's epithelium and lymphangiogenesis in Barrett's adenocarcinoma. J Clin Oncol 2002, 20: 2971–79.
12. Balch CM, Buzaid AC, Soong SJ, et al. Final version of the American Joint Committee on Cancer staging system for cutaneous melanoma. J Clin Oncol 2001, 19: 3635–48.

13. Baldewijns MM, Roskams T, Ballet V, et al.A low frequency of lymph node metastasis in clear-cell renal cell carcinoma is related to low lymphangiogenic activity. BJU Int 2009, Jan 12. [Epub ahead of print]
14. Bando H, Brokelmann M, Toi M, et al. Immunodetection and quantification of vascular endothelial growth factor receptor-3 in human malignant tumor tissues. Int J Cancer 2004, 111: 184–91.
15. Bando H, Weich HA, Horiguchi S, et al. The association between vascular endothelial growth factor-C, its corresponding receptor, VEGFR-3, and prognosis in primary breast cancer: a study with 193 cases. Oncol Rep 2006, 15: 653–9.
16. Banerji S, Ni J, Wang SX, et al. LYVE-1, a new homologue of the CD44 glycoprotein, is a lymph-specific receptor for hyaluronan. J Cell Biol 1999, 144: 789–801.
17. Beasley NJP, Prevo R, Banerji S, et al. Intratumoral lymphangiogenesis and lymph node metastasis in head and neck cancer. Cancer Res 2002, 62: 1315–20.
18. Birner P, Schindl M, Obermair A, et al. Lymphatic microvessel density as a novel prognostic factor in early-stage invasive cervical cancer. Int J Cancer 2001, 95: 29–33.
19. Bono AV. The global state of prostate cancer: epidemiology and screening in the second millennium. BJU Int 2004, 94 Suppl 3: 1–2.
20. Bono P, Wasenius VM, Heikkila P, et al. High LYVE-1-positive lymphatic vessel numbers are associated with poor outcome in breast cancer. Clin Cancer Res 2004, 10: 7144–9.
21. Bonsib SM. The renal sinus is the principal invasive pathway: a prospective study of 100 renal cell carcinomas. Am J Surg Pathol 2004, 28: 1594–600.
22. Bonsib SM. Renal lymphatics, and lymphatic involvement in sinus vein invasive (pT3b) clear cell renal cell carcinoma: a study of 40 cases. Mod Pathol 2006, 19: 746–53.
23. Bonsib SM, Gibson D, Mhoon M, Greene GF. Renal sinus involvement in renal cell carcinomas. Am J Surg Pathol 2000, 24: 451–8.
24. Breiteneder-Geleff S, Matsui K, Soleiman A, et al. Podoplanin, novel 43-kd membrane protein of glomerular epithelial cells, is down-regulated in puromycin nephrosis. Am J Pathol 1997, 151: 1141–52.
25. Cannistra SA. Cancer of the ovary. N Engl J Med 1993, 329: 1550–9.
26. Charoenrat P, Rhys-Evans P, Eccles SA. Expression of vascular endothelial growth factor family members in head and neck squamous cell carcinoma correlates with lymph node metastasis. Cancer 2001, 92: 556–68.
27. Chen F, Takenaka K, Ogawa E, et al. Flt-4-positive endothelial cell density and its clinical significance in non-small cell lung cancer. Clin Cancer Res 2004, 10: 8548–53.
28. Cheng L, Jones TD, Lin H, et al. Lymphovascular invasion is an independent prognostic factor in prostatic adenocarcinoma. J Urol 2005, 174: 2181–5.
29. Choi WW, Lewis MM, Lawson D, et al. Angiogenic and lymphangiogenic microvessel density in breast carcinoma: correlation with clinicopathologic parameters and VEGF-family gene expression. Mod Pathol 2005, 18: 143–52.
30. Creasman WT, Kohler MF. Is lymph vascular space involvement an independent prognostic factor in early cervical cancer? Gynecol Oncol 2004, 92: 525–9.
31. Dadras SS, Lange-Asschenfeldt B, Velasco P, et al. Tumor lymphangiogenesis predicts melanoma metastasis to sentinel lymph nodes. Mod Pathol 2005, 18: 1232–42.
32. Dadras SS, Paul T, Bertoncini J, et al. Tumor lymphangiogenesis: a novel prognostic indicator for cutaneous melanoma metastasis and survival. Am J Pathol 2003, 162: 1951–60.
33. de la Torre NG, Buley I, Wass JA, Turner HE. Angiogenesis and lymphangiogenesis in thyroid proliferative lesions: relationship to type and tumour behaviour. Endocr Relat Cancer 2006, 13: 931–44.
34. Deans GT, Parks TG, Rowlands BJ, Spence RA. Prognostic factors in colorectal cancer. Br J Surg 1992, 79: 608–13.
35. Delgado G, Bundy B, Zaino R, et al. Prospective surgical-pathological study of disease-free interval in patients with stage IB squamous cell carcinoma of the cervix: a Gynecologic Oncology Group study. Gynecol Oncol 1990, 38: 352–7.

36. Ding MX, Li JC, Tang XH. [Lymphatic metastasis and VEGF-C expression in ovarian epithelial tumors]. Shi Yan Sheng Wu Xue Bao 2003, 36: 445–52.
37. Ding MX, Lin XQ, Fu XY, Zhang N, Li JC. Expression of vascular endothelial growth factor-C and angiogenesis in esophageal squamous cell carcinoma. World J Gastroenterol 2006, 12: 4582–85.
38. Dong X, Qiu XS, Wang EH, Li QC, Gu W. [Expression of vascular endothelial growth factor (VEGF) C and VEGF receptor 3 in non-small cell lung cancer]. Zhonghua Bing Li Xue Za Zhi 2003, 32: 128–32.
39. Duff SE, Li C, Renehan A, O'Dwyer ST, Kumar S. Immunodetection and molecular forms of plasma vascular endothelial growth factor-C. Int J Oncol 2003, 22: 339–43.
40. Duff SE, Saunders M, McCredie V, et al. Pre-operative plasma levels of vascular endothelial growth factor A, C and D in patients with colorectal cancer. Clin Oncol (R Coll Radiol) 2005, 17: 367–71.
41. Duntas L, Grab-Duntas BM. Risk and prognostic factors for differentiated thyroid cancer. Hell J Nucl Med 2006, 9: 156–62.
42. Duque JL, Loughlin KR. An overview of the treatment of superficial bladder cancer. Intravesical chemotherapy. Urol Clin North Am 2000, 27: 125–35, X.
43. Ennis RD, Petrylak DP, Singh P, et al. The effect of cystectomy, and perioperative methotrexate, vinblastine, doxorubicin and cisplatin chemotherapy on the risk and pattern of relapse in patients with muscle invasive bladder cancer. J Urol 2000, 163: 1413–18.
44. Evangelou E, Kyzas PA, Trikalinos TA. Comparison of the diagnostic accuracy of lymphatic endothelium markers: Bayesian approach. Mod Pathol 2005, 18: 1490–7.
45. Fellmer PT, Sato K, Tanaka R, et al. Vascular endothelial growth factor-C gene expression in papillary and follicular thyroid carcinomas. Surgery 1999, 126: 1056–61.
46. Fogt F, Zimmerman RL, Ross HM, Daly T, Gausas RE. Identification of lymphatic vessels in malignant, adenomatous and normal colonic mucosa using the novel immunostain D2-40. Oncol Rep 2004, 11: 47–50.
47. Franchi A, Gallo O, Massi D, Baroni G, Santucci M. Tumor lymphangiogenesis in head and neck squamous cell carcinoma: a morphometric study with clinical correlations. Cancer 2004, 101: 973–8.
48. Frankel WL. Update on pancreatic endocrine tumors. Arch Pathol Lab Med 2006, 130: 963–6.
49. Fujimoto J, Toyoki H, Sato E, Sakaguchi H, Tamaya T. Clinical implication of expression of vascular endothelial growth factor-C in metastatic lymph nodes of uterine cervical cancers. Br J Cancer 2004, 91: 466–9.
50. Fuller CD, Wang SJ, Thomas CR, Jr., et al. Conditional survival in head and neck squamous cell carcinoma: results from the SEER dataset 1973–1998. Cancer 2007, 109: 1331–43.
51. Funaki H, Nishimura G, Harada S, et al. Expression of vascular endothelial growth factor D is associated with lymph node metastasis in human colorectal carcinoma. Oncology 2003, 64: 416–22.
52. Furudoi A, Tanaka S, Haruma K, et al. Clinical significance of vascular endothelial growth factor C expression and angiogenesis at the deepest invasive site of advanced colorectal carcinoma. Oncology 2002, 62: 157–66.
53. George ML, Tutton MG, Janssen F, et al. VEGF-A, VEGF-C, and VEGF-D in colorectal cancer progression. Neoplasia 2001, 3: 420–7.
54. Giorgadze TA, Baloch ZW, Pasha T, Zhang PJ, Livolsi VA. Lymphatic and blood vessel density in the follicular patterned lesions of thyroid. Mod Pathol 2005, 18: 1424–31.
55. Goldhirsch A, Glick JH, Gelber RD, et al. Meeting highlights: international expert consensus on the primary therapy of early breast cancer 2005. Ann Oncol 2005, 16: 1569–83.
56. Gombos Z, Xu X, Chu CS, Zhang PJ, Acs G. Peritumoral lymphatic vessel density and vascular endothelial growth factor C expression in early-stage squamous cell carcinoma of the uterine cervix. Clin Cancer Res 2005, 11: 8364–71.

57. Goydos JS, Gorski DH. Vascular endothelial growth factor C mRNA expression correlates with stage of progression in patients with melanoma. Clin Cancer Res 2003, 9: 5962–7.
58. Gunnigham SP, Currie MJ, Han C, et al.The short form of the alternatively spliced flt-4 but not its ligand vascular endothelial growth factor C is related to lymph node metastasis in human breast cancers. Clin Cancer Res 2000, 6: 4278–86.
59. Gunningham SP, Currie MJ, Han C, et al. Vascular endothelial growth factor-B and vascular endothelial growth factor-C expression in renal cell carcinomas: regulation by the von Hippel-Lindau gene and hypoxia. Cancer Res 2001, 61: 3206–211.
60. Guo X, Chen L, Lang R, et al. Invasive micropapillary carcinoma of the breast: association of pathologic features with lymph node metastasis. Am J Clin Pathol 2006, 126: 740–6.
61. Hall FT, Freeman JL, Asa SL, Jackson DG, Beasley NJ. Intratumoral lymphatics and lymph node metastases in papillary thyroid carcinoma. Arch Otolaryngol Head Neck Surg 2003, 129: 716–19.
62. Hall PN, Javaid M, de Takats PG. Therapeutic trends in cutaneous melanoma. Hosp Med 1999, 60: 39–3.
63. Hanrahan V, Currie MJ, Gunningham SP, et al. The angiogenic switch for vascular endothelial growth factor (VEGF)-A, VEGF-B, VEGF-C, and VEGF-D in the adenoma-carcinoma sequence during colorectal cancer progression. J Pathol 2003, 200: 183–94.
64. Hashimoto I, Kodama J, Seki N, et al. Vascular endothelial growth factor-C expression and its relationship to pelvic lymph node status in invasive cervical cancer. Br J Cancer 2001, 85: 93–7.
65. Heasley DD, Toda S, Mihm MC, Jr. Pathology of malignant melanoma. Surg Clin North Am 1996, 76: 1223–55.
66. Hezel AF, Kimmelman AC, Stanger BZ, Bardeesy N, Depinho RA. Genetics and biology of pancreatic ductal adenocarcinoma. Genes Dev 2006, 20: 1218–49.
67. Hinojar-Gutierrez A, Fernandez-Contreras ME, Gonzalez-Gonzalez R, et al. Intratumoral lymphatic vessels and VEGF-C expression are predictive factors of lymph node relapse in T1-T4 N0 laryngopharyngeal squamous cell carcinoma. Ann Surg Oncol 2007, 14: 248–57.
68. Hoar FJ, Chaudhri S, Wadley MS, Stonelake PS. Co-expression of vascular endothelial growth factor C (VEGF-C) and c-erbB2 in human breast carcinoma. Eur J Cancer 2003, 39: 1698–703.
69. Hsieh CY, Chen CA, Chou CH, et al. Overexpression of Her-2/NEU in epithelial ovarian carcinoma induces vascular endothelial growth factor C by activating NF-kappa B: implications for malignant ascites formation and tumor lymphangiogenesis. J Biomed Sci 2004, 11: 249–59.
70. Hu SE, Zhang YJ, Cui YM, Zhang HQ. [Expression of vascular endothelial growth factor A and C in human breast cancer and their significance]. Ai Zheng 2005, 24: 1076–9.
71. Hu WG, Li JW, Feng B, et al. Vascular endothelial growth factors C and D represent novel prognostic markers in colorectal carcinoma using quantitative image analysis. Eur Surg Res 2007, 39: 229–38.
72. Huang JH, Li Y, Liu L, et al. [Lymphangiogenes and location of tumor lymphatic vessels induced by VEGF-C in primary breast carcer]. Zhong Nan Da Xue Xue Bao Yi Xue Ban 2006, 31: 36–9, 51.
73. Hung CJ, Ginzinger DG, Zarnegar R, et al. Expression of vascular endothelial growth factor-C in benign and malignant thyroid tumors. J Clin Endocrinol Metab 2003, 88: 3694–99.
74. Ichikura T, Tomimatsu S, Ohkura E, Mochizuki H. Prognostic significance of the expression of vascular endothelial growth factor (VEGF) and VEGF-C in gastric carcinoma. J Surg Oncol 2001, 78: 132–7.
75. Ishikawa Y, Akasaka Y, Kiguchi H, et al. The human renal lymphatics under normal and pathological conditions. Histopathology 2006, 49: 265–73.
76. Jackson DG. Biology of the lymphatic marker LYVE-1 and applications in research into lymphatic trafficking and lymphangiogenesis. APMIS 2004, 112: 526–38.
77. Jennbacken K, Vallbo C, Wang W, Damber JE. Expression of vascular endothelial growth factor C (VEGF-C) and VEGF receptor-3 in human prostate cancer is associated with regional lymph node metastasis. Prostate 2005, 65: 110–16.

78. Jia YT, Li ZX, He YT, et al. Expression of vascular endothelial growth factor-C and the relationship between lymphangiogenesis and lymphatic metastasis in colorectal cancer. World J Gastroenterol 2004, 10: 3261–3.
79. Jiang HG, Gao M, Tang WP, Li FH, Cai QZ. [Expression and significance of VEGF, VEGF-C, and VEGF-D in papillary thyroid carcinoma]. Ai Zheng 2005, 24: 1136–9.
80. Johnson LA, Prevo R, Clasper S, Jackson DG. Inflammation-induced uptake and degradation of the lymphatic endothelial hyaluronan receptor lyve-1. J Biol Chem 2007, 282: 33671–80.
81. Juttner S, Wissmann C, Jons T, et al. Vascular endothelial growth factor-d and its receptor VEGFR-3: two novel independent prognostic markers in gastric adenocarcinoma. J Clin Oncol 2006, 24: 228–40.
82. Kabashima A, Maehara Y, Kakeji Y, Sugimachi K. Overexpression of vascular endothelial growth factor C is related to lymphogenous metastasis in early gastric carcinoma. Oncology 2001, 60: 146–50.
83. Kahn HJ, Bailey D, Marks A. Monoclonal antibody D2-40, a new marker of lymphatic endothelium, reacts with Kaposi's sarcoma and a subset of angiosarcomas. Mod Pathol 2002, 15: 434–40.
84. Kahn HJ, Marks A. A new monoclonal antibody, D2-40, for detection of lymphatic invasion in primary tumors. Lab Invest 2002, 82: 1255–7.
85. Kaio E, Tanaka S, Kitadai Y, et al. Clinical significance of angiogenic factor expression at the deepest invasive site of advanced colorectal carcinoma. Oncology 2003, 64: 61–73.
86. Kajita T, Ohta Y, Kimura K, et al. The expression of vascular endothelial growth factor C and its receptors in non-small cell lung cancer. Br J Cancer 2001, 85: 255–60.
87. Kalady MF, White RR, Johnson JL, Tyler DS, Seigler HF. Thin melanomas: predictive lethal characteristics from a 30-year clinical experience. Ann Surg 2003, 238: 528–35.
88. Kato T, Prevo R, Steers G, et al. A quantitative analysis of lymphatic vessels in human breast cancer, based on LYVE-1 immunoreactivity. Br J Cancer 2005, 93: 1168–74.
89. Kaushal V, Mukunyadzi P, Dennis RA, et al. Stage-specific characterization of the vascular endothelial growth factor axis in prostate cancer: expression of lymphangiogenic markers is associated with advanced-stage disease. Clin Cancer Res 2005, 11: 584–93.
90. Kawakami M, Furuhata T, Kimura Y, et al. Expression analysis of vascular endothelial growth factors and their relationships to lymph node metastasis in human colorectal cancer. J Exp Clin Cancer Res 2003, 22: 229–37.
91. Kawakami M, Furuhata T, Kimura Y, et al. Quantification of vascular endothelial growth factor-C and its receptor-3 messenger RNA with real-time quantitative polymerase chain reaction as a predictor of lymph node metastasis in human colorectal cancer. Surgery 2003, 133: 300–8.
92. Kazama S, Kitayama J, Watanabe T, Nagawa H. Expression pattern of vascular endothelial growth factor-C in human colorectal normal mucosa and neoplastic mucosa. Hepatogastroenterology 2004, 51: 391–5.
93. Kazama S, Watanabe T, Kanazawa T, Hatano K, Nagawa H. Vascular endothelial growth factor-C (VEGF-C) is a more specific risk factor for lymph node metastasis than VEGF-D in submucosal colorectal cancer. Hepatogastroenterology 2007, 54: 71–6.
94. Kimura Y, Watanabe M, Ohga T, et al. Vascular endothelial growth factor C expression correlates with lymphatic involvement and poor prognosis in patients with esophageal squamous cell carcinoma. Oncol Rep 2003, 10: 1747–51.
95. Kinoshita J, Kitamura K, Kabashima A, et al. Clinical significance of vascular endothelial growth factor-C (VEGF-C) in breast cancer. Breast Cancer Res Treat 2001, 66: 159–64.
96. Kitadai Y, Amioka T, Haruma K, et al. Clinicopathological significance of vascular endothelial growth factor (VEGF)-C in human esophageal squamous cell carcinomas. Int J Cancer 2001, 93: 662–6.
97. Kitadai Y, Kodama M, Cho S, et al. Quantitative analysis of lymphangiogenic markers for predicting metastasis of human gastric carcinoma to lymph nodes. Int J Cancer 2005, 115: 388–92.

98. Kleespies A, Bruns CJ, Jauch KW. Clinical significance of VEGF-A, -C and -D expression in esophageal malignancies. Onkologie 2005, 28: 281–8.
99. Kojima H, Shijubo N, Yamada G, et al. Clinical significance of vascular endothelial growth factor-C and vascular endothelial growth factor receptor 3 in patients with T1 lung adenocarcinoma. Cancer 2005, 104: 1668–77.
100. Koukourakis MI, Giatromanolaki A, Sivridis E, et al. LYVE-1 immunohistochemical assessment of lymphangiogenesis in endometrial and lung cancer. J Clin Pathol 2005, 58: 202–6.
101. Krzystek-Korpacka M, Matusiewicz M, Diakowska D, et al. Up-regulation of VEGF-C secreted by cancer cells and not VEGF-A correlates with clinical evaluation of lymph node metastasis in esophageal squamous cell carcinoma (ESCC). Cancer Lett 2007, 249: 171–7.
102. Kubo H, Fujiwara T, Jussila L, et al. Involvement of vascular endothelial growth factor receptor-3 in maintenance of integrity of endothelial cell lining during tumor angiogenesis. Blood 2000, 96: 546–53.
103. Kurahara H, Takao S, Maemura K, et al. Impact of vascular endothelial growth factor-C and -D expression in human pancreatic cancer: its relationship to lymph node metastasis. Clin Cancer Res 2004, 10: 8413–20.
104. Kurebayashi J, Otsuki T, Kunisue H, et al. Expression of vascular endothelial growth factor (VEGF) family members in breast cancer. Jpn J Cancer Res 1999, 90: 977–81.
105. Kuroyama S, Kobayashi N, Ohbu M, et al. Enzyme histochemical analysis of lymphatic vessels in colon carcinoma: occurrence of lymphangiogenesis within the tumor. Hepatogastroenterology 2005, 52: 1057–61.
106. Kyzas PA, Geleff S, Batistatou A, Agnantis NJ, Stefanou D. Evidence for lymphangiogenesis and its prognostic implications in head and neck squamous cell carcinoma. J Pathol 2005, 206: 170–7.
107. Kyzas PA, Stefanou D, Agnantis NJ. COX-2 expression correlates with VEGF-C and lymph node metastases in patients with head and neck squamous cell carcinoma. Mod Pathol 2005, 18: 153–60.
108. Lerner SP, Skinner E, Skinner DG. Radical cystectomy in regionally advanced bladder cancer. Urol Clin North Am 1992, 19: 713–23.
109. Li J, Hong M, Pan T. Clinical significance of VEGF-C and VEGFR-3 expression in non-small cell lung cancer. J Huazhong Univ Sci Technolog Med Sci 2006, 26: 587–90.
110. Li Q, Dong X, Gu W, Qiu X, Wang E. Clinical significance of co-expression of VEGF-C and VEGFR-3 in non-small cell lung cancer. Chin Med J (Engl) 2003, 116: 727–30.
111. Li QL, Chen FJ, Zeng ZY, et al. [Expression and clinical significance of VEGF-C and Flt-4 in tongue squamous cell carcinoma]. Ai Zheng 2006, 25: 235–40.
112. Li R, Younes M, Wheeler TM, et al. Expression of vascular endothelial growth factor receptor-3 (VEGFR-3) in human prostate. Prostate 2004, 58: 193–9.
113. Liang P, Hong JW, Ubukata H, et al. Increased density and diameter of lymphatic microvessels correlate with lymph node metastasis in early stage invasive colorectal carcinoma. Virchows Arch 2006, 448: 570–5.
114. Liang QC, Wei QY, Fan SQ. [Expression of VEGF-C and angiogenesis, and lymphangiogenesis in papillary thyroid carcinoma]. Zhong Nan Da Xue Xue Bao Yi Xue Ban 2006, 31: 414–6, 419.
115. Liu F, Zhang YJ. [Roles of VEGF-C and its receptor Flt-4 in proliferation and metastasis of primary breast cancer]. Ai Zheng 2003, 22: 1053–56.
116. Loeb S, Roehl KA, Yu X, et al. Lymphovascular invasion in radical prostatectomy specimens: prediction of adverse pathologic features and biochemical progression. Urology 2006, 68: 99–103.
117. Loges S, Clausen H, Reichelt U, et al. Determination of microvessel density by quantitative real-time PCR in esophageal cancer: correlation with histologic methods, angiogenic growth factor expression, and lymph node metastasis. Clin Cancer Res 2007, 13: 76–80.
118. Maeda K, Yashiro M, Nishihara T, et al. Correlation between vascular endothelial growth factor C expression and lymph node metastasis in T1 carcinoma of the colon and rectum. Surg Today 2003, 33: 736–39.

119. Massi D, Puig S, Franchi A, et al. Tumour lymphangiogenesis is a possible predictor of sentinel lymph node status in cutaneous melanoma: a case-control study. J Clin Pathol 2006, 59: 166–73.
120. Mathur SP, Mathur RS, Gray EA, et al. Serum vascular endothelial growth factor C (VEGF-C) as a specific biomarker for advanced cervical cancer: relationship to insulin-like growth factor II (IGF-II), IGF binding protein 3 (IGF-BP3) and VEGF-B. Gynecol Oncol 2005, 98: 467–83.
121. Matsumoto K, Nakayama Y, Inoue Y, et al. Lymphatic microvessel density is an independent prognostic factor in colorectal cancer. Dis Colon Rectum 2007, 50: 308–14.
122. Matsumoto M, Natsugoe S, Okumura H, et al. Overexpression of vascular endothelial growth factor-C correlates with lymph node micrometastasis in submucosal esophageal cancer. J Gastrointest Surg 2006, 10: 1016–22.
123. Matsuyama H, Hirata H, Korenaga Y, et al. Clinical significance of lymph node dissection in renal cell carcinoma. Scand J Urol Nephrol 2005, 39: 30–5.
124. Maula SM, Luukkaa M, Grenman R, et al. Intratumoral lymphatics are essential for the metastatic spread and prognosis in squamous cell carcinomas of the head and neck region. Cancer Res 2003, 63: 1920–26.
125. May M, Kaufmann O, Hammermann F, Loy V, Siegsmund M. Prognostic impact of lymphovascular invasion in radical prostatectomy specimens. BJU Int 2007, 99: 539–44.
126. McDavid K, Lee J, Fulton JP, Tonita J, Thompson TD. Prostate cancer incidence and mortality rates and trends in the United States and Canada. Public Health Rep 2004, 119: 174–86.
127. Meier F, Will S, Ellwanger U, et al. Metastatic pathways and time courses in the orderly progression of cutaneous melanoma. Br J Dermatol 2002, 147: 62–70.
128. Metts MC, Metts JC, Milito SJ, Thomas CR, Jr. Bladder cancer: a review of diagnosis and management. J Natl Med Assoc 2000, 92: 285–94.
129. Minervini A, Lilas L, Morelli G, et al. Regional lymph node dissection in the treatment of renal cell carcinoma: is it useful in patients with no suspected adenopathy before or during surgery? BJU Int 2001, 88: 169–72.
130. Mitsuhashi A, Suzuka K, Yamazawa K, et al. Serum vascular endothelial growth factor (VEGF) and VEGF-C levels as tumor markers in patients with cervical carcinoma. Cancer 2005, 103: 724–30.
131. Miyata Y, Kanda S, Ohba K, et al. Lymphangiogenesis and angiogenesis in bladder cancer: prognostic implications and regulation by vascular endothelial growth factors-A, -C, and -D. Clin Cancer Res 2006, 12: 800–6.
132. Mobius C, Freire J, Becker I, et al. VEGF-C expression in squamous cell carcinoma and adenocarcinoma of the esophagus. World J Surg 2007, 31: 1768–74.
133. Mohammed RA, Green A, El Shikh S, et al. Prognostic significance of vascular endothelial cell growth factors -A, -C and -D in breast cancer and their relationship with angio- and lymphangiogenesis. Br J Cancer 2007, 96: 1092–100.
134. Mori D, Yamasaki F, Shibaki M, Tokunaga O. Lateral peritumoral lymphatic vessel invasion can predict lymph node metastasis in esophageal squamous cell carcinoma. Mod Pathol 2007, 20: 694–700.
135. Morton DL, Wen DR, Wong JH, et al. Technical details of intraoperative lymphatic mapping for early stage melanoma. Arch Surg 1992, 127: 392–9.
136. Mou JH, Yan XC, Li ZP, et al. [Characteristic and clinicopathologic significance of lymphangiogenesis in colorectal cancer]. Zhonghua Bing Li Xue Za Zhi 2005, 34: 348–52.
137. Mountain CF. Revisions in the international system for staging lung cancer. Chest 1997, 111: 1710–17.
138. Mountain CF. The international system for staging lung cancer. Semin Surg Oncol 2000, 18: 106–15.
139. Mouta CC, Nasser SM, di Tomaso E, et al. LYVE-1 is not restricted to the lymph vessels: expression in normal liver blood sinusoids and down-regulation in human liver cancer and cirrhosis. Cancer Res 2001, 61: 8079–84.

140. Munoz-Guerra MF, Marazuela EG, Martin-Villar E, Quintanilla M, Gamallo C. Prognostic significance of intratumoral lymphangiogenesis in squamous cell carcinoma of the oral cavity. Cancer 2004, 100: 553–60.
141. Nagakawa T, Konishi I, Higashino Y, et al. The spread and prognosis of carcinoma in the region of the pancreatic head. Jpn J Surg 1989, 19: 510–18.
142. Nakamura Y, Yasuoka H, Tsujimoto M, et al. Lymph vessel density correlates with nodal status, VEGF-C expression, and prognosis in breast cancer. Breast Cancer Res Treat 2005, 91: 125–32.
143. Nakamura Y, Yasuoka H, Tsujimoto M, et al. Importance of lymph vessels in gastric cancer: a prognostic indicator in general and a predictor for lymph node metastasis in early stage cancer. J Clin Pathol 2006, 59: 77–82.
144. Nakamura Y, Yasuoka H, Zuo H, et al. Nitric oxide in papillary thyroid carcinoma: induction of vascular endothelial growth factor D and correlation with lymph node metastasis. J Clin Endocrinol Metab 2006, 91: 1582–5.
145. Nakamura Y, Yasuoka H, Tsujimoto M, et al. Flt-4-Positive vessel density correlates with vascular endothelial growth factor-D expression, nodal status, and prognosis in breast cancer. Clin Cancer Res 2003, 9: 5313–17.
146. Nakayama Y, Matsumoto K, Nagato M, et al. Significance of lymphangiogenesis as assessed by immunohistochemistry for podoplanin in patients with esophageal carcinoma. Anticancer Res 2007, 27: 619–25.
147. Niakosari F, Kahn HJ, Marks A, From L. Detection of lymphatic invasion in primary melanoma with monoclonal antibody D2-40: a new selective immunohistochemical marker of lymphatic endothelium. Arch Dermatol 2005, 141: 440–44.
148. Niki T, Iba S, Tokunou M, et al. Expression of vascular endothelial growth factors A, B, C, and D and their relationships to lymph node status in lung adenocarcinoma. Clin Cancer Res 2000, 6: 2431–9.
149. Noguchi T, Takeno S, Shibata T, et al. VEGF-C expression correlates with histological differentiation and metastasis in squamous cell carcinoma of the esophagus. Oncol Rep 2002, 9: 995–9.
150. Obermair A, Wanner C, Bilgi S, et al. Tumor angiogenesis in stage IB cervical cancer: correlation of microvessel density with survival. Am J Obstet Gynecol 1998, 178: 314–19.
151. Ogawa E, Takenaka K, Yanagihara K, et al. Clinical significance of VEGF-C status in tumour cells and stromal macrophages in non-small cell lung cancer patients. Br J Cancer 2004, 91: 498–503.
152. Ohta Y, Nozawa H, Tanaka Y, Oda M, Watanabe Y. Increased vascular endothelial growth factor and vascular endothelial growth factor-c and decreased nm23 expression associated with microdissemination in the lymph nodes in stage I non-small cell lung cancer. J Thorac Cardiovasc Surg 2000, 119: 804–13.
153. Okada K, Osaki M, Araki K, et al. Expression of hypoxia-inducible factor (HIF-1alpha), VEGF-C and VEGF-D in non-invasive and invasive breast ductal carcinomas. Anticancer Res 2005, 25: 3003–9.
154. Omachi T, Kawai Y, Mizuno R, et al. Immunohistochemical demonstration of proliferating lymphatic vessels in colorectal carcinoma and its clinicopathological significance. Cancer Lett 2007, 246: 167–72.
155. Ono H, Kondo H, Gotoda T, et al. Endoscopic mucosal resection for treatment of early gastric cancer. Gut 2001, 48: 225–9.
156. Onogawa S, Kitadai Y, Amioka T, et al. Expression of vascular endothelial growth factor (VEGF)-C and VEGF-D in early gastric carcinoma: correlation with clinicopathological parameters. Cancer Lett 2005, 226: 85–90.
157. Onogawa S, Kitadai Y, Tanaka S, et al. Expression of VEGF-C and VEGF-D at the invasive edge correlates with lymph node metastasis and prognosis of patients with colorectal carcinoma. Cancer Sci 2004, 95: 32–9.

158. Pang XH, Tian H, Liu ZY, et al. [Significance and expression of vascular endothelial growth factor-C (VEGF-C) in esophageal squamous carcinoma and glioma]. Ai Zheng 2003, 22: 1166–9.
159. Parr C, Jiang WG. Quantitative analysis of lymphangiogenic markers in human colorectal cancer. Int J Oncol 2003, 23: 533–9.
160. Partanen TA, Alitalo K, Miettinen M. Lack of lymphatic vascular specificity of vascular endothelial growth factor receptor 3 in 185 vascular tumors. Cancer 1999, 86: 2406–12.
161. Pepper MS, Tille JC, Nisato R, Skobe M. Lymphangiogenesis and tumor metastasis. Cell Tissue Res 2003, 314: 167–77.
162. Quon H, Liu FF, Cummings BJ. Potential molecular prognostic markers in head and neck squamous cell carcinomas. Head Neck 2001, 23: 147–59.
163. Renyi-Vamos F, Tovari J, Fillinger J, et al. Lymphangiogenesis correlates with lymph node metastasis, prognosis, and angiogenic phenotype in human non-small cell lung cancer. Clin Cancer Res 2005, 11: 7344–53.
164. Roma AA, Magi-Galluzzi C, Kral MA, et al. Peritumoral lymphatic invasion is associated with regional lymph node metastases in prostate adenocarcinoma. Mod Pathol 2006, 19: 392–8.
165. Rubbia-Brandt L, Terris B, Giostra E, et al. Lymphatic vessel density and vascular endothelial growth factor-C expression correlate with malignant behavior in human pancreatic endocrine tumors. Clin Cancer Res 2004, 10: 6919–28.
166. Saad RS, Kordunsky L, Liu YL, et al. Lymphatic microvessel density as prognostic marker in colorectal cancer. Mod Pathol 2006, 19: 1317–23.
167. Salven P, Lymboussaki A, Heikkila P, et al. Vascular endothelial growth factors VEGF-B and VEGF-C are expressed in human tumors. Am J Pathol 1998, 153: 103–8.
168. Schacht V, Dadras SS, Johnson LA, et al. Up-regulation of the lymphatic marker podoplanin, a mucin-type transmembrane glycoprotein, in human squamous cell carcinomas and germ cell tumors. Am J Pathol 2005, 166: 913–21.
169. Schietroma C, Cianfarani F, Lacal PM, et al. Vascular endothelial growth factor-C expression correlates with lymph node localization of human melanoma metastases. Cancer 2003, 98: 789–97.
170. Schmelz M, Franke WW. Complexus adhaerentes, a new group of desmoplakin-containing junctions in endothelial cells: the syndesmos connecting retothelial cells of lymph nodes. Eur J Cell Biol 1993, 61: 274–89.
171. Schneider M, Buchler P, Giese N, et al. Role of lymphangiogenesis and lymphangiogenic factors during pancreatic cancer progression and lymphatic spread. Int J Oncol 2006, 28: 883–90.
172. Schoppmann SF, Bayer G, Aumayr K, et al. Prognostic value of lymphangiogenesis and lymphovascular invasion in invasive breast cancer. Ann Surg 2004, 240: 306–12.
173. Schoppmann SF, Schindl M, Breiteneder-Geleff S, et al. Inflammatory stromal reaction correlates with lymphatic microvessel density in early-stage cervival cancer. Anticancer Res 2001, 21: 3419–23.
174. Schoppmann SF, Birner P, Stockl J, et al. Tumor-associated macrophages express lymphatic endothelial growth factors and are related to peritumoral lymphangiogenesis. Am J Pathol 2002, 161: 947–56.
175. Shida A, Fujioka S, Ishibashi Y, et al. Prognostic significance of vascular endothelial growth factor D in gastric carcinoma. World J Surg 2005, 29: 1600–607.
176. Shields JD, Borsetti M, Rigby H, et al. Lymphatic density and metastatic spread in human malignant melanoma. Br J Cancer 2004, 90: 693–700.
177. Shimizu K, Kubo H, Yamaguchi K, et al. Suppression of VEGFR-3 signaling inhibits lymph node metastasis in gastric cancer. Cancer Sci 2004, 95: 328–33.
178. Sipos B, Klapper W, Kruse ML, et al. Expression of lymphangiogenic factors and evidence of intratumoral lymphangiogenesis in pancreatic endocrine tumors. Am J Pathol 2004, 165: 1187–97.

179. Sipos B, Kojima M, Tiemann K, et al. Lymphatic spread of ductal pancreatic adenocarcinoma is independent of lymphangiogenesis. J Pathol 2005, 207: 301–12.
180. Sleeman JP, Krishnan J, Kirkin V, Baumann P. Markers for the lymphatic endothelium: in search of the holy grail? Microsc Res Tech 2001, 55: 61–9.
181. Smigal C, Jemal A, Ward E, et al. Trends in breast cancer by race and ethnicity: update 2006. CA Cancer J Clin 2006, 56: 168–83.
182. Soumaoro LT, Uetake H, Takagi Y, et al. Coexpression of VEGF-C and Cox-2 in human colorectal cancer and its association with lymph node metastasis. Dis Colon Rectum 2006, 49: 392–8.
183. Stacker SA, Achen MG, Jussila L, Baldwin ME, Alitalo K. Lymphangiogenesis and cancer metastasis. Nat Rev Cancer 2002, 2: 573–83.
184. Stearns ME, Wang M, Hu Y, Kim G, Garcia FU. Expression of a flt-4 (VEGFR3) splicing variant in primary human prostate tumors. VEGF D and flt-4t(Delta773-1081) overexpression is diagnostic for sentinel lymph node metastasis. Lab Invest 2004, 84: 785–95.
185. Stessels F, Van den Eynden G, Van der Auwera I, et al. Breast adenocarcinoma liver metastases, in contrast to colorectal cancer liver metastases, display a non-angiogenic growth pattern that preserves the stroma and lacks hypoxia. Br J Cancer 2004, 90: 1429–36.
186. Straume O, Jackson DG, Akslen LA. Independent prognostic impact of lymphatic vessel density and presence of low-grade lymphangiogenesis in cutaneous melanoma. Clin Cancer Res 2003, 9: 250–6.
187. Sturgeon C, Angelos P. Identification and treatment of aggressive thyroid cancers. Part 1: subtypes. Oncology (Williston Park) 2006, 20: 253–60.
188. Sundar SS, Zhang H, Brown P, et al. Role of lymphangiogenesis in epithelial ovarian cancer. Br J Cancer 2006, 94: 1650–7.
189. Suzuki K, Morita T, Tokue A. Vascular endothelial growth factor-C (VEGF-C) expression predicts lymph node metastasis of transitional cell carcinoma of the bladder. Int J Urol 2005, 12: 152–8.
190. Takahashi A, Kono K, Itakura J, et al. Correlation of vascular endothelial growth factor-C expression with tumor-infiltrating dendritic cells in gastric cancer. Oncology 2002, 62: 121–7.
191. Takizawa H, Kondo K, Fujino H, et al. The balance of VEGF-C and VEGFR-3 mRNA is a predictor of lymph node metastasis in non-small cell lung cancer. Br J Cancer 2006, 95: 75–9.
192. Tamura M, Oda M, Matsumoto I, et al. The combination assay with circulating vascular endothelial growth factor (VEGF)-C, matrix metalloproteinase-9, and VEGF for diagnosing lymph node metastasis in patients with non-small cell lung cancer. Ann Surg Oncol 2004, 11: 928–33.
193. Tamura M, Ohta Y. Serum vascular endothelial growth factor-C level in patients with primary nonsmall cell lung carcinoma: a possible diagnostic tool for lymph node metastasis. Cancer 2003, 98: 1217–22.
194. Tamura M, Oda M, Tsunezuka Y, et al. Chest CT and serum vascular endothelial growth factor-C level to diagnose lymph node metastasis in patients with primary non-small cell lung cancer. Chest 2004, 126: 342–6.
195. Tanaka K, Kurebayashi J, Sonoo H, et al. Expression of vascular endothelial growth factor family messenger RNA in diseased thyroid tissues. Surg Today 2002, 32: 761–8.
196. Tang RF, Wang SX, Peng L, et al. Expression of vascular endothelial growth factors A and C in human pancreatic cancer. World J Gastroenterol 2006, 12: 280–6.
197. Tezuka K, Onoda N, Takashima T, et al. Prognostic significance of lymphovascular invasion diagnosed by lymphatic endothelium immunostaining in breast cancer patients. Oncol Rep 2007, 17: 997–1003.
198. Tiffen J, Mahon SM. Cervical cancer: what should we tell women about screening? Clin J Oncol Nurs 2006, 10: 527–31.

199. Trojan L, Michel MS, Rensch F, et al. Lymph and blood vessel architecture in benign and malignant prostatic tissue: lack of lymphangiogenesis in prostate carcinoma assessed with novel lymphatic marker lymphatic vessel endothelial hyaluronan receptor (LYVE-1). J Urol 2004, 172: 103–7.
200. Trojan L, Rensch F, Voss M, et al. The role of the lymphatic system and its specific growth factor, vascular endothelial growth factor C, for lymphogenic metastasis in prostate cancer. BJU Int 2006, 98: 903–6.
201. Tsai CS, Lai CH, Wang CC, et al. The prognostic factors for patients with early cervical cancer treated by radical hysterectomy and postoperative radiotherapy. Gynecol Oncol 1999, 75: 328–33.
202. Tsuchiya N, Sato K, Akao T, et al. Quantitative analysis of gene expressions of vascular endothelial growth factor-related factors and their receptors in renal cell carcinoma. Tohoku J Exp Med 2001, 195: 101–13.
203. Tsurusaki T, Kanda S, Sakai H, et al. Vascular endothelial growth factor-C expression in human prostatic carcinoma and its relationship to lymph node metastasis. Br J Cancer 1999, 80: 309–13.
204. Ueda M, Hung YC, Terai Y, et al. Vascular endothelial growth factor-C expression and invasive phenotype in ovarian carcinomas. Clin Cancer Res 2005, 11: 3225–32.
205. Ueda M, Terai Y, Yamashita Y, et al. Correlation between vascular endothelial growth factor-C expression and invasion phenotype in cervical carcinomas. Int J Cancer 2002, 98: 335–43.
206. Vallbohmer D, Lenz HJ. Predictive and prognostic molecular markers in outcome of esophageal cancer. Dis Esophagus 2006, 19: 425–32.
207. Valtola R, Salven P, Heikkila P, et al. VEGFR-3 and its ligand VEGF-C are associated with angiogenesis in breast cancer. Am J Pathol 1999, 154: 1381–90.
208. Van Dam PA, Hauspy J, Vanderheyden T, et al. Intraoperative sentinel node identification with Technetium-99 m-labeled nanocolloid in patients with cancer of the uterine cervix: a feasibility study. Int J Gynecol Cancer 2003, 13: 182–6.
209. Van den Eynden GG, Van der Auwera I, Van Laere SJ, et al. Distinguishing blood and lymph vessel invasion in breast cancer: a prospective immunohistochemical study. Br J Cancer 2006, 94: 1643–9.
210. Van den Eynden GG, Van der Auwerra I, Van Laere SJ, et al. Induction of lymphangiogenesis in and around axillary lymph node metastases of patients with breast cancer. Br J Cancer 2006, 95: 1362–6.
211. Van der Auwera I, Cao Y, Tille JC, et al. First international consensus on the methodology of lymphangiogenesis quantification in solid human tumours. Br J Cancer 2006, 95: 1611–25.
212. Van der Auwera I, Van den Eynden GG, Colpaert CG, et al. Tumor lymphangiogenesis in inflammatory breast carcinoma: a histomorphometric study. Clin Cancer Res 2005, 11: 7637–42.
213. Van der Auwera I, Van Laere SJ, Van den Eynden GG, et al. Increased angiogenesis and lymphangiogenesis in inflammatory versus noninflammatory breast cancer by real-time reverse transcriptase-PCR gene expression quantification. Clin Cancer Res 2004, 10: 7965–71.
214. Van der Schaft DW, Pauwels P, Hulsmans S, et al. Absence of lymphangiogenesis in ductal breast cancer at the primary tumor site. Cancer Lett 2007, 256: 279–81.
215. Van Trappen PO, Steele D, Lowe DG, et al. Expression of vascular endothelial growth factor (VEGF)-C and VEGF-D, and their receptor VEGFR-3, during different stages of cervical carcinogenesis. J Pathol 2003, 201: 544–54.
216. Vermeulen PB, Gasparini G, Fox SB, et al. Second international consensus on the methodology and criteria of evaluation of angiogenesis quantification in solid human tumours. Eur J Cancer 2002, 38: 1564–79.
217. Vleugel MM, Bos R, Van der Groep P, et al. Lack of lymphangiogenesis during breast carcinogenesis. J Clin Pathol 2004, 57: 746–51.
218. Waggoner SE. Cervical cancer. Lancet 2003, 361: 2217–25.

219. Walgenbach-Bruenagel G, Tolba RH, Varnai AD, et al. Detection of lymphatic invasion in early stage primary colorectal cancer with the monoclonal antibody D2-40. Eur Surg Res 2006, 38: 438–44.
220. Wang TB, Deng MH, Qiu WS, Dong WG. Association of serum vascular endothelial growth factor-C and lymphatic vessel density with lymph node metastasis and prognosis of patients with gastric cancer. World J Gastroenterol 2007, 13: 1794–97.
221. Wang TB, Huang YH, Lan P, Song XM, Wang JP. [Correlation of lymphangiogenesis to progression of colorectal cancer]. Ai Zheng 2005, 24: 1276–79.
222. Wang Y, Sun JG, Ye MF, Chen ZT. [Clinical and prognostic significance of podoplanin-positive lymphatic vessel density in non-small cell lung carcinoma]. Zhonghua Jie He He Hu Xi Za Zhi 2006, 29: 758–61.
223. Warburton G, Nikitakis NG, Roberson P, et al. Histopathological and lymphangiogenic parameters in relation to lymph node metastasis in early stage oral squamous cell carcinoma. J Oral Maxillofac Surg 2007, 65: 475–84.
224. Weich HA, Bando H, Brokelmann M, et al. Quantification of vascular endothelial growth factor-C (VEGF-C) by a novel ELISA. J Immunol Methods 2004, 285: 145–55.
225. Weidner N, Semple JP, Welch WR, Folkman J. Tumor angiogenesis and metastasis – correlation in invasive breast carcinoma. N Engl J Med 1991, 324: 1–8.
226. Wetterwald A, Hoffstetter W, Cecchini MG, et al. Characterization and cloning of the E11 antigen, a marker expressed by rat osteoblasts and osteocytes. Bone 1996, 18: 125–32.
227. White JD, Hewett PW, Kosuge D, et al. Vascular endothelial growth factor-D expression is an independent prognostic marker for survival in colorectal carcinoma. Cancer Res 2002, 62: 1669–75.
228. Wigle JT, Oliver G. Prox1 function is required for the development of the murine lymphatic system. Cell 1999, 98: 769–78.
229. Williams CS, Leek RD, Robson AM, et al. Absence of lymphangiogenesis and intratumoural lymph vessels in human metastatic breast cancer. J Pathol 2003, 200: 195–206.
230. Witte D, Thomas A, Ali N, Carlson N, Younes M. Expression of the vascular endothelial growth factor receptor-3 (VEGFR-3) and its ligand VEGF-C in human colorectal adenocarcinoma. Anticancer Res 2002, 22: 1463–6.
231. Xu T, Chen D. Serum vascular endothelial growth factor-C and vascular endothelial growth factor level in patients with colorectal carcinoma and clinical significance. J Huazhong Univ Sci Technolog Med Sci 2006, 26: 329–31, 355.
232. Xu T, Chen D, Chen J. Expression of vascular endothelial growth factor C and its correlation with lymph node metastasis in colorectal carcinoma. J Huazhong Univ Sci Technolog Med Sci 2004, 24: 596–8.
233. Yamaguchi R, Yano H, Nakashima O, et al. Expression of vascular endothelial growth factor-C in human hepatocellular carcinoma. J Gastroenterol Hepatol 2006, 21: 152–60.
234. Yamao T, Shirao K, Ono H, et al. Risk factors for lymph node metastasis from intramucosal gastric carcinoma. Cancer 1996, 77: 602–6.
235. Yang J, Wu HF, Qian LX, et al. Increased expressions of vascular endothelial growth factor (VEGF), VEGF-C and VEGF receptor-3 in prostate cancer tissue are associated with tumor progression. Asian J Androl 2006, 8: 169–75.
236. Yang W, Klos K, Yang Y, et al. ErbB2 overexpression correlates with increased expression of vascular endothelial growth factors A, C, and D in human breast carcinoma. Cancer 2002, 94: 2855–61.
237. Yasuoka H, Nakamura Y, Zuo H, et al. VEGF-D expression and lymph vessels play an important role for lymph node metastasis in papillary thyroid carcinoma. Mod Pathol 2005, 18: 1127–33.
238. Yokoyama Y, Charnock-Jones DS, Licence D, et al. Vascular endothelial growth factor-D is an independent prognostic factor in epithelial ovarian carcinoma. Br J Cancer 2003, 88: 237–44.

239. Yonemura Y, Endo Y, Fujita H, et al. Role of vascular endothelial growth factor C expression in the development of lymph node metastasis in gastric cancer. Clin Cancer Res 1999, 5: 1823–9.
240. Yonemura Y, Fushida S, Bando E, et al. Lymphangiogenesis and the vascular endothelial growth factor receptor (VEGFR)-3 in gastric cancer. Eur J Cancer 2001, 37: 918–23.
241. Yu XM, Lo CY, Chan WF, et al. Increased expression of vascular endothelial growth factor C in papillary thyroid carcinoma correlates with cervical lymph node metastases. Clin Cancer Res 2005, 11: 8063–9.
242. Yuanming L, Feng G, Lei T, Ying W. Quantitative analysis of lymphangiogenic markers in human gastroenteric tumor. Arch Med Res 2007, 38: 106–12.
243. Zeng Y, Opeskin K, Baldwin ME, et al. Expression of vascular endothelial growth factor receptor-3 by lymphatic endothelial cells is associated with lymph node metastasis in prostate cancer. Clin Cancer Res 2004, 10: 5137–44.
244. Zeng Y, Opeskin K, Horvath LG, Sutherland RL, Williams ED. Lymphatic vessel density and lymph node metastasis in prostate cancer. Prostate 2005, 65: 222–30.
245. Zhang GH, Zeng Y, Yang WT, Shi DR. [Study of mRNA expression of vascular endothelial growth factor-(A, C, D) genes and its effect on prognosis of breast cancer]. Zhonghua Bing Li Xue Za Zhi 2006, 35: 473–7.
246. Zhang HZ, Wei YP, Li HG, et al. [Expressions of vascular endothelial growth factor C and matrix metalloproteinases-2 and prognosis of non-small cell lung carcinoma]. Nan Fang Yi Ke Da Xue Xue Bao 2006, 26: 1307–10.
247. Zu X, Tang Z, Li Y, et al. Vascular endothelial growth factor-C expression in bladder transitional cell cancer and its relationship to lymph node metastasis. BJU Int 2006, 98: 1090–3.

Chapter 7

LYMPHANGIOGENESIS AND IMAGING OF THE LYMPHATICS IN CANCER

Baris Turkbey, Gregory Ravizzini, Peter L. Choyke and Hisataka Kobayashi
Molecular Imaging Program, National Cancer Institute (NCI), Bethesda, MD 20892-1088, USA, e-mail: kobayash@mail.nih.gov

Abstract: Metastatic spread of cancer is one of the major causes of cancer death. The lymphatics contribute to metastatic spread by providing a conduit for the spread of cancer cells. Tumors actively induce new lymphatic formation by deploying growth factors, a process known as lymphangiogenesis. Systemic lymphatic imaging with conventional modalities such as computed tomography, magnetic resonance imaging and ultrasound is limited to morphological evaluation for detection of enlarged lymph nodes; on the other hand, functional lymphatic imaging approaches, including positron emission tomography, dynamic contrast-enhanced MRI, lymphotrophic iron oxide nanoparticle enhanced-MRI have been used to diagnose metastatic cancer in lymph nodes. Recently, new targeted lymphatic imaging techniques including gadolinium-conjugated dendrimer-based MRI, optical imaging using nano-sized molecules based on fluorescence-labeled dendrimers, organic macromolecules, or quantum dots, have been developed. In this chapter, we will explain principles and basic findings of conventional and functional lymphatic imaging and will outline newly developed targeted lymphatic imaging approaches.

Key words: Imaging · Lymphangiogenesis · Cancer · Metastases

7.1 Introduction

Tumor invasion and metastasis are critical steps in the development of lethal cancers [1]. While angiogenesis, the recruitment of new vessels to provide necessary nutrients, has been extensively studied and is the target of numerous molecular therapies, lymphangiogenesis, the process by which the tumor acquires larger and more numerous lymphatics, is less well understood. However, angiogenesis and lymphangiogenesis are required for continued tumor growth. As blood flow increases to the

S.A. Stacker, M.G. Achen (eds.), *Lymphangiogenesis in Cancer Metastasis*, Cancer Metastasis – Biology and Treatment 13, DOI 10.1007/978-90-481-2247-9_7,

tumor, intratumoral pressures increase eventually preventing further perfusion of the tumor. The lymphatics relieve this increased intratumoral pressure allowing the tumor to grow. The lymphatics also provide a conduit for the removal of larger molecules and inflammatory cells and may aid the tumor in evading the immune system. Moreover, the lymphatics provide a non vascular conduit for the spread of cancer cells.

Traditionally, lymphatics were thought to passively drain tumors via pre-existing channels. In this model of tumor physiology, tumor cells from the primary tumor entered the lymphatics which were located in the periphery of the primary tumor and passively drained to the regional lymph nodes, the first of which is known as the "sentinel node" [2]. More recently, however, it has become clear that the lymphatics are active participants in the process of metastases. Primary tumors secrete growth factors and cytokines that specifically target receptors on lymphatic endothelial cells causing them to enlarge, proliferate and become more permeable resulting in the process known as "lymphangiogenesis". This process involves both the lymphatic channels and the lymph nodes themselves [3, 4]. Vascular endothelial growth factor (VEGF)-C and VEGF-D, which activate the endothelium via the VEGFR3 receptor, are the best known growth factor-receptors affecting the lymphatics although other ligand-receptor pairs have been implicated [5]. The expression of lymphatic endothelial growth factors by primary tumor cells is associated with a worse clinical prognosis carrying an increased risk of lymphatic metastases [6, 7].

7.2 Anatomy of Lymphatics and Lymph nodes

The lymphatic system is an extensive network of lymphoid organs which include the thymus, bone marrow, tonsils, spleen, Peyer's patches as well as lymphatic vessels and lymph nodes which are present in practically all vascularized tissues except the brain and retina [8]. A lymphatic vessel is composed of thin walled, single layer, capillaries. Distally, the blind ending collecting sacs are highly permeable with large gaps between the endothelial cells. Indeed, the attachments between the lymphatic endothelium can be sparse and have been likened to the buttons on a shirt, allowing large spaces for the absorption of extracellular fluids containing macromolecules and cells in the gaps between the junctions. More proximally, as the lymphatic channels drain toward the lymph nodes, the channels acquire progressively thicker smooth muscle and pericyte layers which provide peristaltic propulsion but allow less transmural leakage. Elastic fibers, which compromise the so-called anchoring filaments, attach to the anti-luminal side of endothelial cells and anchor the lymphatics to the tissue allowing them to expand their pore size as the interstitial pressure increases [7, 9]. Fluid initially enters the blind permeable lymphatic sacs and is initially propagated by pressure and then by peristaltic force to the lymph nodes [7].

A lymph node is composed of a fibrous capsule and a subcapsular sinus which surrounds the islands of T and B cells called germinal centers. The cellular islands can be divided into three components: cortex, which contains primary follicles including B cells and follicular dendritic cells, paracortex which is composed of T and dendritic cells, and the medulla consisting of medullary cords which are separated by medullary sinuses filled with lymph fluid. Lymph fluid, including dendritic cells and antigens, enters the periphery of the lymph node through afferent lymphatic vessels [10]. The vasculature of a lymph node is provided by venules, which enable the constant delivery of T and B cells to the lymph node to process the antigens and develop an immune response. Once processed in the lymph node, such cells leave the lymph node via the efferent lymphatics located in the hilum. Lymph nodes draining tumors undergo early and extensive growth of the lymphatic channels within the sinuses of the node, even before identification of tumor cells within lymph node itself [11].

The first draining lymph node from a tumor is known as the sentinel lymph node (SLN) [7, 12, 13]. The concept of the SLN was first introduced by Cabanas for penile carcinoma in 1977, but gained momentum after Morton et al. applied SLN techniques for malignant melanoma patients in 1992 [14, 15]. The theory behind the SLN technique is that metastases progress in a predictable manner beginning with the SLN rather than spreading randomly to any node in the lymphatic basin. Detection of metastasis within an SLN will affect both the therapeutic approach and prognosis [12]. Detection of a positive SLN usually signals a worse prognosis and results in reclassification of the patient into a group with a higher risk of recurrence; conversely a negative SLN usually connotes disease confined to the organ with a commensurately better prognosis [5]. Moreover, the patient can be spared extensive lymph node surgery. The ability of tumor cells to migrate to the lymph nodes implies that they are biologically more motile than tumor cells that do not have this ability.

7.3 Physiology and Receptors of Lymphatics

The lymphatic system is important for maintaining homeostasis because it transports excess extracellular fluid and macromolecules from tissues into the systemic circulation, thus conserving a fluid balance in the tissues [16]. Normally, the lymphatics safeguard tissue from infection by draining macrophages and pathogens to lymph nodes where immune cells reside, thus facilitating an immune response [7]. In tumors, a variety of cytokines are produced to help the tumor evade the normal immune system. The lymphatics draining the tumor play an important role in this immunomodulation. Damage to the lymphatics due to trauma or infection results in edema, swelling and an attendant risk of infection [17].

The tissue surrounding tumors is often complex, containing normal cells, vascular endothelium, inflammatory cells and extracellular matrix. Identifying lymphatic cells can be difficult without relying on specific markers such as the transcription factor Prox-1, which is necessary for lymphatic development, and which is upregulated in dividing lymphatics and which has been implicated as the "lymphangiogenic switch" [18, 19]. Podoplanin, a mucin-type transmembrane glycoprotein, is another lymphatic marker which is required for lymphatic development and endothelial cell adhesion and migration. Podoplanin is also found in the podocytes of renal tubular cells as well as, osteoblasts, and type 1 alveolar cells [7]. However, the main advantage of podoplanin as a marker of the lymphatics is that it is not expressed in the endothelium of vascular channels enabling the discrimination of lymphangiogenesis from angiogenesis [7]. Lymphatic vessel hyaluronan receptor 1 (LYVE-1) is a homologue of CD44, and is another reliable marker of the lymphatics. LYVE-1 has been implicated in tumor cell trafficking to lymph nodes [20, 21]. VEGFR-3 is an endothelial cell surface receptor but its use for marking the lymphatics is limited since it can also be found on some blood vessels [22]. Finally, molecules such as EphrinB2 and EphrinB4 are differentially expressed in lymphatic vessel subtypes [23]. Thus, a combination of immunohistochemical stains is available to identify the lymphatics.

7.4 Role of Lymphatics in Cancer; Implications for Treatment and Prognosis

When cancers develop in a tissue, normal lymphatic functions are subverted to the needs of the growing primary tumor. The lymphatics are an existing tissue infrastructure that permits the tumor to drain excess macromolecular fluids and shed tumor cells with greater efficiency than is possible with the extracellular tissue space alone [24, 25]. The increased intratumoral pressure within tumors, forces the extracellular fluid to flow outward by convection but the "sump effect", caused by the drainage of the lymphatic channels, directs much of the fluid toward the lymphatic "drain" thus improving the drainage efficiency [26]. For larger particulates such as cellular components or cellular clusters (e.g. tumor cells and macrophages) lymphatic invasion is aided by peristalsis generated by the thin layer of smooth muscle which ensure propagation of cell clusters by contraction [27]. Lymphangiogenesis aids these processes by developing new lymphatic channels, enlarging the lumens of existing lymphatics and providing more lymphatic flow to the lymph nodes [28, 29]. Tumor induced lymphangiogenesis promotes metastasis through increased peri and/or intratumoral lymphatics [24]. The most well documented tumor-lymphangiogenesis promoter system is the VEGF-C/VEGF-D/VEGFR-3 axis in which VEGF-C and VEGF-D, activates VEGFR-3, a receptor expressed on lymphatic endothelium [22]. In several types of cancers, significant correlation exists between VEGF-C and/or VEGF-D expression and lymphatic metastasis and prognosis [7]. Tumor cells are the main source of VEGF-C/D, but perivascular stromal cells such as tumor-associated macrophages can also secrete these growth factors; platelets also contain considerable amounts of VEGF-C [30–32]. VEGF-C mediates distant lymph node and organ metastases [33]. Beside their role in lymphangiogenesis proteolytically activated VEGF-C and VEGF-D can also act on VEGFR-2, which is also found on the endothelium of blood vessels and is a signal for angiogenesis [34].

VEGF-A is also shown to play significant role in peritumoral lymphangiogenesis and metastases, though it is better known as a mediator of angiogenesis [35]. VEGF-A binds to VEGFR-2 that is also expressed on lymphatic vessels and it promotes lymphangiogenesis [36]. Moreover, VEGF-A indirectly activates the VEGF-C/VEGF-D/VEGFR-3 pathway through inflammatory cells which carry VEGF-C and VEGF-D [36].

The success of Bevacizumab, a monoclonal antibody against VEGF-A in reducing tumor angiogenesis in colorectal cancer has encouraged agents directed against tumor lymphangiogenesis. The VEGF-C/VEGF-D/VEGFR-3 axis is a useful target for treating lymphangiogenesis. Theoretically, there are three ways of blocking VEGFR-3 dependent tumor induced lymphangiogenesis [3]: binding to and hence, blocking the effects of VEGF-C and VEGF-D, blocking VEGFR-3 using antibodies or small molecules, sequestering VEGF-C and VEGF-D via soluble dimeric fusion proteins with extracellular ligands for VEGFR-3, and small molecules that interfere with VEGFR-3 signaling through kinase activity blockade [28, 37–39]. Although appealing in concept, the development of agents that block VEGF-C and VEGF-D, has been stymied by a lack of understanding of the whole range of binding characteristics of these growth factors and their possible off-target effects. Moreover, it is likely that lymphangiogenesis is under the control of more than just the known growth factors and these alternate pathways could overcome the effect of the blockade of only one growth factor.

7.5 Role of Imaging of Lymphatics and Lymphangiogenesis in Cancer

7.5.1 Systemic Lymphatic Imaging

Most systemic imaging of the lymphatics is limited to detection of enlarged lymph nodes on cross sectional imaging such as CT, MRI or PET. Normal lymph nodes, despite the large flow of lymph through them, are tightly regulated in size likely because of their fibrous capsule. When they become infected or become a site for the growth of metastases, they enlarge and become visible on imaging studies. CT/MRI rely on size criteria to differentiate benign lymph nodes from metastastic ones; this clearly limits the ability to detect early metastatic nodes, i.e. the node must achieve a short axis diameter ≥ 1 cm to be considered malignant. The long axis of normal nodes is typically parallel to lymphatic vessels. Characteristically, a normal lymph node is elliptical and has a horseshoe-like shape with a hilum containing central fat, smooth outline and homogeneous CT density [40]. Pathologic nodes are usually enlarged, irregular in shape with less central fat (Fig. 7.1). However, normal sized and shaped lymph nodes can often shelter micro-metastases which do not distort the external contour of the node [41]. While these findings are useful in day-to-day clinical practice they are inherently non-specific. Small

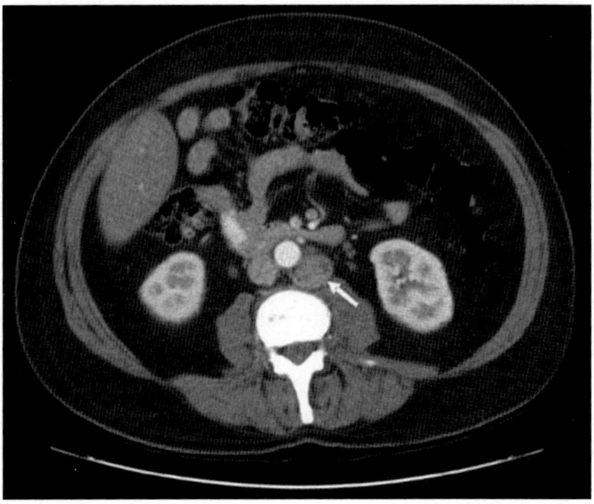

Fig. 7.1 60-year-old male with prostate cancer. Axial contrast enhanced computed tomography image shows metastatic retroperitoneal lymphadenopathy (*arrow*)

nodes may contain microfoci of disease and not be enlarged or distorted in shape; conversely, enlarged nodes may simply be caused by hyperplasia and not malignancy. Thus, additional "functional" systemic imaging methods have been introduced to characterize lymph nodes, including positron emission tomography (PET), dynamic contrast-enhanced MRI (DCE-MRI), ultra-small particles of iron oxide (USPIO)-MRI and color Doppler ultrasound (CDUS). Further developments in spatial resolution, cross-sectional imaging and three-dimensional reconstructions may allow further assessment of morphological features of the nodal cortex and sinus for diagnosis.

7.5.2 Direct Lymphangiography

One of the early attempts to image the lymphatic system was oil-based iodinated dyes that were injected directly into the lymphatics of the feet. The agent then traveled through the leg lymphatics to the abdominal lymphatics whereupon direct radiographs or CT images could be obtained. This technique, known as "Lymphangiography" was classically used in the assessment of nodal metastases in lymphoma and a limited number of other malignancies until the 1980's. Lymphangiography requires great skill and expertise. Initially, a blue dye is injected intradermally into the interdigital space in order to stain and localize the deep lymphatic vessels. Then an incision is made in the skin of the foot to locate lymphatic vessels large enough to allow cannulation with a fine needle. Following this an oily iodinated dye was injected over a period of 60–90 minutes into the lymphatic vessels using a mechanical pump. The initial imaging depicted the lymphatic vessels and follow-up scans obtained after 24 h demonstrated the lymph nodes. However, lymphangiography, in addition to being invasive and requiring skilled practitioners, can cause life-threatening complications such as pulmonary embolization, pulmonary edema and adult respiratory distress syndrome if the oily dye inadvertently enters the veins [42, 43]. The invasiveness of this procedure, its expense and inconvenience for the patient, together with risk of severe complications, has all but eliminated lymphangiography from the modern imaging armory.

7.6 Lymphoscintigraphy and Sentinel Lymph Node Biopsy in Cancer Management

The basic principle for sentinel lymph node biopsy (SLNB) was described in 1907 when Jamieson demonstrated the significance of neoplastic cells spreading to the so-called "primary gland" by which he meant the draining lymph node [44]. The term "sentinel node"— the first lymph node in a regional lymphatic basin, which receives flow from a primary tumor—was proposed by Gould et al. in 1960. The procedure gained prominence only in the early nineties after Morton described the technique in primary cutaneous melanoma using isosulfan blue dye. Intradermal injections were performed circumferentially around the primary lesion and the blue-stained lymphatics were followed surgically until a blue-stained lymph node was identified. In 1993, Alex and Krag, reported the first gamma-probe localization of sentinel lymph nodes utilizing Tc-99 m sulfur colloid. They found good correlation between gamma-probe guided lymphoscintigraphy and the isosulfan blue dye procedure for detecting SLNs and lymphoscintigraphy became more widely used [45]. In a study of the same group, axillary sentinel lymph nodes were identified in 18 out of 22 patients with breast cancer [46]. In 1994, Giuliano et al. extended Morton's technique to breast cancer, injecting isosulfan blue dye into breast masses and surrounding breast parenchyma [47]. In 1996, Albertini performed intraoperative lymphatic mapping using a combination of a vital blue dye and Tc-99 m sulfur colloid for SLN identification in a prospective study [48].

Tc-99 m sulfur colloid is approved for imaging of the liver and spleen by the United States Food and Drug Administration and is widely used for lymphoscintigraphy in filtered and unfiltered forms. Unfiltered sulfur-colloid has variable range of sizes due to aggregation and their size can range from 1–3 μm depending on the preparation technique [49]. The smaller filtered particles are micro-filtered (pore size 220 nm Millipore, Bedford, MA). Particle size determines the intralymphatic kinetics [50]. Filtered particles migrate faster and have been noted to delineate a greater number of echelon or higher order lymph nodes. Worldwide, additional radiotracers are utilized. For example, Tc-99 m antimony trisulfide is commercially available in Australia and Canada and Tc-99 m-HSA nanocolloid is used in Europe and Japan [51].

Radiotracer dose, injection techniques and imaging protocols vary extensively according to the type of tumor being investigated. For example, in patients with melanoma, dynamic or sequential images are usually performed immediately after the intradermal injection of radiotracer around the primary tumor and continued for 30–60 minutes [51]. After the images are acquired, the patient is taken directly to the operating room. In reality, imaging during SLNB is being performed less frequently in breast cancer but continues to be more important in melanoma SLNB. In breast cancer, the most popular approach is to perform an intradermal injection in the skin overlying the primary tumor or a subdermal injection in the periareolar region [52,53] (Fig. 7.2). Dynamic or sequential images are not routinely performed and images can be obtained, if at all, on the same day or the day before surgery [54]. For the two-day protocol, unfiltered sulfur colloid is employed and, due to the slower kinetics of the larger particle, imaged later in the day. The patient is then operated on the following day. Comparable results are achieved with a one day protocol using filtered sulfur colloid in which the imaging and intraoperative localization are performed on the same day. Proponents of the two-day protocol claim greater efficiency in scheduling with less operating room time wasted in waiting for patients to arrive from the Nuclear Medicine Department [54].

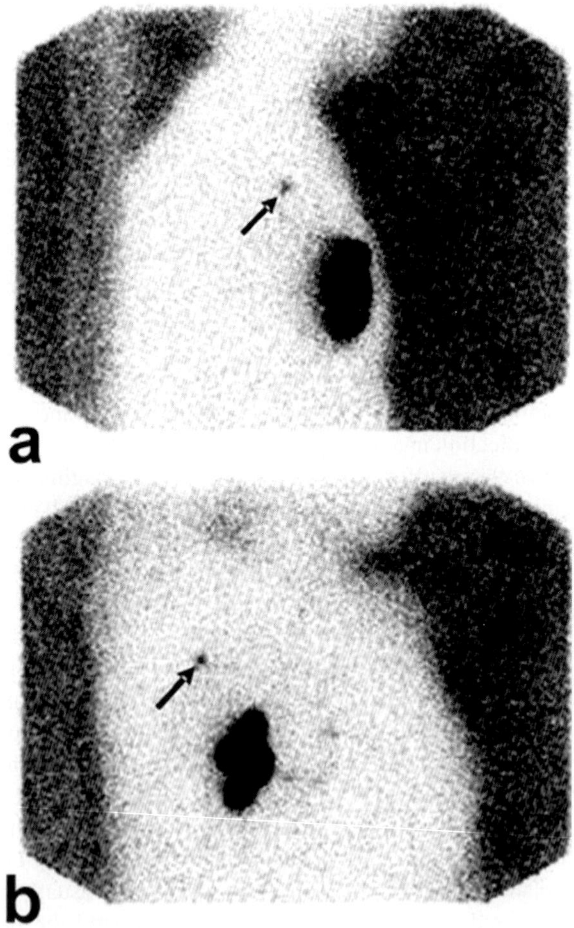

Fig. 7.2 45-year-old female with breast cancer. Lateral (**a**) and right anterior oblique (**b**) planar images obtained after the subdermal injection of unfiltered Tc-99m sulfur-colloid in the periareolar region demonstrating a sentinel lymph node in the right axilla (*arrows*). (Courtesy of Dr. Raghuveer K. Halkar, Emory University)

The main disadvantage of lymphoscintigraphy is the poor spatial resolution and lack of detailed anatomy to guide surgery team during operation. Recently, the introduction of a new imaging instrumentation such as SPECT/CT, promises more accurate depiction of lymphatic channels and draining lymph nodes [55]. SPECT/CT will probably have its highest impact in tumors located in body parts with ambiguous lymph node drainage.

Using the pre-operative images as a guide, the surgeon uses a small handheld probe to detect gamma-rays emissions from the radiotracer. By placing the probe over the region of highest counts, an incision can be made directly over the sentinel lymph node. A SLN usually has at least 10 times the background counts, taken at a location remote from the injection site [51]. After removal of the SLN, residual activity and additional lymph nodes can also be detected via the probe.

In breast cancer patients, the presence of axillary lymph node metastases represents an important prognostic indicator; it has been shown that the presence of regional metastases within the axillary basin decreases a patient's 5-year survival by approximately 28–40% [56, 57]. Axillary lymph nodes receive 85% of the lymphatic drainage from the breast; the remainder drains to the internal mammary chain. The likelihood of axillary LN involvement is related to histologic grade of the primary tumor and the presence of lymphatic or vascular invasion in the specimen [58]. Tumor size and location also play an important role. For example, there is a 16–19% chance of axillary node metastases in very early breast cancers (T1a-b, tumor size $\leq$ to 1 cm) while in T1c lesions (size 1–2 cm), this rate increases up to 30–40% [59, 60].

Axillary lymph node metastases are more common in patients where the primary tumor is located in the outer quadrants of breast [61]. Physical examination has poor predictive value for determination of nodal involvement. Up to one-third of women with non-palpable axillary lymph nodes will have metastases, while one-third of those with palpable lymph nodes will be disease free. In the past, radical axillary lymph node dissection (ALND) was used for staging and prevention of locoregional recurrence. However, the procedure leads to considerable morbidity in the form of lymphedema and sensory motor disturbances in the ipsilateral extremity and its benefit is unproven. For these reasons, the use of ALND for staging purposes in clinically low risk patients is declining with the adoption of the SLNB technique. In a meta-analysis by Kim et al. including sixty nine trials of SLNB in patients with early-stage breast carcinoma between 1970 and 2003, of the 8059 patients studied, 7765 patients (96%) had successfully mapped SLNs. Lymph node involvement was found in 3132 patients (42%) and ranged from 17% to 74% across studies. The false-negative rate (FNR) ranged from 0% to 29%, averaging 7.3%. Eleven trials (15.9%) reported an FNR of 0.0, whereas 26 trials (37.7%) reported an FNR $>$ 10%. Significant inverse correlations were observed between the FNR and the number of patients studied ($r = -0.42$; $P < 0.01$) and the proportion of patients who had successfully mapped SLNs nodes ($r = -0.32$; $P = 0.009$) [62]. Results of SLNB in breast cancer are clearly operator dependent and the importance of experience has been confirmed in multicenter trials [63]. In experienced hands the false negative rate for SLNB is 5% or less. [64].

Metastasis to regional lymph nodes is the most important prognostic factor in early-stage melanoma [65–67]. Before the development of SLNB, the only method to identify regional nodal metastases and stage the nodal basin was elective complete lymph node dissection (CLND). However, only 20% of patients with intermediate-thickness lesions will have metastases to regional nodes [68]. Therefore, the large majority of patients with melanoma underwent extensive surgery without benefit. SLNB has been proposed as a minimally invasive alternative to CLND. The American Joint Committee on Cancer (AJCC) has incorporated the tumor status of the sentinel node into its staging system for melanoma [65]. The Multicenter Selective Lymphadenectomy Trial (MSLT-I) is the largest trial to address the role of lymphatic mapping with SLNB in determining prognosis and its impact on survival [68]. Initial SLN identification rate was 95.3% overall: 99.3% for the groin, 95.3% for the axilla, and 84.5% for the neck basins. The rate of false-negative SLNB during the trial phase, as measured by nodal recurrence in a tumor-negative dissected SLN basin, decreased with increasing case volume at each center: 10.3% for the first 25 cases versus 5.2% after 25 cases. There were no operative mortalities. The low (10.1%) complication rate after SLNB rose to 37.2% if CLND was needed; CLND also increased the severity of complications. Thus, SLNB is a safe, low-morbidity procedure for staging the regional nodal basin in early melanoma. Even after a 30-case learning phase and 25 additional SLNB cases, the accuracy continues to increase with a center's experience. The authors concluded that SLNB should become the standard of care for staging the regional lymph nodes of patients with primary cutaneous melanoma (Fig. 7.3).

^{99m}Tc-sulfur colloid is not the only agent used for lymphoscintigraphy (LS) [69, 70]. In addition to its many applications in oncology, lymphoscintigraphy was studied in the localization of internal mammary lymph nodes for parasternal radiation therapy [71–75] and in staging and treatment planning of patients with lymphoma [76]. However, it is in sentinel lymph node biopsy that the technique has flourished.

The limitations of lymphoscintigraphy and blue dye, combined with advances in imaging technology and contrast media development, have prompted a search for better lymphatic imaging methods. These novel techniques are minimally invasive and potentially offer higher spatial resolution that enables demonstration of lymphatic channels, higher temporal resolution, cross-sectional imaging capabilities, and three-dimensional image reconstruction. Nanoparticle sized contrast agents can access the lymphatic system by three different routes of administration: intravenous, intra-lymphatic (direct lymphatic injection) or interstitial. Contrast agents for direct intra-lymphatic injection are not being developed due to the inherent difficulties in finding and cannulating the lymphatic vessels; thus the newer contrast agents tend to use the other two routes. Contrast agents further increase the potential to provide functional imaging- a particular advantage for cancer imaging, where anatomical demonstration not alone provides sufficient information about disease status any more.

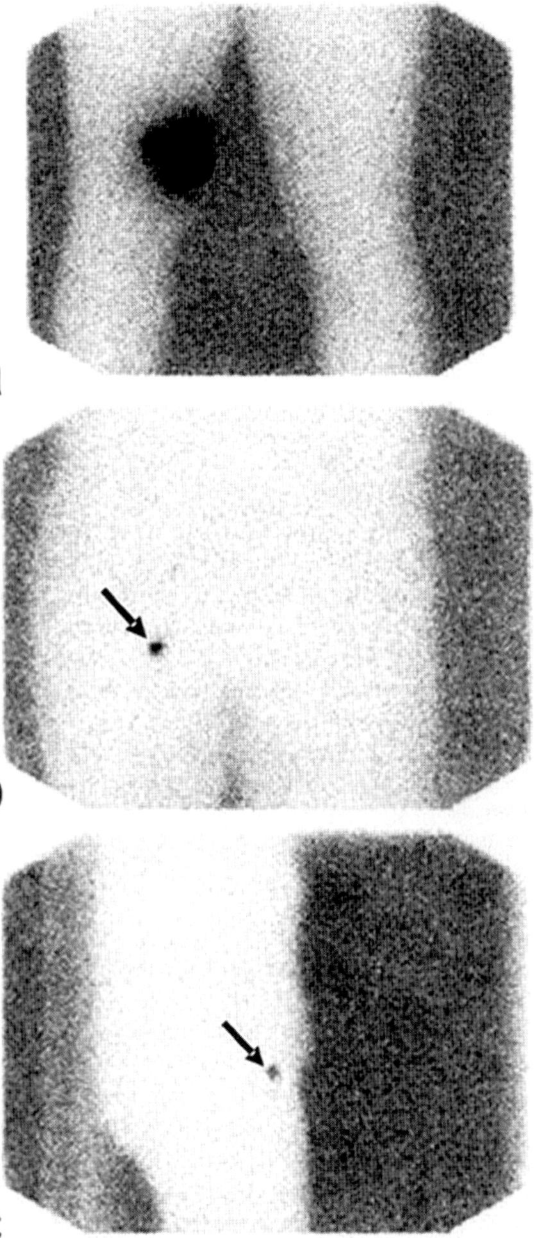

a b c

Fig. 7.3 40-year-old male with cutaneous malignant melanoma. Images obtained after the intra-dermal injection of microfiltered Tc-99 m sulfur-colloid in the right distal thigh demonstrate intense radiotracer activity at the injection site (**a**). Anterior (**b**) and right lateral (**c**) images of the pelvis demonstrate a sentinel lymph node in the right groin (*arrows*). (Courtesy of Dr. Raghuveer K. Halkar, Emory University)

7.7 Novel imaging Techniques

Lymphatic imaging across a number of modalities, including CT, PET, US, MRI and optical imaging are minimally invasive and are becoming more available. They potentially offer higher spatial and temporal resolution, three-dimensional image reconstruction and sensitivity/specificity.

7.7.1 Ultrasound

Ultrasound (US) has long been used to investigate lymphadenopathy [77]. Specific features supporting nodal metastases include loss of central hilum, hypo-echogenicity, irregular borders, and enlargement [78]. The main disadvantages of US are the poor spatial resolution, its limited use in the thorax and deep retroperitoneum, and the fact that interpretation is highly operator dependent. Doppler US can offer functional imaging of the lymph node and may aid diagnosis but relies on lymph node angiogenesis. Color Doppler ultrasound (CDUS) studies can demonstrate differences in vascularity and, as such, may be able to classify node as being reactive, metastatic or neoplastic but there is an overlap among these (Fig. 7.4). Ultrasound (US) using microbubbles, which are gas filled liposomes 2–10 μm in diameter, have been adapted for sentinel lymph node imaging. Choi et al., injected different microbubble agents subcutaneously into rabbits [79]. The agents, such as 'AF0150' were small enough in diameter to enter the lymphatic system, presumably via gaps between the lymphatic endothelial cells. All agents rapidly and markedly enhanced the popliteal lymph node following foot pad injection and hind limb massage. In addition, the lymphatic ducts were visible after interstitial injection of the microbubbles.

7.7.2 Computed Tomography (CT)

Recently, multi-slice computed tomography has been attempted for sentinel node imaging using a low molecular weight iodinated contrast agent (iopamidol) injected interstitially. Suga et al., described seventeen patients with breast cancer who underwent thin-section, three-dimentional CT after subcutaneous injection of iopamidol in the peri-tumoral and peri-areolar areas [80]. Peri-operative blue-dye injection was performed for comparison. CT imaging allowed localization of SLNs in all patients by opacifying the lymphatic vessel draining the injection site and an SLN. Minato et al. were also able to predict SLNs in 13/15 patients with breast cancer, either by enhancement of the lymphatic vessels draining into the SLN, or enhancement in the SLN itself, correlating well with blue-dye detection [81]. The problem with iopamidol-based sentinel lymph node imaging is that the low molecular weight of the agent leads to rapid enhancement and early wash out from the nodes, therefore only a small temporal window remains for imaging. Other disadvantages of CT are

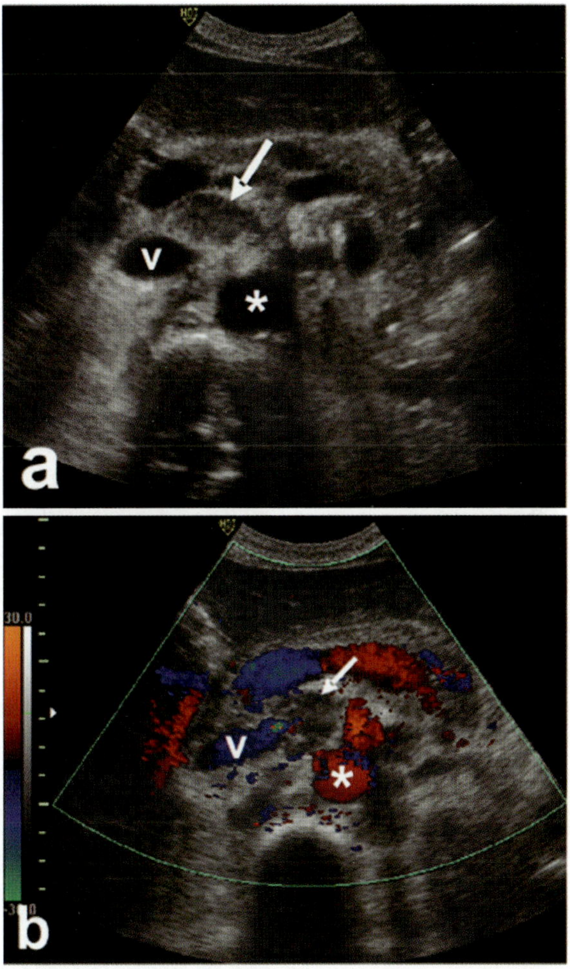

Fig. 7.4 63-year-old male with lymphoma. Gray scale abdominal ultrasound image (**a**) depicts a round hypoechoic lesion consistent with malignant lymphadenopathy (*arrow*) anterior to inferior vena cava (v) and aorta (asterix); color Doppler image (**b**) shows diminished vascularity within the involved lymph node (*arrow*)

the exposure to ionizing radiation and lack of real time guidance to the operating surgeon during identification and resection of the sentinel node.

7.7.3 Magnetic Resonance (MR) Imaging

MRI, which has been used in a similar manner to CT for lymph node imaging, offers good spatial resolution, functional information, and a range of new contrast media, without exposure to ionizing radiation (Fig. 7.5). A number of different contrast agents including gadolinium-diethynetriamine pentaacetic acid (Gd-DTPA), iron oxide particles and Gd (III)-containing macromolecular agents (liposomes, dendrimers) have been specifically developed for and applied to imaging

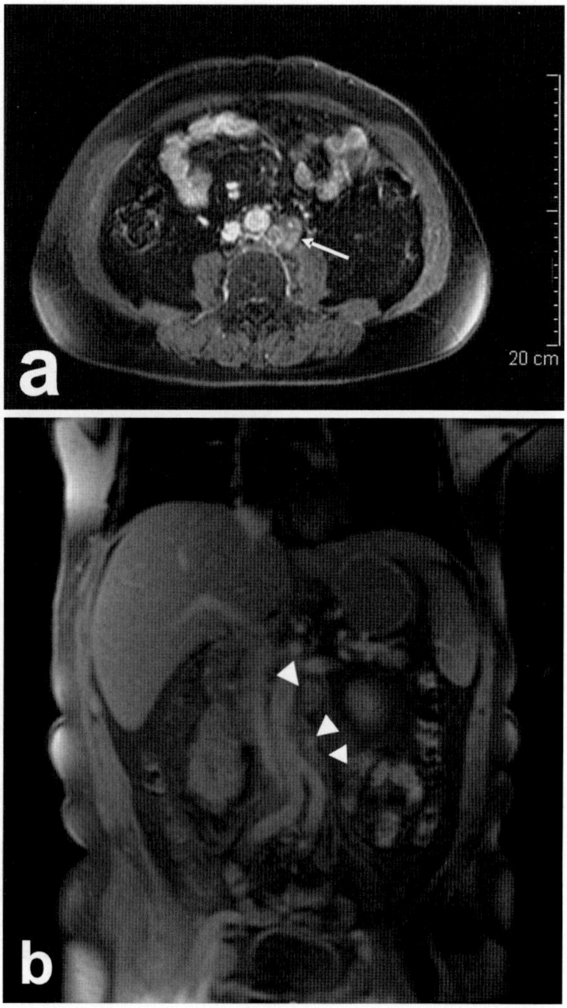

Fig. 7.5 74-year-old male with prostate cancer. Gadolinum enhanced axial T1-weighted magnetic resonance image (**a**) shows malignant lymphadenopathies with slight enhancement (*arrow*); coronal T1-weighted image (**b**) shows conglomerated metastatic nodes (*arrowheads*) along the course of abdominal aorta

of the lymphatic system. Un-enhanced MRI is equivalent to CT since it relies predominantly on nodal size in order to distinguish benign from malignant lymph nodes; whereas, dynamic contrast-enhanced MR imaging is a new functional tool that is readily available in the clinical setting. DCE-MRI acquires serial images following intravenous injection of a low-molecular weight Gd-DTPA including contrast agent. Wash-in and wash-out curves can be derived from designated regions of interest (ROIs) for direct comparison; moreover, pharmacokinetic models can be applied in order to derive permeability parameters. Resulting parameters reflect differences of microvascularity in terms of blood flow and permeability which have been shown to correlate with the degree of angiogenesis within tumors [82] (Fig. 7.6).

Lymphotrophic nanoparticle enhanced MRI has emerged as a useful imaging modality for lymph node characterization. Superparamagnetic iron oxide (SPIO) and ultra-small SPIO (USPIO) nanoparticles used as lymphotrophic nanoparticle MRI contrast agents have the potential to evaluate the reticuloendothelial system (liver, spleen, bone marrow, lymph nodes) with unique relaxation mechanisms and physiologic distribution [83]. SPIO nanoparticles slowly move to the interstitial space where they are transported to lymph nodes through lymph vessels. In lymph nodes they are captured by macrophages and this uptake results in loss of signal within normal nodes. Due to the strong magnetic susceptibility effect as well as T2 shortening effect, normal uptake is seen as a loss of signal whereas metastases are seen as residual signal. SPIO enhanced MRI contains two sessions; the initial one in which patient is scanned just before contrast injection and a 24–36 hours post-contrast follow-up scan which is important to ensure sufficient accumulation of nanoparticle within lymph nodes. A report by Saokar et al. on overall performance of lymphotrophic nanoparticle enhanced MRI in differentiation of benign

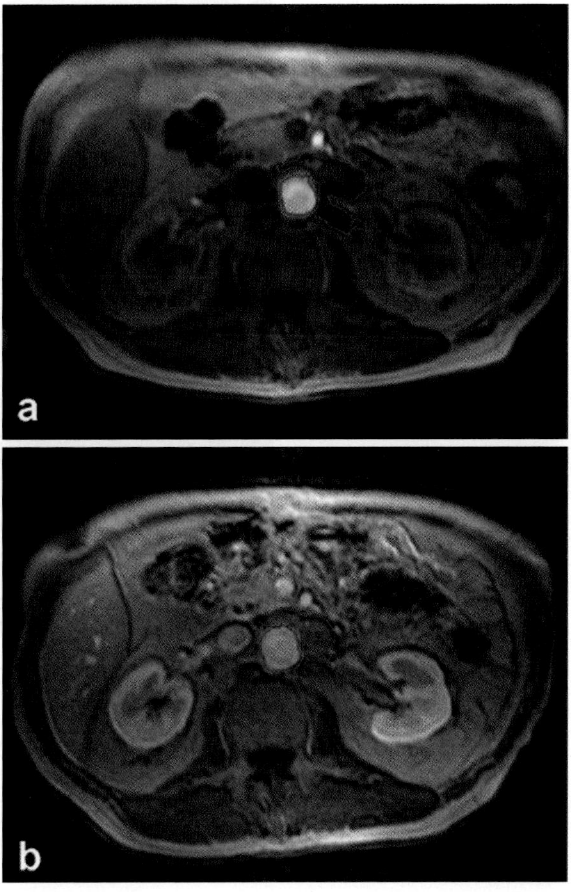

Fig. 7.6 (continued)

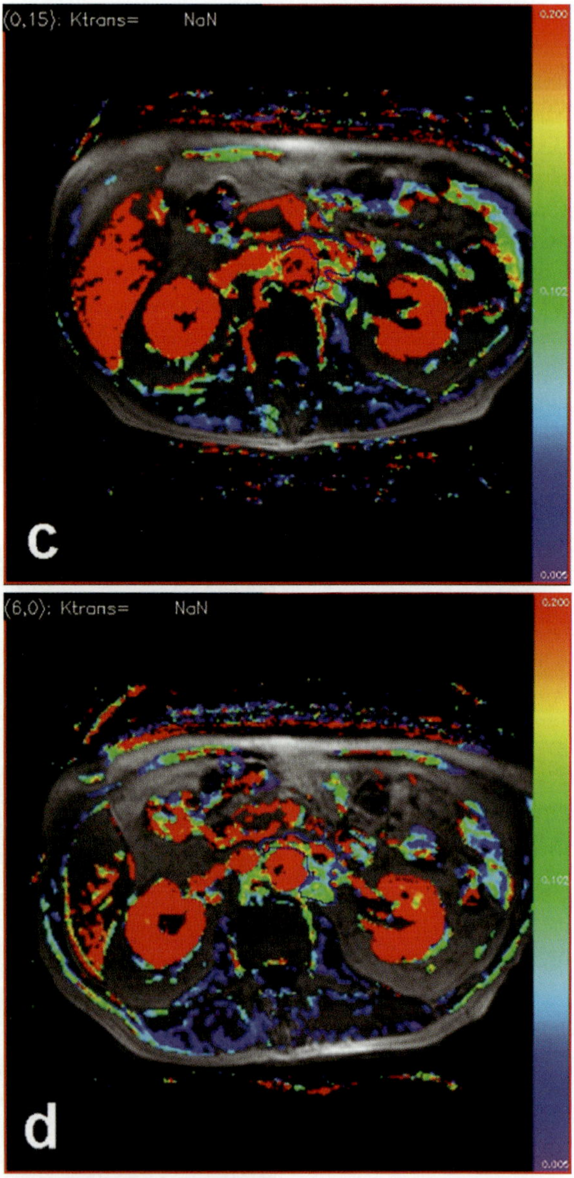

Fig. 7.6 72-year-old male with prostate cancer. Axial contrast enhanced T1-weighted magnetic resonance images before (**a**) and one-month after (**b**) experimental anti-angiogenic therapy show retroperitoneal lymphadenopathy with minimal change in size after therapy (*arrow*); K^{trans}, a vascular leakage parameter, reductions consistent with response to anti-angiogenic therapy are seen on the K^{trans} maps (*arrowheads*) before (**c**) and one-month after therapy (**d**)

lymph nodes from metastatic ones reveals sensitivity and specificity values varying between 33–100% and 37.5–100%, respectively [84]. Though the utility of SPIO-enhanced MRI in the detection of nodal metastases varies for different regions of the body, it is most accurate in identifying metastases to normal sized lymph nodes (Fig. 7.7).

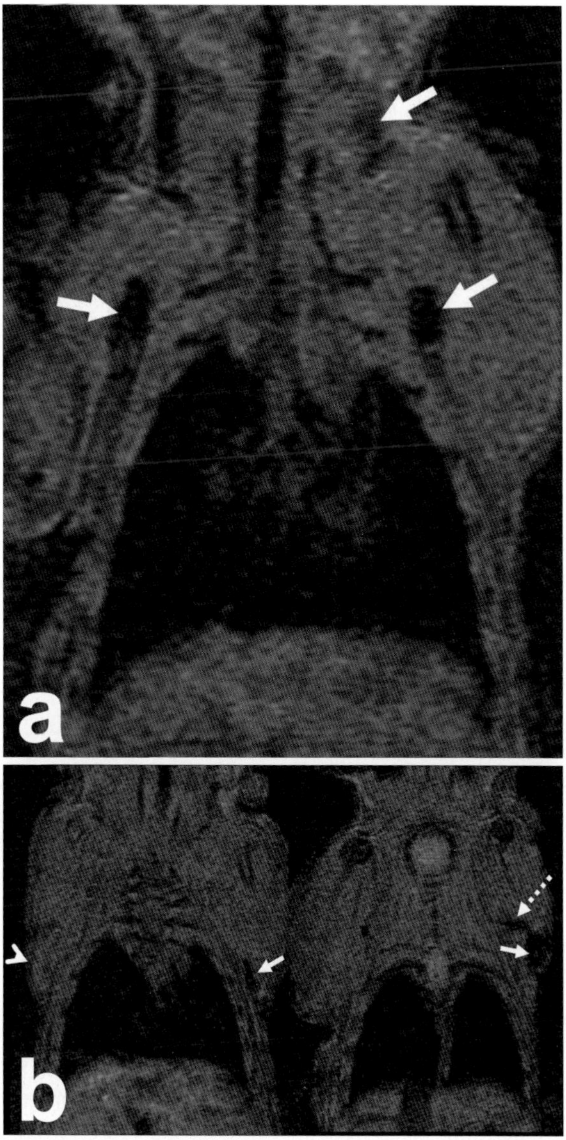

Fig. 7.7 USPIO lymphography of a mouse 24 hrs after intravenous injection from the tail vein (**a**) and 10 min after subcutaneous injection through the finger (**b**). All LN are seen in *black* after intravenous injection (*arrows*) (a and b); lymph node on right (*arrowhead*) is not enhanced since injection was made from left subcutaneously, whereas both draining lymph nodes (*short arrows*) and lymphatic vessels (*long arrow*) are demonstrated after subcutaneous injection from left side (b)

The use of macromolecular agents employing Gadolinium labeling is at a relatively early stage of research. Some groups have demonstrated the feasibility of magnetic resonance lymphography (MRL) with Gd-containing liposomes in animal models [85, 86]. Good uptake was demonstrated in regional lymph nodes following subcutaneous injection, which is likely related to trapping of the liposomes by macrophages. Misselwitz et al. used the macromolecular contrast medium Gadomer-17 to image the inguinal and iliac nodes in dogs following hind limb injection [87].

Enhancement was seen as early as 15 min post-injection, but was maximal 60–90 min after injection, with signal enhancement increasing by as much as 450–960%, depending on the initial dose.

It is also possible to use MRI to image in vivo lymphatic-convective transport. Pathak et al. selected two murine breast cell lines, known to have differences in invasiveness [88]. Using albumin-Gd-DTPA as a contrast agent they were able to classify ROIs as "pooling" if the macromolecular contrast media (MMCM) concentration increased over time, or "draining" if it decreased relative to early phase images. The more invasive tumor line had a significantly higher MRI-detected number of 'draining' voxels. Thus, the lymphatic drainage pattern correlates with the metastasis rate and lymphangiogenesis. Drainage may be dependent on both the 'invasiveness' of the tumor and the extracellular matrix integrity which, if reduced, can facilitate passage of tumor cells, along with extracellular fluid.

Dendrimers are monodispersed synthetically produced organic polymers. They can be produced at precise but, chemically identical sizes. Two forms of dendrimers are commercially available: polyamidoamine (PAMAM) and diaminobutane core polypropylimine (DAB or PPI). Different generations (sizes) of dendrimers have been investigated as MRL macromolecular contrast agents. Kobayashi et al. used interstitially injected generation-6 (G6) PAMAM dendrimers loaded with Gadolinium chelates (Gd) to image the lymphatic system and the sentinel nodes of normal mice and mice with xenografted breast tumors [89]. Gd-G6 dendrimer was injected directly into the mammary gland or peri-tumorally, imaged by T1 weighted MRI and 3D reconstruction was used to aid anatomical localization. They were able to differentiate normal and abnormal lymphatics and distinguish intralymphatic from extralymphatic disease in a mouse lymphoma model [90]. Kobayashi et al. also compared MRL with either Gd-PAMAM dendrimers of different generations, or the less hydrophilic Gd-DAB generations in murine models [91]. Gd-PAMAM-G8 was retained in the fine lymphatic vessels without major leakage, resulting in excellent imaging of the lymphatic channels. However, Gd-PAMAM-G4 provided better localization of lymph nodes that were close to the liver, due to a reduced background signal (Fig. 7.8).

Another interesting advance in lymph node imaging is the use of dual-modality contrast agents. Talanov et al. synthesized a PAMAM G6 dendrimer conjugated to Gadolinium for MR lymphography and Cy5.5, a near infrared dye for optical imaging [92]. The agent was injected into the mammary fat pad of mice and sentinel lymph nodes were successfully imaged on MRI, followed by optical imaging. MR has a number of advantages, including good spatial resolution and lack of ionizing radiation exposure while optical imaging offers portability to the operating theatre and real time imaging. Further advances in contrast agent development may eventually lead to substantial progress in this field. One concern that places the future in some doubt is the possibility of gadolinium leaching from interstitially injected macromolecules and dissociating in the interstitial tissues. A new syndrome, found mainly in patients with renal failure who have received intravenous injections of Gadolinium-chelates, produces severe interstitial fibrosis and is known as Nephrogenic sclerosing fibrosis (NSF) [93, 94]. Its origins have been traced to

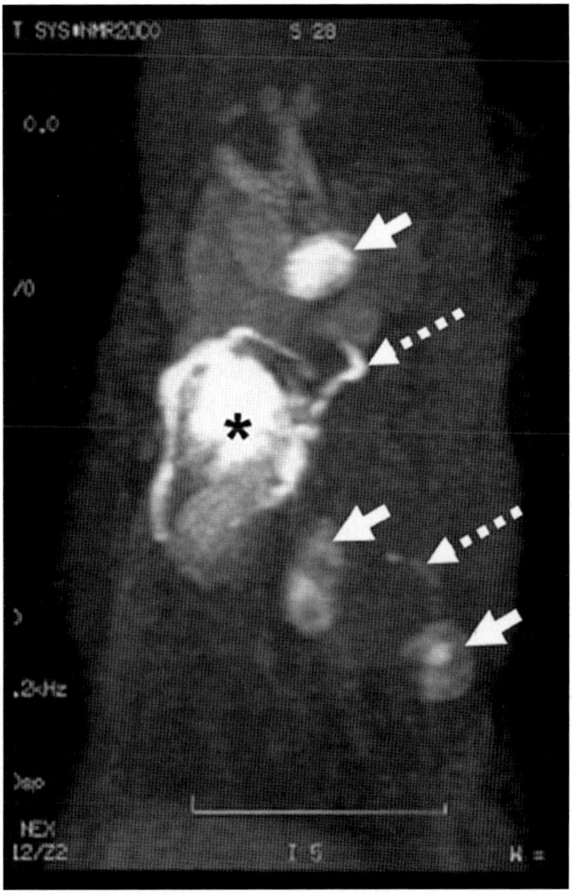

Fig. 7.8 12 min post-injection gadolinum-G6 dendrimer enhanced magnetic resonance lymphangiography image of a mouse with breast cancer demonstrates tumor tissue (*black asterix*), sentinel lymph nodes in superficial cervical, lateral thoracic and axillary regions from the top in order (*arrows*) and lymphatic vessels (*broken arrow*)

free Gadolinium in the interstitial tissues due to the slow clearance of the chelate in patients with renal failure undergoing dialysis. The exact pathogenesis of NSF remains obscure; however, its recognition has had a chilling effect on the development of new Gadolinium containing compounds, particularly those that are to be injected interstitially. Certainly, there is the theoretical risk of Gadolinium leaching from interstitially injected macromolecules with the potential for unknown toxic events. For this reason, the field is currently progressing cautiously and potential side effects are being evaluated in animal trials.

7.7.4 Optical Imaging

Optical imaging is a rapidly advancing branch of medical imaging that does not require ionizing radiation exposure and utilizes relatively low cost and portable equipment. It can be easily incorporated into the operating theatre for SLN biopsies. The near infra-red (NIR) spectrum is often used for in vivo imaging because hemoglobin, muscle and fat are least absorbent in this light range, allowing light arising from deeper tissue planes to be imaged. [95]. Fluorescent probes are safe and can be bound to various macromolecules, including antibodies. The main disadvantage remains the poor depth sensitivity of the technique, and penetration beyond 1–2 cm is currently unrealistic, however, in many cases this may be adequate for clinical use, especially in the context of identifying superficial SLNs in melanoma surgery (e.g. sentinel node imaging).

A special kind of optical fluorophore is known as a quantum dot. Quantum dots (Qdots) are semiconductor crystals in the nanometer size ranging between 5–20 nm in diameter. They have high quantum yield and thus are very bright. The wavelength of emitted energy of Qdots can be controlled by changing their size and shape; therefore they can emit light in the near infrared spectrum [82]. In addition to their high light output, Qdots can be excited with broadband light below the emission wavelength. Moreover, due to their narrow emission bandwidth, multiple Qdots can be used simultaneously with the same excitation light. In comparison, organic fluorophores usually require specific excitation light near to the emission wavelength limiting the number of dyes that can be simulantenously excited and detected [96]. On the other hand, the main disadvantage of Qdots is their potential toxicity related to their cadmium-selenium or cadmium-tellurium core [97]. Additionally, Qdots may induce cell death by formation of reactive oxygen species since in some cases Qdots can transfer absorbed optical energy to oxygen radicals which may cause DNA damage, potentially resulting in apoptosis and necrosis [98]. However, their clinical feasibility is still to be determined since the necessary dose is extremely low and Qdots can be designed to avoid the creation of reactive oxygen species.

The range of maximum fluorescence of Qdots is determined by their elemental composition extending over the whole visible and near infrared spectrum [98]. Their size makes them ideal agents for lymphatic imaging via interstitial injection. In order to make Qdots suitable for lymphatic imaging, their surfaces have to be modified [98].

Kim et al. reported successful in vivo imaging of the lymphatics using a near infrared Qdots in order to detect sentinel lymph nodes arising from breast tissue [99]. Retention of Qdots in lymph nodes was shown by several studies in animal models [100–102]. Parungo et al. reported use of Qdots in the detection of lymphatic drainage of peritoneal and pleural spaces in animal models [103, 104]. Knapp et al. used near infrared Qdots to demonstrate sentinel lymph nodes in invasive urinary bladder cancer in animal models [105]. Additionally, utility of using Qdots in sentinel lymph node mapping of gastrointestinal tract and lung was reported by Soltesz et al. [106, 107].

Hama et al. have shown utility of fluorescence lymphangiography using two near infrared Qdots with different emission spectra to simultaneously detect two lymphatic basins [108]. Recently, Kobayashi et al. demonstrated simultaneous multicolor fluorescence imaging of five different lymphatic basins using five near infrared Qdots [109] (Fig. 7.9). This is only possible with Qdots because of their narrow emission wavelengths and broad excitation tolerance. Fluorescence imaging enables improved sensitivity for those lymph nodes close enough to surface to be

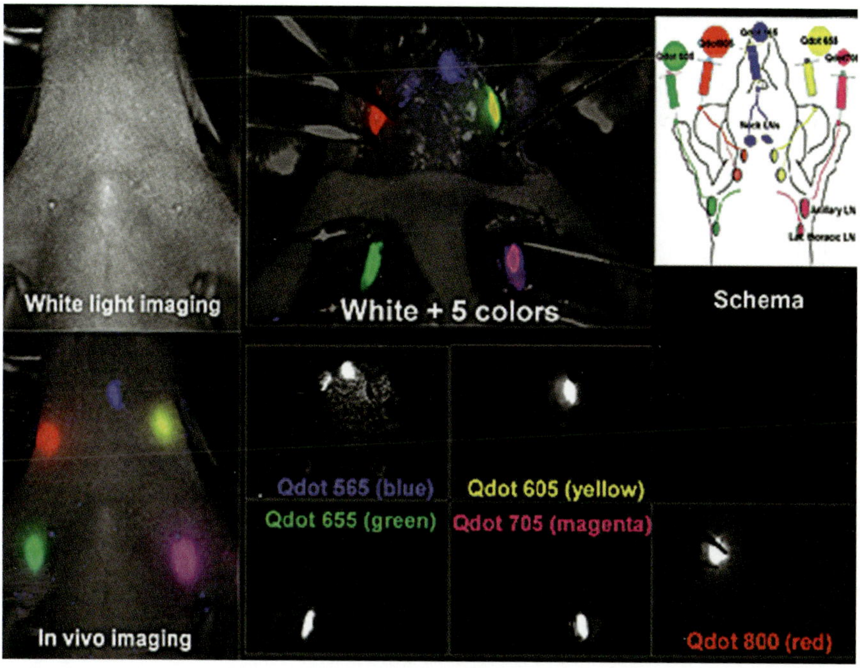

Fig. 7.9 Simultaneous 5-color (*blue, green, yellow, magenta and red*) lymphatic drainage imaging of the head and neck region using 5 quantum dots (Qdots 605, 805, 565, 655 and 705) injected intracutaneously from 5 different parts of the body in a mouse

imaged; moreover the combination of Qdots fluorescence imaging with other imaging modalities such as magnetic resonance imaging, computed tomography, scintigraphy and positron emission tomography may allow mapping of deeper lymph nodes, [99].

7.8 Summary and Outlook

Metastatic spread of cancer is the major cause of cancer related death. The lymphatics contribute to metastatic spread by allowing the primary tumor to grow larger and by providing a conduit for the spread of shed cancer cells. Evidence is accumulating from clinical and experimental studies that inhibition of the VEGF-C/VEGF-D/VEGFR-3 axis and other growth factors (VEGF-A, PDGF-BB, angiopoetins, FGF, HGF, IGF), which promote lymphatic spread of tumors through lymphangiogenesis, could lead to improvements in prognosis and survival. Systemic lymphatic imaging with computed tomography, magnetic resonance imaging and ultrasound is limited to the detection of enlarged lymph nodes; on the other hand, functional systemic lymphatic imaging methods, including positron emission tomography, dynamic contrast-enhanced MRI, lymphotrophic nanoparticle enhanced-MRI have been introduced to differentiate benign lymph nodes from malignant ones. The concept of sentinel lymph node biopsy under imaging guidance of Tc-99 m sulfur colloid lymphoscintigraphy and isosulfan blue dye procedure has had a profound influence on cancer management. Moreover, new targeted lymphatic imaging techniques including dendrimer conjugated gadolinium-MRI, quantum dot-fluorescence lymphangiography and macromolecule labeled optical imaging have been intensely studied in several trials. Advances in both systemic and targeted lymphatic imaging techniques may open up the possibility of delivery of intralymphatic treatment to tumor burden in addition to diagnostic imaging.

References

1. Su JL, Yen CJ, Chen PS, et al. The role of the VEGF-C/VEGFR-3 axis in cancer progression. Br J Cancer 2007, 96: 541–5.
2. Pepper MS. Lymphangiogenesis and tumor metastasis: myth or reality? Clin Cancer Res 2001, 7: 462–8.
3. Achen MG, McColl BK, Stacker SA. Focus on lymphangiogenesis in tumor metastasis. Cancer Cell 2005, 7: 121–7.
4. Ferrara N, Kerbel RS. Angiogenesis as a therapeutic target. Nature 2005, 438: 967–74.
5. Shayan R, Achen MG, Stacker SA. Lymphatic vessels in cancer metastasis: bridging the gaps. Carcinogenesis 2006, 27: 1729–38.
6. He Y, Karpanen T, Alitalo K. Role of lymphangiogenic factors in tumor metastasis. Biochim Biophys Acta 2004, 1654: 3–12.
7. Thiele W, Sleeman JP. Tumor-induced lymphangiogenesis: a target for cancer therapy? J Biotechnol 2006, 124: 224–41.
8. Tobler NE, Detmar M. Tumor and lymph node lymphangiogenesis – impact on cancer metastasis. J Leukoc Biol 2006, 80: 691–6.
9. Saharinen P, Tammela T, Karkkainen MJ, Alitalo K. Lymphatic vasculature: development, molecular regulation and role in tumor metastasis and inflammation. Trends Immunol 2004, 25: 387–95.
10. Drayton DL, Liao S, Mounzer RH, Ruddle NH. Lymphoid organ development: from ontogeny to neogenesis. Nat Immunol 2006, 7: 344–53.
11. Harrell MI, Iritani BM, Ruddell A. Tumor-induced sentinel lymph node lymphangiogenesis and increased lymph flow precede melanoma metastasis. Am J Pathol 2007, 170: 774–86.
12. Allan R. Sentinel node localization: do or dye alone? Br J Radiol 2001, 74: 475–7.
13. Hsueh EC, Hansen N, Giuliano AE. Intraoperative lymphatic mapping and sentinel lymph node dissection in breast cancer. CA Cancer J Clin 2000, 50: 279–91.
14. Cabanas RM. An approach for the treatment of penile carcinoma. Cancer 1977, 39: 456–66.
15. Morton DL, Wen DR, Wong JH, et al. Technical details of intraoperative lymphatic mapping for early stage melanoma. Arch Surg 1992, 127: 392–9.
16. Witte MH, Bernas MJ, Martin CP, Witte CL. Lymphangiogenesis and lymphangiodysplasia: from molecular to clinical lymphology. Microsc Res Tech 2001, 55: 122–45.
17. Baldwin ME, Stacker SA, Achen MG. Molecular control of lymphangiogenesis. Bioessays 2002, 24: 1030–40.
18. Wigle JT, Harvey N, Detmar M, et al. An essential role for Prox1 in the induction of the lymphatic endothelial cell phenotype. EMBO J 2002, 21: 1505–13.
19. Oliver G. Lymphatic vasculature development. Nat Rev Immunol 2004, 4: 35–45.
20. Prevo R, Banerji S, Ferguson DJ, Clasper S, Jackson DG. Mouse LYVE-1 is an endocytic receptor for hyaluronan in lymphatic endothelium. J Biol Chem 2001, 276: 19420–30.

21. Jackson DG, Prevo R, Clasper S, Banerji S. LYVE-1, the lymphatic system and tumor lymphangiogenesis. Trends Immunol 2001, 22: 317–21.
22. Valtola R, Salven P, Heikkila P, et al. VEGFR-3 and its ligand VEGF-C are associated with angiogenesis in breast cancer. Am J Pathol 1999, 154: 1381–90.
23. Makinen T, Adams RH, Bailey J, et al. PDZ interaction site in ephrinB2 is required for the remodeling of lymphatic vasculature. Genes Dev 2005, 19: 397–410.
24. Sundlisaeter E, Dicko A, Sakariassen PO, Sondenaa K, Enger PO, Bjerkvig R. Lymphangiogenesis in colorectal cancer-Prognostic and therapeutic aspects. Int J Cancer 2007, 121: 1401–9.
25. Tuttle TM. Technical advances in sentinel lymph node biopsy for breast cancer. Am Surg 2004, 70: 407–13.
26. Farnsworth RH, Achen MG, Stacker SA. Lymphatic endothelium: an important interactive surface for malignant cells. Pulm Pharmacol Ther 2006, 19: 51–60.
27. Ji RC, Kato S. Intrinsic interrelation of lymphatic endothelia with nerve elements in the monkey urinary bladder. Anat Rec 2000, 259: 86–96.
28. Skobe M, Hawighorst T, Jackson DG, et al. Induction of tumor lymphangiogenesis by VEGF-C promotes breast cancer metastasis. Nat Med 2001, 7: 192–8.
29. Stacker SA, Caesar C, Baldwin ME, et al. VEGF-D promotes the metastatic spread of tumor cells via the lymphatics. Nat Med 2001, 7: 186–91.
30. Stacker SA, Williams RA, Achen MG. Lymphangiogenic growth factors as markers of tumor metastasis. APMIS 2004, 112: 539–49.
31. Schoppmann SF, Birner P, Stockl J, et al. Tumor-associated macrophages express lymphatic endothelial growth factors and are related to peritumoral lymphangiogenesis. Am J Pathol 2002, 161: 947–56.
32. Wartiovaara U, Salven P, Mikkola H, et al. Peripheral blood platelets express VEGF-C and VEGF which are released during platelet activation. Thromb Haemost 1998, 80: 171–5.
33. Hirakawa S, Brown LF, Kodama S, Paavonen K, Alitalo K, Detmar M. VEGF-C-induced lymphangiogenesis in sentinel lymph nodes promotes tumor metastasis to distant sites. Blood 2007, 109: 1010–7.
34. Achen MG, Jeltsch M, Kukk E, et al. Vascular endothelial growth factor D (VEGF-D) is a ligand for the tyrosine kinases VEGF receptor 2 (Flk1) and VEGF receptor 3 (Flt4). Proc Natl Acad Sci U S A 1998, 95: 548–53.
35. Roy H, Bhardwaj S, Yla-Herttuala S. Biology of vascular endothelial growth factors. FEBS Lett 2006, 580: 2879–87.
36. Bjorndahl MA, Cao R, Burton JB, et al. Vascular endothelial growth factor-a promotes peritumoral lymphangiogenesis and lymphatic metastasis. Cancer Res 2005, 65: 9261–8.
37. Persaud K, Tille JC, Liu M, et al. Involvement of the VEGF receptor 3 in tubular morphogenesis demonstrated with a human anti-human VEGFR-3 monoclonal antibody that antagonizes receptor activation by VEGF-C. J Cell Sci 2004, 117: 2745–56.
38. Krishnan J, Kirkin V, Steffen A, et al. Differential in vivo and in vitro expression of vascular endothelial growth factor (VEGF)-C and VEGF-D in tumors and its relationship to lymphatic metastasis in immunocompetent rats. Cancer Res 2003, 63: 713–22.
39. Wilhelm SM, Carter C, Tang L, et al. BAY 43-9006 exhibits broad spectrum oral antitumor activity and targets the RAF/MEK/ERK pathway and receptor tyrosine kinases involved in tumor progression and angiogenesis. Cancer Res 2004, 64: 7099–109.
40. Luciani A, Itti E, Rahmouni A, Meignan M, Clement O. Lymph node imaging: basic principles. Eur J Radiol 2006, 58: 338–44.
41. Cserni G. Metastases in axillary sentinel lymph nodes in breast cancer as detected by intensive histopathological work up. J Clin Pathol 1999, 52: 922–4.
42. Derveaux L, Demedts M, Lijnen P, Amery A. Severe pulmonary thromboembolism following lymphography. Acta Clin Belg 1983, 38: 119–21.
43. Silvestri RC, Huseby JS, Rughani I, Thorning D, Culver BH. Respiratory distress syndrome from lymphangiography contrast medium. Am Rev Respir Dis 1980, 122: 543–9.

44. Jamieson JK, Dobson JF. Lectures on the lymphatic system of the stomach. Lancet 1907, 1: 1061–2.
45. Alex JC, Krag DN. Gamma-probe localization of lymph nodes. Surg Oncol 1993, 2: 137–43.
46. Krag DN, Weaver DL, Alex JC, Faibank JT. Surgical resection and radiolocalization of the sentinel lymph node in breast cancer using gamma probe. Surg Oncol 1993, 2: 335–9.
47. Giuliano AE, Kirgan DM, Guenter JM, Morton DL. Lymphatic mapping and sentinel lymphadenectomy for breast cancer. Ann Surg 1994, 220: 391–8.
48. Albertini JJ, Lyman GH, Cox C, et al. Lymphatic mapping and sentinel node biopsy in the patient with breast cancer. JAMA 1996; 276: 1818–22.
49. Larson SM, Nelp WB. Radiopharmacology of a Simplified Technetium-99 m-Colloid Preparation fir Photoscanning. J Nucl Med 1966, 7: 817–26.
50. Bergqvist L, Strand S-E, Perssom BRR. Particle sizing and biokinetics of interstitial lymphoscintigraphic agents. Semin Nucl Med 1983; 12: 9–19.
51. Alazraki N, Glass EC, Castronovo F, Olmos RA, Podoloff D. Society of Nuclear Medicine. Procedure guideline for lymphoscintigraphy and the use of intraoperative gamma probe for sentinel lymph node localization in melanoma of intermediate thickness 1.0. J Nucl Med 2002, 43: 1414–8.
52. Maza S, Valencia R, Geworski L, et al. Peritumoral versus subareaolar administration of technetium-99 m nanocolloid for sentinel lymph node detection in breast cancer: preliminary results of a prospective intra-individual comparative study. Eur J Nucl Med Mol Imaging 2003, 30: 651–6.
53. Maza S, Thomas A, Winzer KJ, et al. Subareolar injection of technetium-99 m nanocolloid yields reliable data on the axillary lymph node tumor status in breast cancer patients with previous manipulations on the primary tumour: a prospective study of 117 patients. Eur J Nucl Med Mol Imaging 2004, 31: 671–5.
54. Yeung HW, Cody III HS, Turlakow A, et al. Lymphoscintigraphy and sentinel node localization in breast cancer patients: a comparison between 1-day and 2-day protocols. J Nucl Med 2001; 42: 420–3.
55. Mar MV, Miller SA, Kim EE, Macapinlac HA. Evaluation and localization of lymphatic drainage and sentinel lymph nodes in patients with head and neck melanomas by hybrid SPECT/CT lymphoscintigraphy imaging. J Nucl Med Technol 2007, 35: 10–6.
56. Nemoto T, Vana J, Bedwani RN, Baker HW, McGregor FH, Murphy GP. Management and survival of female breast cancer: results of a national survey by the American College of Surgeons. Cancer 1980, 45: 2917–24.
57. Carter CL, Allen C, Henson DE. Relation of tumor size, lymph node status, and survival in 24,740 breast cancer cases. Cancer 1989, 63: 181–7.
58. Mustafa IA, Cole B, Wanebo HJ, Bland KI, Chang HR. Prognosis analysis of survival in small breast cancers. J Am Coll Surg 1998, 186: 562–9.
59. Rivadeneira DE, Simmons RM, Christos PJ; Hanna K, Daly JM, Osborne MP. Predictive factors associated with axillary lymph node metastases in T1a and T1b breast carcinomas: analysis in more than 900 patients. J Am Coll Surg 2000, 191: 1–6.
60. Mariani G, Moresco L, Viale G, et al. Radioguided sentinel lymph node biospsy in breast cancer. J Nucl Med 2001, 42: 1198–215.
61. Martin C, Cutuli B, Velten M. Predictive model of axillary lymph node involvement in women with small invasive breast carcinoma: axillary metastases in breast carcinoma. Cancer 2002, 94: 314–22.
62. Kim T, Giuliano AE, Lyman GH. Lymphatic mapping and sentinel lymph node biospsy in early-stage breast carcinoma: a metaanalysis. Cancer 2006, 106: 4–16.
63. Krag D, Weaver D, Ashikaga T, et al. The sentinel node in breast cancer – a multicenter validation study. N Engl J Med 1998, 339: 941–6.
64. Lyman GH, Giuliano AE, Somerfield MR, et al. American Society of Clinical Oncology guideline recommendations for sentinel lymph node biopsy in early-stage breast cancer. J Clin Oncol 2005, 23: 7703–20.

65. Balch CM, Buzaid AC, Soong SJ, et al. Final version of the American Joint Committee on cancer staging system for cutaneous melanoma. J Clin Oncol 2001, 19: 3635–48.
66. Cascinelli N. Margin of resection in the management of primary melanoma. Semin Surg Oncol 1998, 14: 272–5.
67. Balch CM, Soong SJ, Gershenwald J, et al. Prognostic factors analysis of 17,600 melanoma patients: validation of the AJCC melanoma staging system. J Clin Oncol 2001, 19: 3622–34.
68. Morton DL, Cochran AJ, Thompson JF, et al. Sentinel node biopsy for early-stage melanoma: accuracy and morbidity in MSLT-I, an international multicenter trial. Ann Surg 2005, 242: 302–11.
69. Sherman AI, Ter-Pogossian M. Lymph-node concentration of radioactive colloidal gold following interstitial injection. Cancer 1953, 6: 1238–40.
70. Zum Winkel K: Lymphologie mit Radionukliden. Verlag Hildergard Hoffman, Berlin, 1972, 30–32.
71. Juma N, Andrey T, Ege GN. Comparison as a lymphoscintigraphic agent between 99Tcm dextran and 99Tcm antimony sulphide colloid. Br J Radiol 1985, 58: 325–30.
72. Ege GN, Warbick A. Lymphoscintigraphy: a comparison of 99Tc(m) antimony sulphide colloid and 99Tc(m) stannous phytate. Br J Radiol 1979, 52: 124–9.
73. Strand SE, Persson BR. Quantitative lymphoscintigraphy I: basic concepts for optimal uptake of radiocolloids in the parasternal lymph nodes of rabbits. J Nucl Med 1979, 20: 1038–46.
74. Allen JF. An improved technetium-99 m generator for medical applications. Int J Appl Radiat Isot 1965, 16: 332–4.
75. Dufresne EN, Kaplan WD, Zimmerman RE, Rose CM. The application of internal mammary lymphoscintigraphy to planning of radiation therapy. J Nucl Med 1980, 21: 697–9.
76. Pearlman AW. Abdominal lymph node scintiscanning with radioactive gold (Au198) for evaluation and treatment of patients with lymphoma. Am J Roentgenol Radium Ther Nucl Med 1970, 109: 780–92.
77. Vassallo P, Wernecke K, Roos N, Peters PE. Differentiation of benign from malignant superficial lymphadenopathy: the role of high-resolution US. Radiology 1992 Apr; 183(1): 215–20.
78. Tregnaghi A, De Candia A, Calderone M, et al. Ultrasonographic evaluation of superficial lymph node metastases in melanoma. Eur J Radiol 1997 May, ;4(3): 216–21.
79. Choi SH, Kono Y, Corbeil J, Lucidarme O, Mattrey RF. Model to quantify lymph node enhancement on indirect sonographic lymphography. AJR Am J Roentgenol 2004, 183: 513–7.
80. Suga K, Yuan Y, Okada M, et al. Breast sentinel lymph node mapping at CT lymphography with iopamidol: preliminary experience. Radiology 2004, 230: 543–52.
81. Minato M, Hirose C, Sasa M, Nishitani H, Hirose Y, Morimoto T. 3-dimensional computed tomography lymphography-guided identification of sentinel lymph nodes in breast cancer patients using subcutaneous injection of nonionic contrast medium: a clinical trial. J Comput Assist Tomogr 2004, 28: 46–51.
82. Barrett T, Choyke PL, Kobayashi H. Imaging of the lymphatic system: new horizons. Contrast Media Mol Imaging 2006, 1: 230–45.
83. Anzai Y. Superparamagnetic iron oxide nanoparticles: nodal metastases and beyond. Top Magn Reson Imaging 2004, 15: 103–11.
84. Saokar A, Braschi M, Harisinghani MG. Lymphotrophic nanoparticle enhanced MR imaging (LNMRI) for lymph node imaging. Abdom Imaging 2006, 31: 660–7.
85. Fujimoto Y, Okuhata Y, Tyngi S, Namba Y, Oku N. Magnetic resonance lymphography of profundus lymph nodes with liposomal gadolinium-diethylenetriamine pentaacetic acid. Biol. Pharm. Bull 2000, 23: 97–100.
86. Misselwitz B, Sachse A. Interstitial MR lymphography using Gd-carrying liposomes. Acta Radiol 1997, 412: 51–55.
87. Misselwitz B, Schmitt-Willich H, Michaelis M, Oellinger JJ. Interstitial magnetic resonance lymphography using a polymeric t1 contrast agent: initial experience with Gadomer-17. Invest Radiol 2002, 37: 146–51.

88. Pathak AP, Artemov D, Neeman M, Bhujwalla ZM. Lymph node metastasis in breast cancer xenografts is associated with increased regions of extravascular drain, lymphatic vessel area, and invasive phenotype. Cancer Res 2006, 66: 5151–8.
89. Kobayashi H, Kawamoto S, Sakai Y, et al. Lymphatic drainage imaging of breast cancer in mice by micromagnetic resonance lymphangiography using a nano-size paramagnetic contrast agent. J Natl Cancer Inst 2004, 96: 703–8.
90. Kobayashi H, Kawamoto S, Brechbiel MW, et al. Detection of lymph node involvement in hematologic malignancies using micromagnetic resonance lymphangiography with a gadolinum-labeled dendrimer nanoparticle. Neoplasia 2005, 7: 984–91.
91. Kobayashi H, Kawamoto S, Choyke PL, et al. Comparison of dendrimer-based macromolecular contrast agents for dynamic micro-magnetic resonance lymphangiography. Magn Reson Med 2003, 50: 758–66.
92. Talanov VS, Regino CA, Kobayashi H, Bernardo M, Choyke PL, Brechbiel MW. Dendrimer-based nanoprobe for dual modality magnetic resonance and fluorescence imaging. Nano Lett 2006, 6: 1459–63.
93. Grobner T. Gadolinium – a specific trigger for the development of nephrogenic fibrosing dermopathy and nephrogenic systemic fibrosis? Nephrol Dial Transplant 2006, 21: 1104–8.
94. Marckmann P, Skov L, Rossen K, et al. Nephrogenic systemic fibrosis: suspected causative role of gadodiamide used for contrast-enhanced magnetic resonance imaging. J Am Soc Nephrol 2006, 17: 2359–62.
95. Carrington C. Optical imaging sheds light on cancer's signature—regional blood flow and tissue oxygenization measures may permit earlier breast cancer detection. Diagnost Imag Accessed June 2004; www.diagnosticimaging.com.
96. Dahan M, Laurence T, Pinaud F, et al. Time-gated biological imaging by use of colloidal quantum dots. Opt Lett 2001, 26: 825–7.
97. Derfus AM, Chan WCW, Bhatia SN. Probing the cytotoxicity of semiconductor quantum dots. Nano Lett 2004, 4: 11–8.
98. Hild WA, Breunig M, Goepferich A. Quantum dots – nano-sized probes for the exploration of cellular and intracellular targeting. Eur J Pharm Biopharm 2008, 68: 153–68;
99. Kim S, Lim YT, Soltesz EG, De Grand AM, et al. Near-infrared fluorescent type II quantum dots for sentinel lymph node mapping. Nat Biotechnol 2004, 22: 93–7.
100. Gopee NV, Roberts DW, Webb P, et al. Migration of intradermally injected quantum dots to sentinel organs in mice. Toxicol Sci 2007, 98: 249–57.
101. Ballou B, Ernst LA, Andreko S, et al. Sentinel lymph node imaging using quantum dots in mouse tumor models. Bioconjug Chem 2007, 18: 389–96.
102. Frangioni JV, Kim SW, Ohnishi S, Kim S, Bawendi MG. Sentinel lymph node mapping with type-II quantum dots. Methods Mol Biol 2007, 374: 147–60.
103. Parungo CP, Soybel DI, Colson YL, et al. Lymphatic drainage of the peritoneal space: a pattern dependent on bowel lymphatics. Ann Surg Oncol 2007, 14: 286–98.
104. Parungo CP, Colson YL, Kim SW, et al. Sentinel lymph node mapping of the pleural space. Chest 2005, 127: 1799–804.
105. Knapp DW, Adams LG, Degrand AM, et al. Sentinel lymph node mapping of invasive urinary bladder cancer in animal models using invisible light. Eur Urol 2007; 52: 1700–8.
106. Soltesz EG, Kim S, Kim SW, et al. Sentinel lymph node mapping of the gastrointestinal tract by using invisible light. Ann Surg Oncol 2006, 13: 386–96.
107. Soltesz EG, Kim S, Laurence RG, et al. Intraoperative sentinel lymph node mapping of the lung using near-infrared fluorescent quantum dots. Ann Thorac Surg 2005, 79: 269–77.
108. Hama Y, Koyama Y, Urano Y, Choyke PL, Kobayashi H. Two-color lymphatic mapping using Ig-conjugated near infrared optical probes. J Invest Dermatol 2007, 127: 2351–6.
109. Kobayashi H, Hama Y, Koyama Y, et al. Simultaneous multicolor imaging of five different lymphatic basins using quantum dots. Nano Lett 2007, 7: 1711–16.

Chapter 8

LYMPHANGIOLEIOMYOMATOSIS

Kuniaki Seyama[1], Toshio Kumasaka[2], Teruhiko Sato[1] and Keiko Mitani[2]
[1]*Department of Respiratory Medicine, Juntendo University, School of Medicine, Bunkyo-Ku, Tokyo 113-8421 Japan, e-mail: kseyama@juntendo.ac.jp*
[2]*Department of Human Pathology, Juntendo University School of Medicine, Bunkyo-Ku, Tokyo 113-8421, Japan, e-mail: kumasaka toshio@med.jrc.or.jp*

Abstract: Lymphangioleiomyomatosis (LAM) is a rare, neoplastic disease in which abnormal smooth muscle-like cells (LAM cells) proliferate in the lungs and along the axial lymphatic systems including the lymph nodes and thoracic ducts. LAM cells are transformed cells due to loss-of-function type mutations of either the *TSC1* or *TSC2* gene, which are tumor suppressor genes originally identified to be the genetic cause for tuberous sclerosis complex. LAM shows an extreme gender predilection and it usually occurs in women of reproductive age. Its pathological findings are characterized by the existence of abundant lymphatic vessels resulting from LAM-associated lymphangiogenesis since LAM cells produce potent lymphangiogenic growth factors, VEGF-C and VEGF-D. Consequently its clinical manifestations include the symptoms and signs related with abnormalities in the lymphatic system, such as lymphangioleiomyomas, chylous leaks into body cavities and urine, from the airways or even the vagina, or lymphedema of the lower extremities as well as a progressive cystic destruction of the lungs, thus resulting in respiratory failure. The extent of LAM-associated lymphangiogenesis correlates with the histologic severity of LAM. The mechanism for the progression of LAM is now hypothesized to be a unique invasion-independent mechanism mediated with LAM-associated lymphangiogenesis. LAM cells are considered to disseminate and form a metastatic lesion in the lungs and axial lymphatic systems through the lymphangiogenesis-mediated fragmentation of LAM foci and followed by the subsequent shedding of LAM cell clusters into the lymphatic circulation.

Key words: Estrogen · LAM cell cluster · Lymphangiogenesis · Tuberous sclerosis complex · VEGF-C · VEGF-D

S.A. Stacker, M.G. Achen (eds.), *Lymphangiogenesis in Cancer Metastasis*, Cancer Metastasis – Biology and Treatment 13, DOI 10.1007/978-90-481-2247-9_8,

8.1 Introduction

Lymphangioleiomyomatosis (LAM) is a rare, neoplastic disease in which abnormal smooth muscle-like cells (LAM cells) proliferate in the lungs and along axial lymphatic systems, including the lymph nodes and thoracic ducts. LAM is a disease showing an extreme gender predilection for women. LAM usually occurs in women of reproductive age, but some occurrences in post-menopausal women have also been reported. LAM can be grouped into patients with a sporadic occurrence of LAM (sporadic LAM) or those associated with tuberous sclerosis complex (TSC) (TSC-LAM). TSC is an autosomal-dominant neurocutaneous syndrome resulting from the mutations of either the *TSC1* or *TSC2* gene. TSC-LAM is rarely seen in men [1, 2].

In the literature, the first description of the cases with LAM seems to be back in 1937, when patients with chylous pleural effusion, a proliferation of abnormal smooth muscle cells in the lymph node, and honeycomb lungs were reported under the diagnosis of leiomyosarcoma [3] or muscular cirrohosis of the lungs [4]. Subsequently, Frack et al. used the term, "the lymphangiomyomatosis syndrome", for the first time to describe a case with progressive dyspnea culminating in pneumothrax, chylothorax, chylous ascites, chyluria, and lymphedema of the left lower extremity [5]. A comprehensive evaluation of this disease from the viewpoints of physiologic, pathologic and radiologic aspects, was reported in 1975 under the name of "pulmonary lymphangioleiomyomatosis" [6], and the name of lymphangiomyomatosis or lymphangioleiomyomatosis thereafter became widely used. It was generally considered that the lungs were involved in lymphangioleiomyomatosis as the term "pulmonary" was added as an adjective, for example in such term as, "pulmonary lymphangioleiomyomatosis" and the patients with this disease tended to die of respiratory failure approximately 10 years after being diagnosed with this disease [7]. However, extrapulmonary involvement of lymph nodes in the retroperitoneum or pelvic cavity was thereafter recognized as a part of the clinical picture of lymphangioleiomyomatosis and renal angiomyolipoma in which the identical genetic abnormalities were demonstrated to frequently coexist with lymphangioleiomyomatosis [8, 9], the disease is now recognized to be a systemic disease affecting the lungs and axial lymphatic system and, as a result, it has since that time been simply called either as lymphangioleiomyomatosis or its abbreviation, LAM. Significant advances in the understanding of its clinical features and molecular mechanisms have been obtained over the past 10 years and to date LAM is still considered to be somewhat peculiar in terms of its gender predilection, the heterogeneity of the clinical manifestations and disease course, and also in regard to its intimate association with lymphatic systems which are considered to play a significant role in both its clinical manifestations and the progression of the disease.

8.2 Clinical Features

8.2.1 Epidemiology

True incidence and prevalence of LAM are not known. There reported to be approximately 400 known patients with LAM in North America [10]. A nation-wide survey recently performed in Japan reported and estimated prevalence of 1.2–2.3 cases per million in the Japanese population [11] which was quite similar to the reported prevalence of 0.9 cases in per million in the population of United Kingdom [12], 1.3 cases per million in the population of France [13]. These figures suggest that there is no apparent racial difference in the prevalence of LAM. However, the LAM Foundation, a patient organization of LAM in the United States, has identified over 1,000 patients with LAM in the United States and approximately 1,200 patients with LAM in other countries, suggesting a minimum prevalence of 2–6 per million women [14].

8.2.2 Symptoms and Signs

Several retrospective clinical studies and epidemiological studies including a sufficient number of patients with LAM have been published from different countries and there appear to be no significant racial differences in the clinical manifestations of LAM [11–14]. For convenience sake, the numbers in parentheses refer to the frequency of symptoms at the time of LAM diagnosis according to Hayashida et al. [11] in the following description.

LAM can be asymptomatic in its early stages. Due to advances in modern imaging techniques and the popularity of regular health checkups based on concerns or an increased awareness of heath promotion, a certain number of patients have been identified as individuals suspected of having abnormalities on chest X-ray examination and thereafter have been eventually been found to have LAM without pulmonary symptoms. However, LAM often demonstrates such respiratory symptoms as exertional dyspnea (74%), chest pain accompanying pneumothorax (53%), cough (32%), sputum (usually small in amount) (21%), or hemosputum (8%). The most common symptoms and signs at presentation are insidiously progressive dyspnea on exertion or pneumothrax. LAM is an important cause of the underlying diseases of women with pneumothrax. Pneumothorax frequently recurs and it rarely occurs concurrently in both lungs. Cough, sputum, and hemosputum are less frequent. However, hemosputum is a peculiar symptom and LAM is listed as one of underlying diseases of pulmonary hemorrhage, but in general massive hemoptysis is rare.

LAM occasionally occurs with extrathoracic symptoms. Symptoms and signs due to extrathoracic LAM lesions include the feeling of abdominal distension, abdominal discomfort, or pain which may be attributable to lymphangioleiomyomas in the retroperitoneum and the pelvic cavity or renal angiomyolipomas (AMLs). Severe abdominal pain or hematuria due to intratumoral hemorrhage of renal AMLs may be an episode leading to disclose the existence of LAM. Lymphatic edema of lower extremities may be the occasion of seeking medical attention and subsequently explored to have LAM in the pelvic cavity, although it is rare.

The most peculiar complication frequently seen in LAM is the chyle leakage into body cavities such as pleural, abdominal, or pericardial spaces, into urine (chylouria), or from the vagina, and the expectoration of chyle (chylous sputum). Chylous pleural effusion (7%) or chylous ascites (5%) are the most common types among LAM-associated chyle leakage. Some patients have chyle leakage into several body cavities concurrently. For example, some are complicated with both chylous ascites and pleural effusion, and even with chyle leakage from the vagina.

8.2.3 Radiographic and Physiological Findings

The plain chest radiographs may demonstrate increased bilateral reticulonodular interstitial markings with a normal or increased lung volume. Pneumothorax or pleural effusion may also be demonstrated. In patients with severe disease or at an advanced stage, cysts and bullae may also be visualized. However, chest X-rays are often normal and generally less sensitive for the identification of LAM than computed tomography (CT) of the chest since parenchymal involvement is mild in many patients at presentation.

High resolution CT of the chest, the images obtained at the slice thickness of 1–2 mm, is the most sensitive to identify LAM and hence the modality of choice for patients with LAM or those who are suspected of having LAM. Thin-walled cysts scattered evenly over all normal lung fields (Fig. 8.1). The cysts measure mostly between a few mm and 1 cm in diameter, have thin walls with clear borders from the underlying normal parenchymal image. As LAM gradually progresses, the number of cysts tends to increase over the time (Fig. 8.1) with an insidious deterioration of the pulmonary function. In some patients, especially in patients with TSC-LAM, small nodular shadows with hazy edges consistent with multifocal micronodular pneumocyte hyperplasia (MMPH) may be seen [16, 17]. The enlargement of mediastinal lymph nodes or the dilatation of the thoracic duct may also rarely be seen. CT of the abdominopelvic cavity may demonstrate the existence of lymphangioleiomyomas, renal AMLs, or chylous ascites (Fig. 8.2). The frequency of AML may differ according to race or ethnicity, with a reported frequency of 37.8% in the study of a NIH cohort [15] but 27% in a study based on a Japanese population [11].

Cystic destruction of lung parenchyma has devastating physiological consequences, including the progressive impairment of pulmonary function. The representative abnormalities found in LAM is the impairment in gas transfer and airflow (Fig. 8.3). The lowered diffusing capacity (DLco) is the most characteristic abnormality found in LAM and it is often demonstrated at an early stage of disease when no airflow limitation is recognized [11–13, 15, 18]. An insidious decline of DLco may also be demonstrated while the forced expiratory volume in 1 s (FEV1)

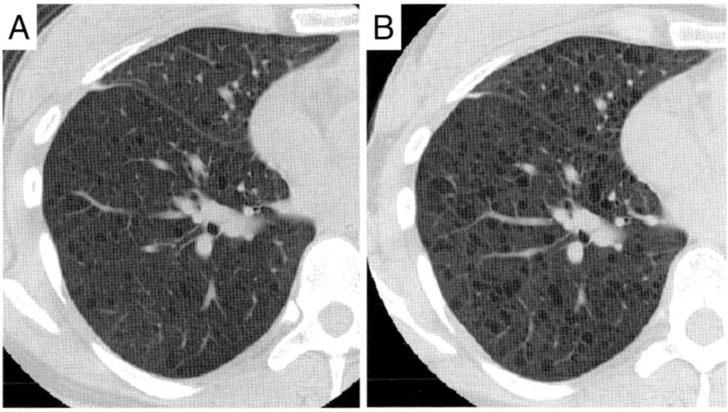

Fig. 8.1 Progression of pulmonary LAM on HRCT images of the chest. The HRCT images of the chest was obtained when the patient was at 26 (**A**) and 29 year olds (**B**), respectively, thus demonstrating an increasing number of cysts in right lower lung field during the following 3 years. This case presented with pneumothorax and was subsequently diagnosed to have LAM. Clearly demarcated thin-wall-cysts are recognized

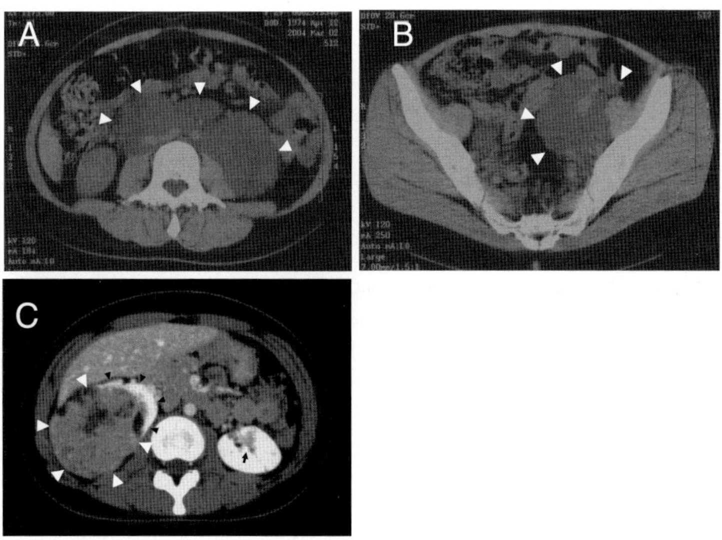

Fig. 8.2 The CT images of the extrapulmonary LAM and renal angiomyolipoma. Lymphangioleiomyomas, the cystic dilatation of lymph nodes involved by LAM, are demonstrated in the retroperitoneum (**A**) and the pelvic cavity (**B**) (indicated by *arrowheads*). Some patients with LAM are complicated with renal angiomyolipoma (AML) (**C**). Contrast-enhanced CT clearly delineates the AML composing of solid and fat density in both kidneys (*white arrowheads* in the *right* and an *arrow* in the *left* kidney, respectively). Normal *right* kidney tissue is compressed by huge AML as a crescent shape (*black arrowheads*)

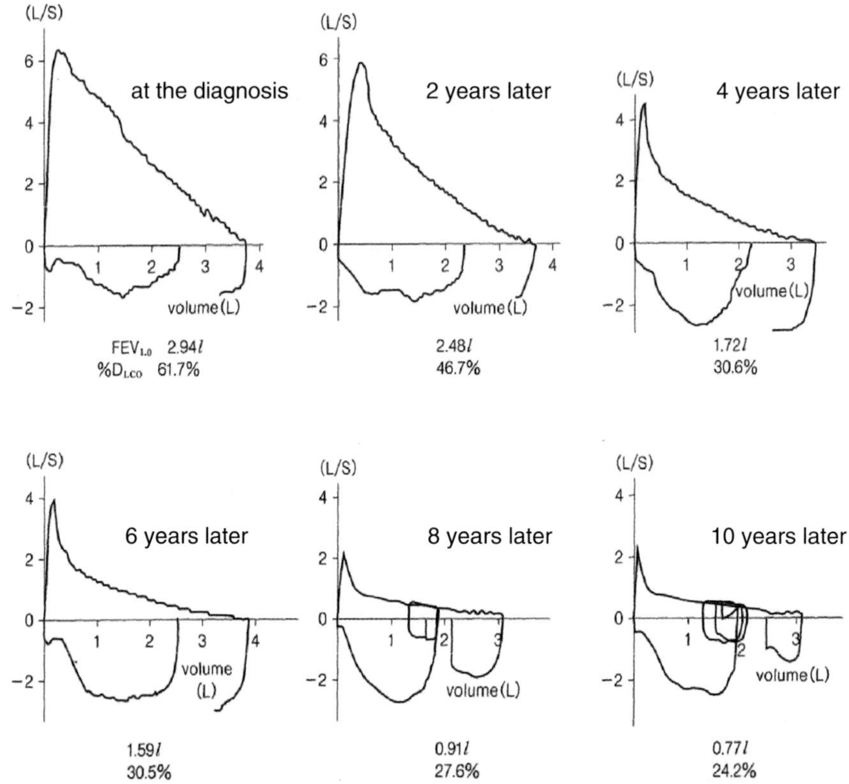

Fig. 8.3 Representative serial changes of pulmonary function demonstrated in a case with sporadic LAM. As a consequence of the progressive increase of cysts in the lung parenchyma, the pulmonary function gradually becomes impaired over time in LAM patients. This figure shows the serial change in the flow-volume curve obtained from a 31-year-old woman who presented with abdominal discomfort related to lymphangioleiomyomas in the pelvic cavity. At the diagnosis of a sporadic LAM made by surgical resection of lymphangioleiomyomas in the pelvic cavity, there was no airflow limitation but moderately impaired diffusing capacity was demonstrated. However, the gradual progression of airflow limitation, indicated by scooping of the forced expiratory flow-volume curve, was revealed during her 10-year course, together with consistent decline of the diffusing capacity

tend to remain stable [18]. The obstructive ventilatory impairment, as demonstrated with lowered FEV1 and FEV1/forced vital capacity (FVC) ratio, usually accompanies an increased residual volume (RV), total lung capacity (TLC) and RV/TLC ratio, thus indicating the hyperinflation of the lungs due to gas trapping [11–13,15,18].

8.2.4 Pathological Findings

The characteristic microscopic finding in LAM is the proliferation of smooth muscle-like cells (LAM cells) (Fig. 8.4A). The proliferation of LAM cells can be demonstrated in the lungs (cyst wall, pleura, bronchioles, and peripheral vessels,

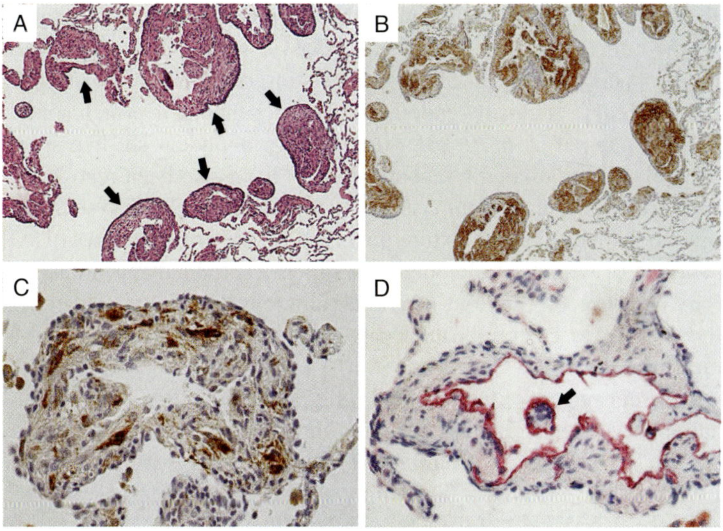

Fig. 8.4 Histopathological findings of LAM-affected lung tissues. The nodular proliferation of smooth muscle-like cells (LAM cells) (*arrows*) was demonstrated in the lungs (**A**, HE stain, original magnification ×50). Most LAM cells are immunopositive for α-smooth muscle actin (α-SMA) (**B**, indicated by *brown*, original magnification ×50) and some of them are immunopositive for one of melanoma-related antigens, HMB45 (**C**, indicated by *brown*, original magnification ×100). Within LAM cell nodules, slit-like or dilated spaces were frequently recognized (**A**). They are lymphatic vessels since they are lined with lymphatic endothelial cells (LEC), as clearly demonstrated by the immnuopositivity for VEGFR-3, a specific marker for LEC (**D**, indicated by *red*, original magnification ×100). Note that LAM cell cluster (LCC) (*arrow*), tightly packed LAM cells enveloped with LEC, is floating within LAM-associated lymphatic vessel (**D**)

etc.), and axial lymph nodes (supraclavicular region, pulmonary hilum and mediastinum, and, retroperitoneum, pelvic cavity, etc.). In the lung specimens stained with ordinary hematoxylin-esoin (HE stain), LAM cells proliferate in clusters and form distinct nodules or foci in lung parenchyma (Fig. 8.4A). LAM cells vary in form from spindle-shaped to epithelioid, and their nuclei are oval to spindle-shaped, with either no nucleolus or only one nucleolus and very fine chromatin. Their cytoplasm is eosinophilic or foamy. Immunohistochemical examinations are useful to confirm the proliferation of LAM cells and to also disclose their cellular features. Most LAM cells are immunopositive for muscular antigens such as α-smooth muscle actin (α-SMA) (Fig. 8.4B) and desmin. Some LAM cells react with anti-HMB45 antibodies (granular staining pattern in the cytoplasm surrounding the nucleus) that is a monoclonal antibody recognizing gp100, a premelanosomal protein presenting in certain melanoma cells (Fig. 8.4C). Although HMB45 is considered to be a specific marker for LAM cells [19], the positive staining rate for anti-HMB45 antibody is low in LAM cells, and sometimes no HMB45-positive LAM cells can be detected. Some LAM cells show positive staining for sex steroid hormone receptors, estrogen receptor and progesterone receptor [20, 21].

Another peculiar pathologic feature is the existence of abundant lymphatic vessels in LAM-affected tissues (Fig. 8.4D). Classic studies have characteristically described the histopathological features consisting of an anastomosing meshwork of cellular trabeculae enclosing endothelial-lined channels, in which some contain proteinaceous material [5, 6, 22, 23]. Alternatively, abundant slit-like spaces lined by endothelial cells or dilated lymphatic vessels have also been recognized. Lymphangioleiomyomas, usually demonstrated along axial lymphatic system, are cystic, soft lymph nodes with spongy texture, involved by the proliferation of LAM cells. The cut surface of lymphangioleiomyomas shows a shredded appearance [5]. They contain chylous or bloody chylous fluid and numerous particles which are composed of proliferating LAM cells trabeculated by lymphatic endothelial cells (LEC) and lymphatic channels [24].

LAM cells not only proliferate in the lung parenchyma but also invade the airways, pulmonary artery, diaphragm, aorta and retroperitoneal fat tissue [24, 25] (Fig. 8.5). We have previously examined the explanted lungs from 7 sporadic LAM patients who underwent lung transplantation with special reference to the airway involvement by LAM cells. When the airways in LAM-affected lung were analyzed from the central to peripheral direction by preparing tissue block every 10 mm along lower lobe bronchus and B9 bronchus, it was demonstrated that LAM cells invaded into the mucosal tissues in 5 of 7 patients with LAM (Fig. 8.5A and B). Abundant dilated lymphatic vessels existed with proliferating LAM cells (Fig. 8.5A and B). In the lower lobe bronchus or segmental bronchus, airway cartilages was often destroyed by proliferating LAM cells (Fig. 8.5C). Pulmonary arterioles can also be invaded by LAM cells. Pulmonary arterioles may be totally occluded by LAM cells or the wall of arterioles may be destroyed (Fig. 8.5D and E), thus suggesting that such these direct destruction of pulmonary arteries is likely to be the cause of pulmonary hemorrhage, hemosputum, or hemoptysis that are symptoms frequently seen in patients with LAM. In some patients, especially in those with chylous effusion, the nodular proliferation of LAM cells has been demonstrated in the diaphragm [24] (Fig. 8.5F). The transmural involvement by LAM cells with abundant lymphatic vessels may also be observed [24].

8.2.5 Molecular Pathogenesis of LAM

A similarity in the pulmonary lesions between sporadic LAM and TSC-LAM has been recognized since the early phase in the investigation of LAM [22, 26]. This led to the hypothesis that sporadic LAM might have some genetic abnormalities in common with TSC. TSC is an autosomal dominant neurocutaneous syndrome characterized with the development of hamartomas in multiple organ including central nervous system, skin, eye, heart, kidney, and lung. The two causative genes for TSC have been identified. The *TSC2* gene was identified on chromosome 16p13.3 in 1993 [27] followed by the identification of the *TSC1* gene on chromosome 9q34 in 1997 [28]. Since the loss of heterozygosity (LOH) for either *TSC1* or *TSC2* has been

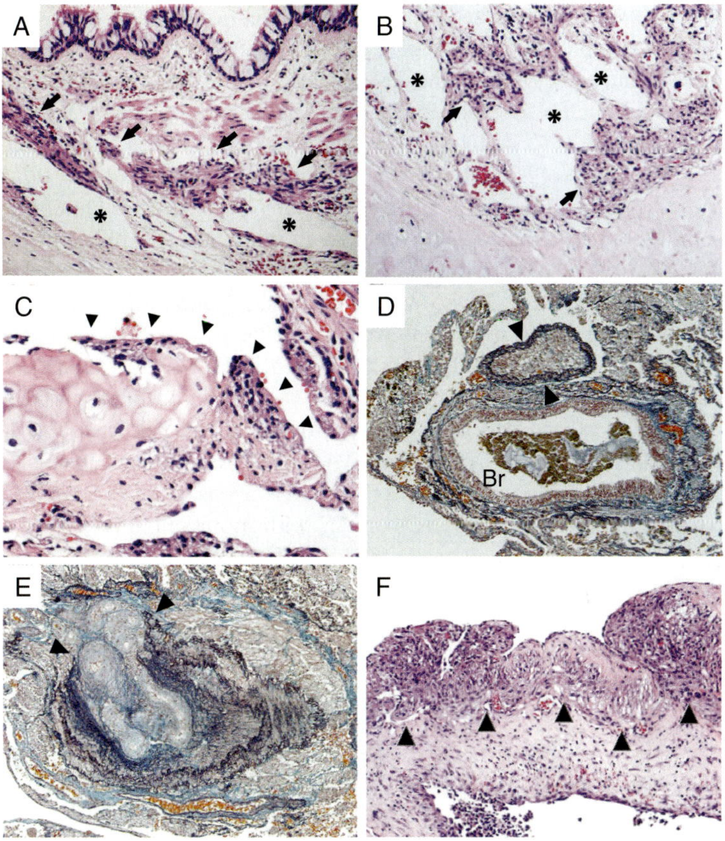

Fig. 8.5 LAM cells invade into airways, pulmonary arteries, and diaphragm. The proliferation of LAM cells with eosinophilic cytoplasm were identified in the bronchial wall (**A** and **B**, *arrows*) where abundant dilated lymphatic vessels (*) were also demonstrated (**A** and **B**, HE stain of the bronchial wall of explanted lung tissues, original magnification ×100). In Fig. 8.5A, bronchial cartilage was demonstrated on the *left side*. A part of bronchial cartilage was destroyed by proliferating LAM cells, which was clearly demonstrated in magnified view (**C**, HE stain of explanted lung tissues, original magnification ×100: *arrowheads* indicate LAM cells with eosinophilic cytoplasm). Pulmonary arterioles may be affected by proliferating LAM cells. Arterioles may be totally occluded by LAM cells (**D**, Masson-Trichrome stain of lung tissue specimens from an autopsy case, original magnification ×25: *arrowheads*, an arteriole occluded by LAM cells and Br, bronchiole). Proliferating LAM cells may encompass and directly invade into the arteriolar wall (**E**, Masson-Trichrome stain of lung tissue specimens from an autopsy case, original magnification ×25: *arrowheads* indicate the disruption and lacking of elastic fibers of arteriolar wall by LAM cells). Nodular proliferation of LAM cells (*arrowheads*) with cleft-like lymphatic spaces was identified in the fibrous stroma of the diaphragm (*top*, thoracic side; *bottom*, abdominal side) (**F**, HE staining of the diaphragm obtained by a surgical resection, original magnification ×35)

demonstrated in TSC-associated hamartomas [29, 30], both genes are considered to function as a tumor suppressor gene and a classic two-hit model proposed by Knudson [31] can be applied for the development of TSC-associated tumors.

The first evidence for the involvement of the *TSC2* gene in LAM was provided by the identification of *TSC2* LOH in renal AMLs or lymph nodes from patients with sporadic LAM [9]. No *TSC1* LOH was demonstrated in their study. In addition, none of these patients had clinical manifestations of TSC. In addition, no *TSC2* germline mutation was found in 41 patients with sporadic LAM [32]. Subseqeuntly, *TSC2* mutations were demonstrated in 5 AML tissue specimens obtained from 7 patients with sporadic LAM who concurrently had renal AML [8]. In addition, LAM-affected lung tissue specimens were available in 4 of 5 sporadic LAM patients for a *TSC2* mutation analysis and the identical *TSC2* mutations with renal AMLs have been demonstrated in all patients [8]. Based on these findings, it has been demonstrated that LAM cells are transformed, neoplastic cells due to *TSC2* mutations and might spread from renal AML via metastatic mechanisms [33]. We also confirmed the same findings from a mutation analysis of the *TSC1* or *TSC2* gene in 6 Japanese patients with TSC-LAM and 22 with sporadic LAM [34]. In addition to the mutation analysis, we demonstrated the clonal proliferation of transformed

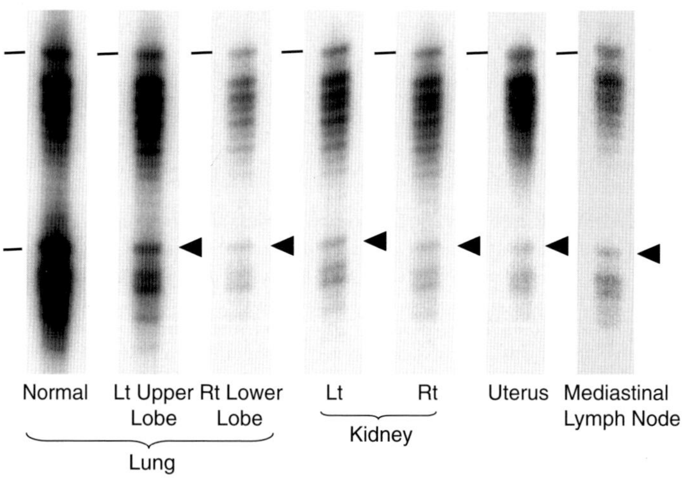

Fig. 8.6 LAM cells at different anatomical sites had the identical genetic abnormality. From an autopsy case with TSC-LAM who had a *TSC1* germline mutation, LAM cells were microdissected at different anatomical locations including *left* (lt) *upper lobe* or *right* (rt) *lower lobe* lung tissues, *left* (lt) or *right* (rt) kidney, uterus, and mediastinal lymph node, and then LOH was analyzed using D9S149, a commonly used microsatellite marker for the *TSC1* gene. The two alleles of D9S149 are indicated with *lines* and the lost allele is indicated with an *arrow*. As compared with two alleles of the marker D9S149 detected in the normal lung tissues, the identical pattern of *TSC1* LOH (loss of the lower allele) was demonstrated in all LAM cells obtained from various locations. As compared with two alleles detected in the normal lung tissues, the identical pattern of *TSC1* LOH was demonstrated in all LAM cells obtained from various locations. (Modified from Sato et al. [34]. By copyright permission of The Japan Society of Human Genetics and Springer Japan)

LAM cells carrying the identical LOH or *TSC2* mutations by microdissecting LAM cells from more than one anatomical location (Fig. 8.6), thus suggesting a metastatic spread of LAM cells [34]. Interestingly, the hypothesis has been raised with an early insightful observation that cells with a smooth muscle phenotype in LAM lesion may represent a clonal population despite the lack of any malignant features [22].

8.3 LAM-Associated Lymphangiogenesis

How do LAM cells metastasize and cause a progressive cystic destruction of the lungs? Do LAM cells truly metastasize even though LAM cells lack the morphological features of malignancy seen in most cancers? Do LAM cells invade blood vessels in a manner similar to that of cancer cells and perform a complex biological process comprised of multiple steps, such as intravasation, transport via the blood, extravasation, and secondary growth in the target organ?

Obviously, angiogenesis is an important process for tumor growth and blood borne- metastasis in cancers. The *TSC* gene products, hamartin for *TSC1* and tuberin for *TSC2*, have been demonstrated to be associated together in cytoplasm and they function as a negative regulator of mTOR signaling pathway, which control protein synthesis, cell size and cell growth [35]. Several studies have reported that mutations of the *TSC* genes may be associated with angiogenesis in TSC-associated tumors of the brain, kidney and skins [36, 37]. However, LAM-affected lesions are abundant with lymphatic vessels and LAM cells appear to have intimate association with LEC (Figs. 8.4 and 8.5).

8.3.1 Lymphatic Vessels in LAM-Affected Tissues

As the name of disease, "lymphangioleiomyomatosis" is self-explanatory regarding its important pathological features, the proliferation of LAM cells (abnormal smooth muscle-like cells) and the involvement of abundant lymphatic vessels in LAM-affected tissues. In addition, peculiar complications of chyle leakage into various body cavities, urine, and from vagina or intestine (manifested as protein-loosing enteropathy) [38] as well as lymphedema of lower extrimeties [39, 40], clearly remind us that LAM is a disease involving the lymphatic system.

Recent advances in lymphatic research identified several markers for LEC including LYVE-1, VEGFR-3, podoplanin (D2-40), and prox-1 [41], while also providing an opportunity to re-evaluate the pathology of LAM and for opening up a new field in regard to LAM research. With immunohistochemical examinations using CD31 as a marker for vascular endothelial cells and VEGFR-3 as a marker for LEC, the existence of abundant lymphatic vessels in LAM lesions were clearly demonstrated while a few CD31-positive blood vessels have been identified [25]. LAM-affected tissue specimen including the lungs (14 cases), lymph nodes (4 cases), uterus (2 cases), and ovary (a single case) were obtained from a total of 15 cases with

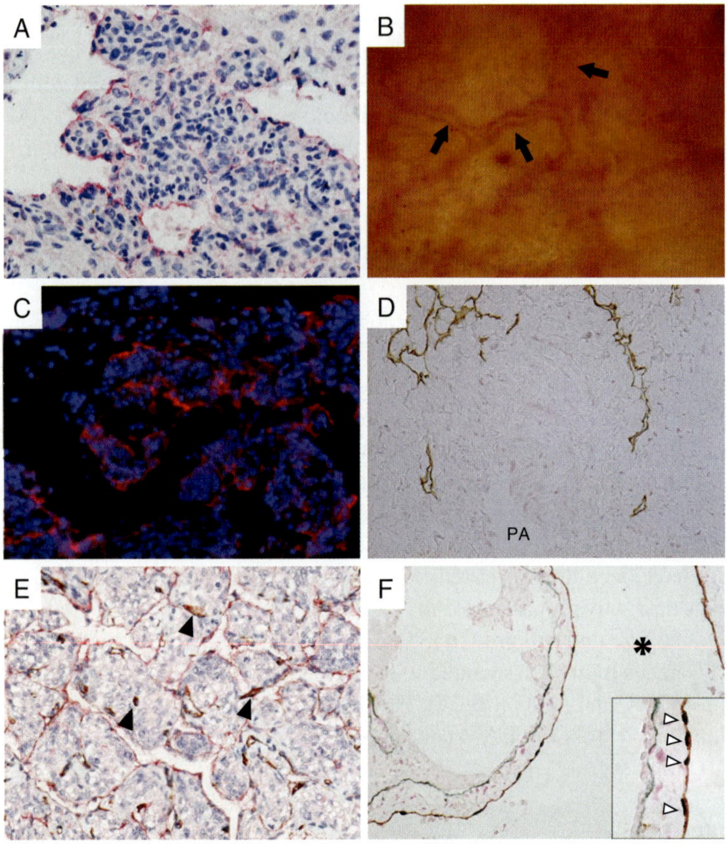

Fig. 8.7 LAM-associated lymphangiogenesis. Immunohistochemistry for VEGFR-3 demonstrates that slit-like or dilated space surrounded by LAM cells are lymphatic vessels with VEGFR-3-immunopositive LEC lining (**A**, original magnification ×100). Note that LEC was identified within the foci of proliferating LAM cells and LEC appear to demarcate the nodular proliferation of LAM cells and fragment it into LAM cell cluster. In the whole-mount lung tissue, lymphatic vessels (indicated by *arrows*) in LAM lesion were demonstrated by immunohistochemistry for VEGFR-3 (**B**, original magnification ×50). Immunofluorescent labeling of VEGFR-3 delineated a fine lymphatic network lined with LEC in a tissue specimen obtained from the inner marginal tissues of retroperitoneal lymphangioleiomyoma (**C**, smear preparation, original magnification ×100). Lymphatic vessels associated with the proliferation of LAM cells were seen in the wall of the large vessels in the lung. Lymphangiogenesis associated with proliferating LAM cells was demonstrated even in the wall of the large pulmonary artery (indicated with PA) by immunohistochemistry for podoplanin (**D**, original magnification ×100). In contrast, double Immunostaining for CD31 (*brown*, VEC, indicated by *arrowheads*) and VEGFR-3 (*red*, LEC) clearly revealed that blood vessels were much less abundant than lymphatic vessels in LAM foci (**E**, original magnification ×100). Double immunostaining for podoplanin (*brown*, LEC) and MIB-1 (*green color*, a cell cycle marker) revealed many LEC to proliferate in the dilated LAM-associated lymphatic vessels (*) of which nuclei were immunopositive for MIB-1 (*black*, double positive for podoplanin and MIB-1, indicated by *arrowheads* in the inset) (**F**, original magnification ×25; Inset, ×400). Note that anti-MIB-1 antibody (Ki67) has cross-reactivity for alveolar epithelial cells

LAM (6 autopsy cases, a single lung transplant case, and 8 surgical cases) to examine both vascular endothelial cells (VEC) and LEC. Immunostaining with VEGFR-3 clearly demonstrated that lymphatics were extremely abundant in both pulmonary and extrapulmonary LAM lesions (Fig. 8.7A and B). LEC not only outlined the inside of slit-like or dilated spaces within or adjacent LAM foci but also infiltrated within LAM foci. In lymphangioleiomyomas, the lymph nodes involved in the proliferation of LAM cells, a fine lymphatic network immunopositive for VEGFR-3 developed well to separate LAM foci into clusters were clearly delineated (Fig. 8.7C and E). LAM-associated lymphangiogenesis has even been demonstrated within the vascular wall and its adjacent interstitium where LAM cells proliferate (Fig. 8.7D). In contrast, immunostaining with CD31 confirmed that VEC were scanty in the LAM foci (Fig. 8.7E). Are these LAM-associated lymphatic vessels the ones that have already existed in the lungs and lymph nodes and have LAM cells simply proliferated and migrated along the pre-existing lymphatic vessels as considered in an early observation? [6, 22]. This would not be the case since the normal lungs or lymph nodes do not have as many lymphatic vessels as LAM-affected tissues. Some LEC in the LAM foci showed immunopositivity for Ki67 (Fig. 8.7F), a protein related to cell proliferation and expressed in cell nuclei throughout the entire cell cycle except for the G0 phase, thus indicating LAM-associated lymphangiogenesis rather the pre-existing lymphatic vessels.

8.3.2 LAM Cell Clusters

In the pathologic examination of LAM-affected tissue, a part of the LAM foci were observed to apparently protrude into the lymphatic lumen (Fig. 8.7A) and thereafter fragment into cell clusters. Immunostaining for VEGFR-3 revealed that cell clusters were enveloped by a monolayer of VEGFR-3-positive LEC and were confirmed to be free in the lymphatic lumen as examined with based on serial sections. These cell clusters could be demonstrated in intra-LAM foci lymphatic lumen of the lungs (Figs. 8.4A and 8.7A) or in the lymphatic lumen of lymph nodes. Chyle leakage such as chylous pleural effusion and ascites are the peculiar complications of patients with LAM, which approximately 10% of all such patients are reported to have [11–13, 15]. If these cell clusters are floating in the lymphatic stream, then they would be detected in LAM-associated chylous fluids. This is truly the case and so far these cell clusters have been successfully demonstrated in all LAM-associated chylous fluid specimens that we have so far examined. A total of 17 sample of chylous fluids of body cavities, including 9 of pleural effusion, seven of ascites and one of pericardial effusion, were collected from 13 patients with LAM. Using microscopy, cell clusters were able to be easily recognized as a sphere of three-dimensional structure and to have cellular overlapping with an apparent cellular border with a spool-like appearance. In a Papanicolaou stain, LAM cells in LCC had light green-stained cytoplasm and oval and/or elliptic nuclei with fine-granular hyperchromatin (Fig. 8.8A). The cell cluster always consisted of two distinct cell components, a

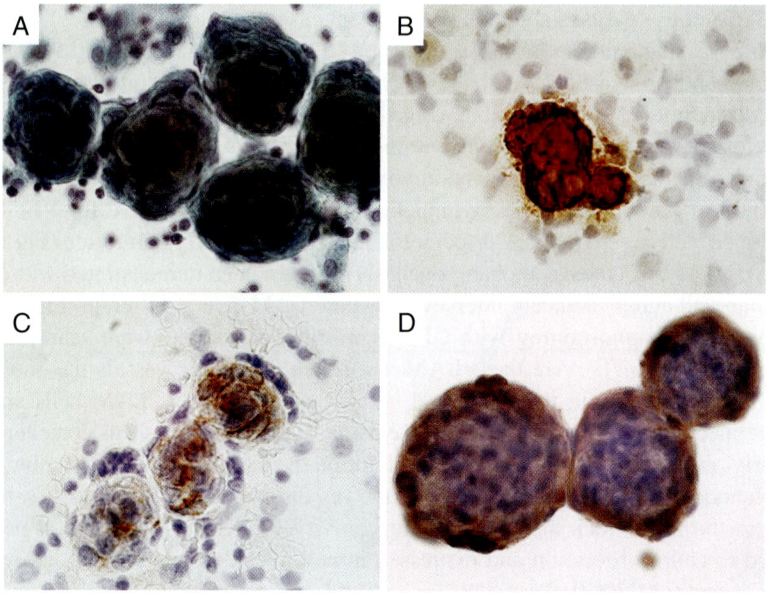

Fig. 8.8 Representative cytological and immunocytological features of LCC in LAM-associated chylous effusion. Five LCC are loosely connected with slit-like lymphatic vessels intervening among LCC (**A**, Papanicolaou stain, original magnification ×400). LCC has spherically-shaped, whorled, and well-demarcated structure, in which spindle-shaped cells are tightly packed and its surface is covered by flat cells with small nucleus. LAM cells had oval nuclei with moderate hyperchromatism (Papanicolaou stain, original magnification ×400). The inner cells of LCC were strongly immunopositive for α-SMA (**B**, smear, original magnification ×150). The perinuclear area in the cytoplasm of the inner cells was granularly immunopositive for the melanoma-associated antigens, HMB45 (**C**, smear, original magnification ×150). Podoplanin was expressed on the flat cells enveloping LCC (**D**, smear, original magnification like ×400)

superficial monolayer of LEC and tightly packed LAM cells inside, and accordingly were named LAM cell clusters (LCC) [24]. Immunocytochemistry demonstrated the cells inside to consist of LAM cells immunopositive for α-SMA (Fig. 8.8B) and HMB45 (Fig. 8.8C) and they were enveloped by a monolayer of flattened lymphatic endothelial cells immunopositive for VEGFR-3 or podoplanin (Fig. 8.8D). On smear preparation, LCC measure from approximately 25 to 125 μm in diameter. When LAM-associated chylous effusion contained many LCC, a concatenation of LCC, loosely connecting to the slit-like space, was often demonstrated (Fig. 8.8A). The background in smear preparation showed many lymphocytes and macrophages. However, neither tumor diathesis nor solitary LAM cell can be detected in contrast to the smear of cancer-associated effusions such as breast adenocarcinomas and malignant mesotheliomas in which tumor cell clusters are sometimes observed. The number of LCC appears to vary significantly from case to case and differ in the source of LAM-associated chylous effusion even in the same case with effusion in several body cavities concurrently. Although we could not evaluate the number of LCC quantitatively in all samples of chylous effusion, ascites tended to have more LCC than pleural effusion.

The detection of globular LAM cell nests of which surface was covered by a monolayer of endothelial cells has already been demonstrated and suggested to demonstrate diagnostic significance for LAM [42, 43]. Solid and globular cell clusters were recognized to be pathognomonic for LAM in two reports although the nature of endothelial cells was not identified as LEC [42, 43]. We have recently reported a case whose diagnosis of LAM was made based on the demonstration of *TSC2* LOH in cultured cells derived from LCC as well as the detection of LCC with these cytological and immunocytochemical features described above [44]. Accordingly, the cytological and immunocytochemical confirmations that the surface of cell cluster is LEC and the inside is LAM cells together with characteristic clinical settings, thus allow us to avoid the need to perform invasive diagnostic examinations such as the biopsy of the lungs or lymphangioleiomyomas.

8.3.3 Expression of VEGF-C and VEGF-D by LAM Cells

There are several growth factors regulating lymphangiogenesis, such as VEGF-C and VEGF-D, both of which bind to their cognate receptor VEGFR-3 on LEC [41]. The VEGF-C/-D and VEGFR-3 signaling system seems to be very important for tumor-associated lymphangiogenesis and metastasis since both VEGF-C-transduced and VEGF-D-transduced tumor cells promote tumor-associated lymphangiogenesis in both peri- and intra-tumoral area when explanted to an experimental animal [45, 46]. From the points of view that LAM cells are transformed, thus neoplastic cells through mutations of the *TSC* genes, lymphatic vessels in the LAM lesion may be a form of tumor-associated lymphangiogenesis if LAM cells produce lymphangiogenic growth factors.

Immunostaining with anti-VEGF-C and ant-VEGF-D antibodies demonstrated that LAM cells did express both VEGF-C and VEGF-D at varying intensities [25, 47] (Fig. 8.9). When the VEGF-C expression was examined in LAM-affected tissue specimens including the lungs (14 cases), lymph nodes (4 cases), uterus (2 cases), and ovary (a single case), the immunoreactivity for VEGF-C was detected in the cytoplasm of LAM cells with high reactivity in 11 (6 in lung, 3 in lymph node,

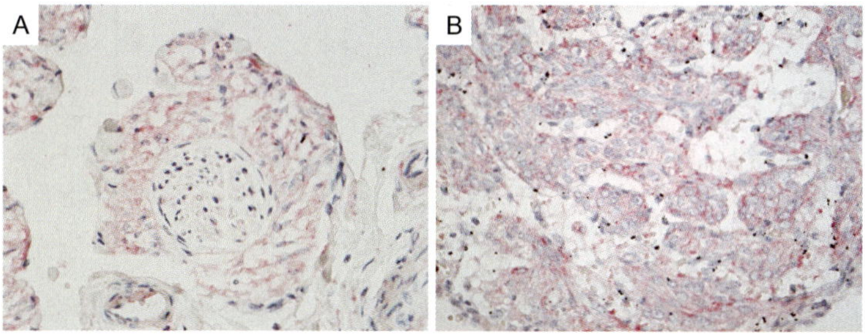

Fig. 8.9 Expression of VEGF-C and VEGF-D by LAM cells. The representative results of immunohistochemistry of LAM cells for VEGF-C (**A**) and VEGF-D (**B**) were presented (both, original magnification $\times 100$)

1 each in uterus and ovary), moderate in 1 (lung), and weak in 9 (7 in lung and 1 each in lymph node and uterus) [25]. Based on the semi-quantification of VEGF-C expression and lymphangiogenesis in LAM-affected lung tissues, a statistically significant correlation ($p<0.001$) was noted among the degree of lymphangiogenesis, the VEGF-C expression by LAM cells, and the histologic score representing the histologic severity of LAM [25, 48]. On the other hand, the VEGF-D expression by LAM cells was examined in 7 LAM-affected lung tissues. Similar to VEGF-C expression, VEGF-D immunoreactivity was demonstrated in the cytoplasm of LAM cells at varying intensities, including a high level in 2, a moderate level in 4, and a weak level in 1 [47].

VEGF-D appears to play a greater role in LAM-associated lymphangiogenesis than VEGF-C. When VEGF-A, VEGF-C, and VEGF-D were measured in the serum of 44 patients with LAM and 24 age-matched control women, only VEGF-D was demonstrated to increase in the serum of patients with LAM {LAM vs. control, geometric mean (95% confidence interval); 1,069.3 pg/ml (809.4 ~ 1,412.6) vs. 295.9 pg/ml (262.6 ~ 333.5)} (Fig. 8.10) [47]. The serum VEGF-D level thus appears to be higher in severe cases than in mild cases since a cross-sectional analysis found a statistically significant negative correlation between the serum VEGF-D level and an impairment in the pulmonary function [47]. The higher the serum VEGF-D level is, the lower the FEV1/FVC ($r = -0.365$, $p<0.05$) and %DLCO/VA ($r = -0.560$, $p<0.01$) are demonstrated to be although the correlation of airflow obstruction with serum VEGF-D is not as compelling as that of %DLCO/VA (Fig. 8.11). Accordingly, both LAM-associated lymphangiogenesis and an increased

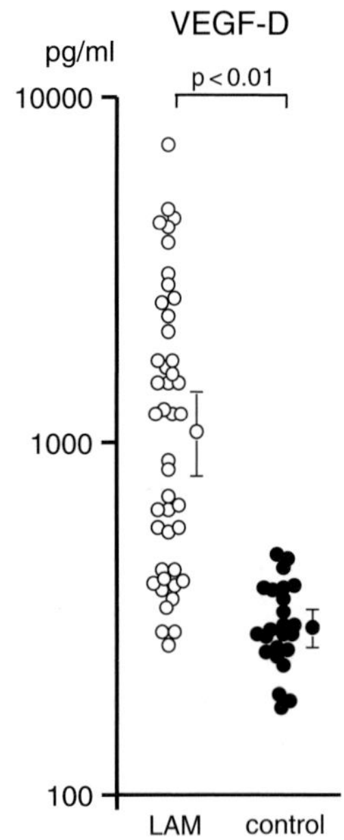

Fig. 8.10 The VEGF-D level is increased in the serum of patients with LAM. Data are plotted on a logarithmic scale with geometric mean ± 95% confidence interval. (Data from Seyama et al. [47]. By copyright permission of Lymphatic Research and Biology)

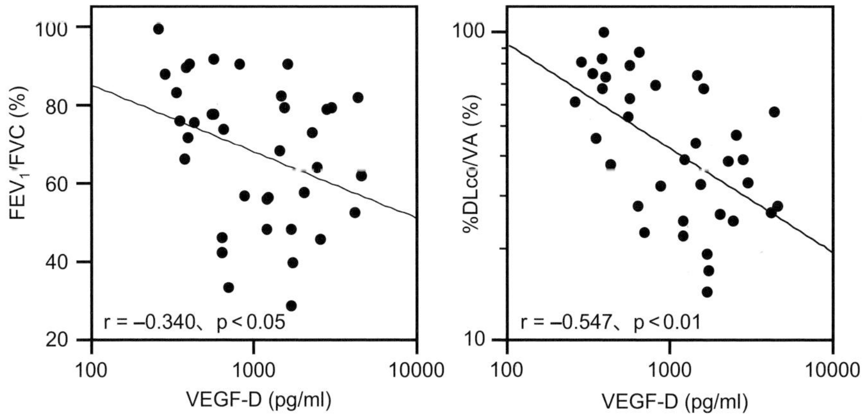

Fig. 8.11 Serum VEGF-D level negatively correlated with the pulmonary function. The serum VEGF-D level is plotted on the abscissa with a logarithmic scale and the correlation with FEV_1/FVC (*left panel*, *n* = 38) or %DLco/VA (*right panel*, *n* = 37), respectively, on the ordinate was evaluated. (Data from Seyama et al. [47]. By copyright permission of Lymphatic Research and Biology)

VEGF-D level in the serum seem to be implicated in the progression and disease severity of LAM. On the other hand, VEGF-A and VEGF-C were not elevated in LAM [47]. However, the platelet granules contain both VEGF-A and VEGF-C and which are then released upon platelet activation [49]. Accordingly, the determining the serum concentration may therefore be inappropriate to precisely evaluate their role in LAM and further analysis on this subject is thus called for.

8.3.4 Lymphangiogenesis and mTOR Signaling

Some of TSC-associated lesions, including renal anigomyolipoma, skin tumors (facial angiofibroma and subungual fibroma) and subependimal giant cell astrocytoma, are angiogenic and characterized by abundant vascular vessels [36, 37]. There have been many studies reported that loss of *TSC1* or *TSC2* function induce VEGF-A through mTOR-dependent pathways (induction of hypoxia-inducible factor 1a) and mTOR-independent pathway involving chromatic remodeling [50–52]. Why lymphangiogenesis dominates angiogenesis in LAM? It was recently reported that mTOR regulated the production of not only VEGF-A but also VEGF-C in a lymphatic metastasis-prone rat pancreatic tumor cell line and rapamycin inhibited VEGF-C expression [53]. In contrast, no study regarding mTOR signaling pathway and VEGF-D production is available in the literature. Both VEGF-C and VEGF-D are important lymphangiogenic growth factors and can transduce growth signal into lymphatic endothelial cells through VEGFR-3, but there seems to exist functional differences between VEGF-C and VEGF-D [54]. In the transgenic mouse model of pancreatic β-cell carcinogenesis, VEGF-D-expressing β-cells resulted in the formation of peri-insular lymphatic lacunae, often with blood-lymphatic vessel shunts, and the frequent development of lymph node and lung metastasis during the tumorigenesis of β-cells while a comparable transgenic expression of VEGF-C also provoked lymphangiogenesis, but without apparent blood-lymphatic shunts, and promoted lymph node metastasis in the absence of lung metastasis. Interestingly, angiogenesis was suppressed in VEGF-D-transgenic system but not in showed VEGF-C-transgenic system while lymphangiogenesis was induced similarly, suggesting that VEGF-D has distinct roles in lymphangiogenesis and metastasis from VEGF-C [54]. Some similarities between clinical manifestations of LAM and these experimental results may be realized. Chylous pleural effusion or ascites, a peculiar complication of LAM patients, are often body and contaminated with red blood cells, suggesting the presence of blood-lymphatic vessel shunts. Abundant lymphatic vessels with few blood vessels in LAM lesions may indicate that VEGF-D-mediated lymphangiogensis suppress angiogenesis in LAM lesions.

8.3.5 Lymphangiogenesis-Mediated Shedding of LCC as a Mechanism for the Metastasis and the Progression of LAM

The existence of tumor cell clusters enveloped by endothelial cells of sinusoidal vasculature, similar to LCC in which lymphatic rather than vascular endothelial cells envelop cell clusters, have been reported and their role in the mechanism for invasion-independent metastasis in multiple human cancers has been thus demonstrated [55, 56]. LCC were able to decompose into LAM cells and LEC and LAM cells proliferate thereafter once LCC were cultured in vitro on collagen-coated dishes [24]. Since LCC were detected in all samples of the LAM-associated chylous effusion that we have so far examined, these findings suggest that the generation of LCC is one of common pathophysiologic mechanisms in LAM and this phenomenon is also implicated in the metastasis of LAM cells such as in the case of vascular endothelial cell-coated tumor cells.

When the axial lymphatic system was retrospectively analyzed in the archived samples of 5 autopsy cases with LAM according to the notion mentioned above, the dissemination of LAM lesions was identified along the thoracic duct and axial lymph nodes in various regions [24]. The walls of the thoracic ducts in all cases were involved with LAM lesions and in some cases, not only the wall of thoracic ducts but also area of adjacent fat tissue had LAM lesions (Fig. 8.12). LAM lesions were demonstrated along the axial lymphatic system from the distal (pelvic region) to the proximal regions, finally at the left jugulosubclavian angle where the thoracic dust drains into the venous blood circulation (Fig. 8.12). The region most frequently and extensively affected by LAM was both the retroperitoneal and left supraclavicular region. Moreover, the conglomerated lymph nodes due to the extension of LAM lesion to adjacent fat tissues were frequently found in the retroperitoneal region. In contrast, no LAM lesion was demonstrated in the tributary lymphatic regions such as the mesenteric, axillary, or cervical region.

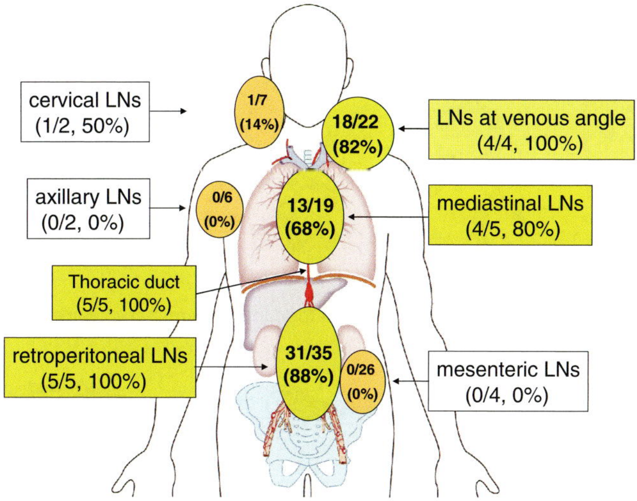

Fig. 8.12 Summary of the histopathological examinations of the lymphatic system and the detection of LAM lesions. Five autopsy cases with LAM were retrospectively analyzed to detect LAM lesions in the lymphatic system. The thoracic duct was cut at 5-mm intervals to prepare tissue block and then examined. The *circle* indicates the detection rate of LAM lesions among the total number of lymph node examined at mesenteric, retroperitoneal, mediastinal, axillary, supraclavicular, and cervical regions. The *square* indicates the detection rate on an individual basis

Based on in-depth pathologic analyses of LAM lesion, while paying special attention to LAM-associated lymphangiogenesis as well as the analysis of lymphangiogenic growth factors, a hypothesis regarding the mechanism for the metastasis of LAM cells and the progression of LAM has thus emerged (Fig. 8.13) [24]. At the site where LAM cells proliferate, LAM cells produce lymphangiogenic factors VEGF-C and VEGF-D and induce LAM-associated lymphangiogenesis which thus demarcate the LAM foci to form LCC. LCC can then become implanted inside the lymphatic vessels from where LCC expose to the extracellular matrix through the LEC–LEC interaction between LEC of lymphatic vessel and of LCC. Once LAM cells form a new lesion, then both LAM-associated lymphangiogenesis and lymphangiogenesis-mediated shedding of LCC into lymphatic circulation occur. Due to an anatomic connection with the venous circulation system at the jugulosubclavian angle, LCC are then delivered to the pulmonary circulation, thus resulting in the development of a new LAM lesion in the lung. A continuous series of events consisting of the proliferation of LAM cells, LAM-associated lymphangiogenesis, and the resultant shedding of LCC into the lymphatic circulation will enable LAM cells to communicate between pulmonary lymphatic and venous circulations, and thus resulting in progressive cystic formation in the lungs (Fig. 8.14). Although it still unknown where LAM cells originally occur within the body, this hypothesis may help to explain the progressive development of cysts and the deterioration of the lung function even in LAM cases who do not have renal AML. Other group has postulated that renal AML was the source for the dissemination and progression of

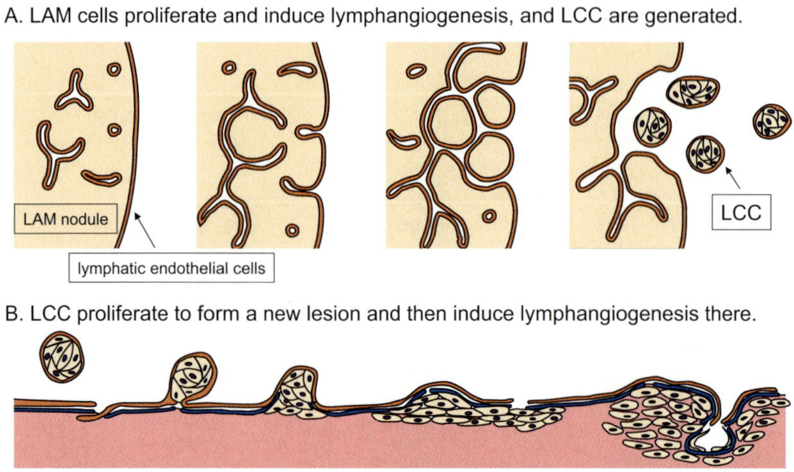

Fig. 8.13 Schematic illustration of a postulated lymphangiogenesis-mediated fragmentation of LAM lesion, shedding of LCC into lymphatic circulation, and the subsequent formation of a new LAM lesion at the site where LCC implanted. (Modified from Kumasaka et al. [24]. By copyright permission of Lippincott Williams & Wilkins)

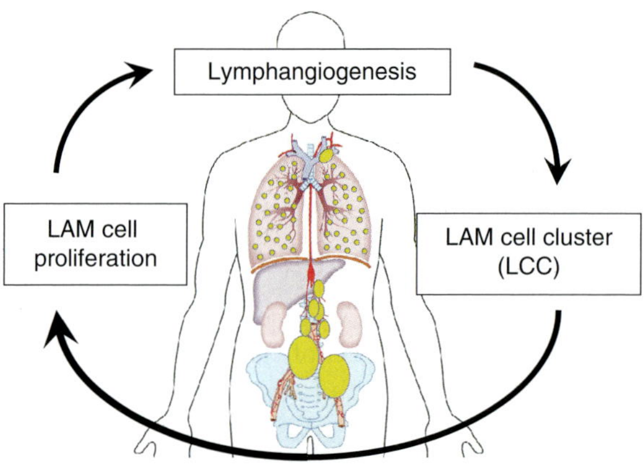

Fig. 8.14 Hypothesis regarding the mechanism for the progression and metastasis of LAM. Once transformed LAM cells are generated somewhere within the body, then such LAM cells proliferate and induce lymphangiogenesis. LAM-associated lymphangiogenesis demarcates LAM foci and LCC are then eventually shed into the lymphatic circulation. As a consequence, a new LAM lesion is formed along the axial lymphatic system and these series of events continue to occur. As a consequence, LAM cells disseminate along axial lymphatic system and also into the lungs. Once LAM cells form a pulmonary lesion, then LAM cells travel back and forth between the pulmonary lymphatic circulation and the pulmonary blood circulation via LAM-associated lymphangiogenesis and the shedding of LCC, thus eventually destroying the lungs by generating multiple cysts

LAM [33] because approximately 50% of all LAM patients have been reported to be complicated with renal AML [57]. Supporting this hypothesis, it has been demonstrated that LAM cells and AML shared the common *TSC2* mutations [8,34] and that *TSC* mutation caused a dysfunction of cytoskeletons, thus leading to aberrant cell migration [33, 58, 59]. However, this hypothesis is not pertinent to the patients with LAM who have no renal AML. Indeed, sporadic LAM patients have been reported in the literature who had no renal AML but who did experience metastatic LAM lesions in the donor lung after unilateral lung transplantation [60]. In addition, in our 5 autopsy cases who showed the extensive involvement of axial lymphatic system with the proliferation of LAM cells and LAM-associated lymphangiogenesis, no AML was demonstrated in the kidneys or other organs [24].

Why do LAM cells generate a metastatic new lesion only in the lungs and along the axial lymphatic systems? How do LCC implant themselves into the lung parenchyma and thus form metastatic lesions after draining into pulmonary venous circulation from the lymphatic circulation? Although we carefully examined lung specimens, we have never identified the occlusion of pulmonary vasculature with LCC. The existence of a solitary LAM cells circulating in the systemic blood circulation has been reported using the density gradient method and a subsequent cell sorting analysis [61]. Are these circulating LAM cells derived from LCC or are they due to the direct invasion into the vasculature and subsequent extravasation from the site of proliferation of LAM cells? Further studies on the interaction between LAM cells and LEC will thus be needed to verify the pathophysiologic significance of LAM-associated lymphangiogenesis in LAM.

8.4 Future Directions

LAM-associated lymphangiogenesis is likely to be a potential therapeutic target in LAM since it apparently associated with disease progression and appears to be the basis on metastatic spread [24,25]. Several experimental systems targeting lymphangiogenesis have been established [62, 63]. Adenovirus-medicated transduction and expression of the fusion protein consisting of extracellular domain of VEGFR-3 and Fc fragment of immunoglobulin (VEGFR-3-Ig) was reported to abolish lymphangiogenesis very efficiently [62]. VEGF-Ig induced apoptosis of proliferating lymphatic endothelial cells and regression of pre-formed lymphatic vessels in the embryonic skin and inhibited tumor-related lymphangiogenesis and regional lymph node metastasis. Neutralizing antibody against VEGF-C and VEG-D were reported to successfully block VEGFR-3 signaling pathway [63]. VEGFR-3-Ig may have an advantage since it can trap both VEGF-C and VEGF-D.

Rapamycin is the most attractive drug in terms of theoretical consideration since constitutively activated mTOR due to the loss of *TSC1/TSC2* function is the pathogenic basis of the disease. Recently the result of the Cincinnati Angiomyolipoma Sirolimus Trial (CAST) was published that sirolimus treatment shrunk renal AML and improved pulmonary function in patients with LAM [64].

Accordingly, the Multicenter International Lymphangioleiomyomatosis Efficacy of Sirolimus (MILES) Trial was designed to determine if sirolimus truly improves pulmonary function in patients with LAM and is currently underway in the United States, Canada, and Japan. Interestingly, rapamycin inhibited VEGF-C-mediated proliferation and migration of human LECs and impedes lymphangiogenesis [65]. Rapamycin also decreased tumor-associated lymphangiogenesis and lymph node metastasis in tumor xenograft experiment [53]. Further studies are needed to investigate precise mechanism for LAM-associated lymphangiogenesis through elucidating the interaction between LAM cells and LECs. In addition, the establishment of a mouse model of LAM is needed to promote the development of a new treatment.

References

1. Aubry MC, Myers JL, Ryu JH, et al. Pulmonary lymphangioleiomyomatosis in a man. Am J Respir Crit Care Med 2000, 162: 749–52.
2. Schiavina M, Di Scioscio V, Contini P, et al. Pulmonary lymphangioleiomyomatosis in a karyotypically normal man without tuberous sclerosis complex. Am J Respir Crit Care Med 2007, 176: 96–8.
3. Burrell LST, Ross JM. A case of chylous effusion due to leiomyosarcoma. Br J Tuberc 1937, 31: 38–9.
4. Von Stossel E. Uber musculare cirrose der lunge. Beitr Klin Tuberk 1937, 90: 432–42.
5. Frack MD, Simon L, Dawson BH. The lymphangiomyomatosis syndrome. Cancer 1968, 22: 428–37.
6. Carrington CB, Cugell DW, Gaensler EA, et al. Lymphangioleiomyomatosis. Physiologic-pathologic-radiologic correlations. Am Rev Respir Dis 1977, 116: 977–95.
7. Silverstein EF, Ellis K, Wolff M, Jaretzki A, III. Pulmonary lymphangiomyomatosis. Am J Roentgenol Radium Ther Nucl Med 1974, 120: 832–50.
8. Carsillo T, Astrinidis A, Henske EP. Mutations in the tuberous sclerosis complex gene TSC2 are a cause of sporadic pulmonary lymphangioleiomyomatosis. Proc Natl Acad Sci U S A 2000, 97: 6085–90.
9. Smolarek TA, Wessner LL, McCormack FX, Mylet JC, Menon AG, Henske EP. Evidence that lymphangiomyomatosis is caused by TSC2 mutations: chromosome 16p13 loss of heterozygosity in angiomyolipomas and lymph nodes from women with lymphangiomyomatosis. Am J Hum Genet 1998, 62: 810–5.
10. Arnold S, Kristof JM. Lymphangioleiomyomatosis. In Interstitial Lung Disease (Marvin I, Schawarz TEKJ, Eds.), 4th ed, pp. 851–64. BC Decker Inc., London, 2003;
11. Hayashida M, Seyama K, Inoue Y, Fujimoto K, Kubo K. The epidemiology of lymphangioleiomyomatosis in Japan: a nationwide cross-sectional study of presenting features and prognostic factors. Respirology 2007, 12: 523–30.
12. Johnson SR, Tattersfield AE. Clinical experience of lymphangioleiomyomatosis in the UK. Thorax 2000, 55: 1052–7.
13. Urban T, Lazor R, Lacronique J, et al. Pulmonary lymphangioleiomyomatosis. A study of 69 patients. Groupe d'Etudes et de Recherche sur les Maladies "Orphelines" Pulmonaires (GERM"O"P). Medicine (Baltimore) 1999, 78: 321–37.
14. Foundation TL. The international LAM symposium 2008, 2008.
15. Ryu JH, Moss J, Beck GJ, et al. The NHLBI lymphangioleiomyomatosis registry: characteristics of 230 patients at enrollment. Am J Respir Crit Care Med 2006, 173: 105–11.
16. Guinee D, Singh R, Azumi N, et al. Multifocal micronodular pneumocyte hyperplasia: a distinctive pulmonary manifestation of tuberous sclerosis. Mod Pathol 1995, 8: 902–6.

17. Maruyama H, Seyama K, Sobajima J, et al. Multifocal micronodular pneumocyte hyperplasia and lymphangioleiomyomatosis in tuberous sclerosis with a TSC2 gene. Mod Pathol 2001, 14: 609–14.
18. Seyama K, Kira S, Takahashi H, et al. Longitudinal follow-up study of 11 patients with pulmonary lymphangioleiomyomatosis: diverse clinical courses of LAM allow some patients to be treated without anti-hormone therapy. Respirology 2001, 6: 331–40.
19. Bonetti F, Chiodera PL, Pea M, et al. Transbronchial biopsy in lymphangiomyomatosis of the lung. HMB45 for diagnosis. Am J Surg Pathol 1993, 17: 1092–102.
20. Brentani MM, Carvalho CR, Saldiva PH, Pacheco MM, Oshima CT. Steroid receptors in pulmonary lymphangiomyomatosis. Chest 1984, 85: 96–9.
21. Graham ML, II, Spelsberg TC, Dines DE, Payne WS, Bjornsson J, Lie JT. Pulmonary lymphangiomyomatosis: with particular reference to steroid-receptor assay studies and pathologic correlation. Mayo Clin Proc 1984, 59: 3–11.
22. Cornog JL, Jr., Enterline HT. Lymphangiomyoma, a benign lesion of chyliferous lymphatics synonymous with lymphangiopericytoma. Cancer 1966, 19: 1909–30.
23. Enterline HT, Roberts B. Lymphangiopericytoma; case report of a previously undescribed tumor type. Cancer 1955, 8: 582–7.
24. Kumasaka T, Seyama K, Mitani K, et al. Lymphangiogenesis-mediated shedding of LAM cell clusters as a mechanism for dissemination in lymphangioleiomyomatosis. Am J Surg Pathol 2005, 29: 1356–66.
25. Kumasaka T, Seyama K, Mitani K, et al. Lymphangiogenesis in lymphangioleiomyomatosis: its implication in the progression of lymphangioleiomyomatosis. Am J Surg Pathol 2004, 28: 1007–16.
26. Capron F, Ameille J, Leclerc P, et al. Pulmonary lymphangioleiomyomatosis and Bourneville's tuberous sclerosis with pulmonary involvement: the same disease? Cancer 1983, 52: 851–5.
27. European Chromosome 16 Tuberous Sclerosis Consortium. Identification and characterization of the tuberous sclerosis gene on chromosome 16. Cell 1993, 75: 1305–15.
28. Van Slegtenhorst M, de Hoogt R, Hermans C, et al. Identification of the tuberous sclerosis gene TSC1 on chromosome 9q34. Science 1997, 277: 805–8.
29. Sepp T, Yates JR, Green AJ. Loss of heterozygosity in tuberous sclerosis hamartomas. J Med Genet 1996, 33: 962–4.
30. Henske EP, Scheithauer BW, Short MP, et al. Allelic loss is frequent in tuberous sclerosis kidney lesions but rare in brain lesions. Am J Hum Genet 1996, 59: 400–6.
31. Knudson AG, Jr. Mutation and cancer: statistical study of retinoblastoma. Proc Natl Acad Sci U S A 1971, 68: 820–3.
32. Astrinidis A, Khare L, Carsillo T, et al. Mutational analysis of the tuberous sclerosis gene TSC2 in patients with pulmonary lymphangioleiomyomatosis. J Med Genet 2000, 37: 55–7.
33. Henske EP. Metastasis of benign tumor cells in tuberous sclerosis complex. Genes Chromosomes Cancer 2003, 38: 376–81.
34. Sato T, Seyama K, Fujii H, et al. Mutation analysis of the TSC1 and TSC2 genes in Japanese patients with pulmonary lymphangioleiomyomatosis. J Hum Genet 2002, 47: 20–8.
35. Inoki K, Corradetti MN, Guan KL. Dysregulation of the TSC-mTOR pathway in human disease. Nat Genet 2005, 37: 19–24.
36. Arbiser JL, Brat D, Hunter S, et al. Tuberous sclerosis-associated lesions of the kidney, brain, and skin are angiogenic neoplasms. J Am Acad Dermatol 2002, 46: 376–80.
37. Nguyen-Vu PA, Fackler I, Rust A, et al. Loss of tuberin, the tuberous-sclerosis-complex-2 gene product is associated with angiogenesis. J Cutan Pathol 2001, 28: 470–5.
38. Koets AP, Rutten VP, de Boer M, Bakker D, Valentin-Weigand P, Van Eden W. Differential changes in heat shock protein, lipoarabinomannan, and purified protein derivative-specific immunoglobulin G1 and G2 isotype responses during bovine Mycobacterium avium subsp. paratuberculosis infection. Infect Immun 2001, 69: 1492–8.
39. Abe R, Kimura M, Airosaki A, et al. Retroperitioneal lymphangiomyomatosis with lymphedema of the legs. Lymphology 1980, 13: 62–7.

40. Van Lith JM, Hoekstra HJ, Boeve WJ, Weits J. Lymphoedema of the legs as a result of lymphangiomyomatosis. A case report and review of the literature. Neth J Med 1989, 34: 310–6.
41. Jussila L, Alitalo K. Vascular growth factors and lymphangiogenesis. Physiol Rev 2002, 82: 673–700.
42. Itami M, Teshima S, Asakuma Y, Chino H, Aoyama K, Fukushima N. Pulmonary lymphangiomyomatosis diagnosed by effusion cytology. A case report. Acta Cytol 1997, 41: 522–8.
43. Yamauchi M, Nakahara H, Uyama K, Tsujimoto A, Tamai M, Aozasa K. Cytologic finding of chyloascites in lymphangioleiomyomatosis. A case report. Acta Cytol 2000, 44: 1081–4.
44. Hirama M, Atsuta R, Mitani K, et al. Lymphangioleiomyomatosis diagnosed by immunocytochemical and genetic analysis of lymphangioleiomyomatosis cell clusters found in chylous pleural effusion. Intern Med 2007, 46: 1593–6.
45. Karpanen T, Egeblad M, Karkkainen MJ, et al. Vascular endothelial growth factor C promotes tumor lymphangiogenesis and intralymphatic tumor growth. Cancer Res 2001, 61: 1786–90.
46. Mandriota SJ, Jussila L, Jeltsch M, et al. Vascular endothelial growth factor-C-mediated lymphangiogenesis promotes tumour metastasis. Embo J 2001, 20: 672–82.
47. Seyama K, Kumasaka T, Souma S, et al. Vascular endothelial growth factor-D is increased in serum of patients with lymphangioleiomyomatosis. Lymphat Res Biol 2006, 4: 143–52.
48. Matsui K, Beasley MB, Nelson WK, et al. Prognostic significance of pulmonary lymphangioleiomyomatosis histologic score. Am J Surg Pathol 2001, 25: 479–84.
49. Wartiovaara U, Salven P, Mikkola H, et al. Peripheral blood platelets express VEGF-C and VEGF which are released during platelet activation. Thromb Haemost 1998, 80: 171–5.
50. Brugarolas J, Kaelin WG, Jr. Dysregulation of HIF and VEGF is a unifying feature of the familial hamartoma syndromes. Cancer Cell 2004, 6: 7–10.
51. Brugarolas JB, Vazquez F, Reddy A, Sellers WR, Kaelin WG, Jr. TSC2 regulates VEGF through mTOR-dependent and independent pathways. Cancer Cell 2003, 4: 147–58.
52. El-Hashemite N, Walker V, Zhang H, Kwiatkowski DJ. Loss of TSC1 or TSC2 induces vascular endothelial growth factor production through mammalian target of rapamycin. Cancer Res 2003, 63: 5173–7.
53. Kobayashi S, Kishimoto T, Kamata S, Otsuka M, Miyazaki M, Ishikura H. Rapamycin, a specific inhibitor of the mammalian target of rapamycin, suppresses lymphangiogenesis and lymphatic metastasis. Cancer Sci 2007, 98: 726–33.
54. Kopfstein L, Veikkola T, Djonov VG, et al. Distinct roles of vascular endothelial growth factor-D in lymphangiogenesis and metastasis. Am J Pathol 2007, 170: 1348–61.
55. Sugino T, Kusakabe T, Hoshi N, et al. An invasion-independent pathway of blood-borne metastasis: a new murine mammary tumor model. Am J Pathol 2002, 160: 1973–80.
56. Sugino T, Yamaguchi T, Ogura G, et al. Morphological evidence for an invasion-independent metastasis pathway exists in multiple human cancers. BMC Med 2004, 2: 9.
57. Bernstein SM, Newell JD, Jr., Adamczyk D, Mortenson RL, King TE, Jr., Lynch DA. How common are renal angiomyolipomas in patients with pulmonary lymphangiomyomatosis? Am J Respir Crit Care Med 1995, 152: 2138–43.
58. Astrinidis A, Cash TP, Hunter DS, Walker CL, Chernoff J, Henske EP. Tuberin, the tuberous sclerosis complex 2 tumor suppressor gene product, regulates Rho activation, cell adhesion and migration. Oncogene 2002, 21: 8470–6.
59. Astrinidis A, Henske EP. Aberrant cellular differentiation and migration in renal and pulmonary tuberous sclerosis complex. J Child Neurol 2004, 19: 710–5.
60. Karbowniczek M, Astrinidis A, Balsara BR, et al. Recurrent lymphangiomyomatosis after transplantation: genetic analyses reveal a metastatic mechanism. Am J Respir Crit Care Med 2003, 167: 976–82.
61. Crooks DM, Pacheco-Rodriguez G, DeCastro RM, et al. Molecular and genetic analysis of disseminated neoplastic cells in lymphangioleiomyomatosis. Proc Natl Acad Sci U S A 2004, 101: 17462–7.
62. Makinen T, Jussila L, Veikkola T, et al. Inhibition of lymphangiogenesis with resulting lymphedema in transgenic mice expressing soluble VEGF receptor-3. Nat Med 2001, 7: 199–205.

63. He Y, Kozaki K, Karpanen T, et al. Suppression of tumor lymphangiogenesis and lymph node metastasis by blocking vascular endothelial growth factor receptor 3 signaling. J Natl Cancer Inst 2002, 94: 819–25.
64. Bissler JJ, McCormack FX, Young LR, et al. Sirolimus for angiomyolipoma in tuberous sclerosis complex or lymphangioleiomyomatosis. N Engl J Med 2008, 358: 140–51.
65. Huber S, Bruns CJ, Schmid G, et al. Inhibition of the mammalian target of rapamycin impedes lymphangiogenesis. Kidney Int 2007, 71: 771–7.

Chapter 9

KAPOSI'S SARCOMA AND THE LYMPHATICS

Victoria Emuss and Chris Boshoff
CR-UK Viral Oncology Group, UCL Cancer Institute, WC1E 6BT London, UK,
e-mail: v.emuss@ucl.ac.uk

Abstract: Kaposi's sarcoma herpesvirus (KSHV) is the etiological agent of Kaposi's sarcoma (KS). KS presents as multifocal, angiogenic lesions involving an inflammatory infiltrate and KSHV-infected spindle cells display characteristic markers of lymphatic endothelia. The precise origin of the spindle cell component of KS lesions is uncertain and may derive from the reprogramming of the transcriptome of endothelial cells or their precursors to adopt a lymphatic-like gene expression profile. The lymphotrophic nature of KSHV corresponds to its pathological association with two further AIDS-related malignancies: primary effusion lymphomas (PEL) and a plasmablastic variant of multicentric Castleman's disease (MCD). KSHV infection of B-cells in lymph node follicles creates a reservoir for the persistence of KSHV infection that may influence the characteristics of the associated lymphomas.

Here we discuss the mechanisms of KSHV infection in the context of KS and KSHV-associated lymphomas and examine the potential for KSHV to determine the fate of cells associated with the lymphatic system.

Key words: Endothelial cells · KSHV · Kaposi's sarcoma · Lymphangiogenesis · Spindle cells

Up to 20% of global incidents of cancer can be attributed to infectious agents, primarily viruses, which influence the genesis of malignancies through a variety of mechanisms. Papillomaviruses (HPV) of types 11, 16 and 32, the human herpesviruses Epstein-Barr virus (EBV) and Kaposi's sarcoma herpesvirus (KSHV, or HHV8), polyomaviruses, hepatitis viruses B and C, human T-cell leukaemia virus-1 and the bacterium *Helicobacter pylori* are causally associated with a variety of malignancies (reviewed in [52, 59, 66]).

S.A. Stacker, M.G. Achen (eds.), *Lymphangiogenesis in Cancer Metastasis*,
Cancer Metastasis – Biology and Treatment 13, DOI 10.1007/978-90-481-2247-9_9,

The majority of tumour-associated viruses establish a latent infection in the host cell following primary infection. This leads to the persistence of the viral genome in the host tissue with infected cells expressing a subset of viral genes. Transformation may result from direct disruption of cell growth and proliferation through dysregulation of host signalling pathways or the induction of growth factors (e.g. EBV and HPV). Alternatively, the infectious agent may not be inherently oncogenic and transformation is achieved through indirect mechanisms (e.g. Hepatitis B and C). Commensurate with the multi-factorial nature of cancer, additional factors may participate in establishing the transformed phenotype in infected cells. Such factors include host immunosuppression, carcinogenic exposure and genetic predisposition or additional somatic mutation. The presence or absence of the infectious agent may serve to define subsets of a given tumour or determine the progression of the disease (reviewed in [33, 52, 59]).

Kaposi's sarcoma-associated herpesvirus (KSHV) is a γ2-herpesvirus (See Box 1) whose global seroprevalence varies from less than 1% in Japan, to over 50% in much of sub-Saharan Africa where KSHV infection is endemic [77]. KSHV has been recognised by the International Agency for Research Against Cancer (IARC) as a class I carcinogen [IARC, 1] and is considered the causative agent of Kaposi's sarcoma (KS) according to the Hill criteria [70]. The global incidence of KS also correlates with the seroprevalence of KSHV in different countries (for review see [3]).

Box 1–Biology of KSHV

Virus Evolution

KSHV was discovered in 1994 following PCR-based Representational Difference Analysis, which identified unique DNA sequences in AIDS-KS that were absent in adjacent skin [18]. KSHV is a double-stranded DNA virus that belongs to the *Rhadinovirus* (γ2) genus of the γ-herpesviridae subfamily of herpesviruses. The structure of the KSHV capsid purified from BCBL-1 cells induced with TPA is shown in Figure a (the image was kindly provided by Z. Hong Zhou. Data were taken from [88]). It is the eighth and most recently identified human herpesvirus (thereby designated HHV-8). Close homologues of KSHV have been identified in chimpanzees, gorillas and rhesus macaques and phylogenetic analysis has indicated that it shares substantial genetic homology with the *Rhadinovirus* Herpesvirus Saimiri (HVS), found in *Saimiri sciureus* [43, 70, 73]. The closest human relative of KSHV is the gammaherpesvirus Epstein-Barr Virus (EBV), genus *Lymphocryptovirus* (γ1) [55].

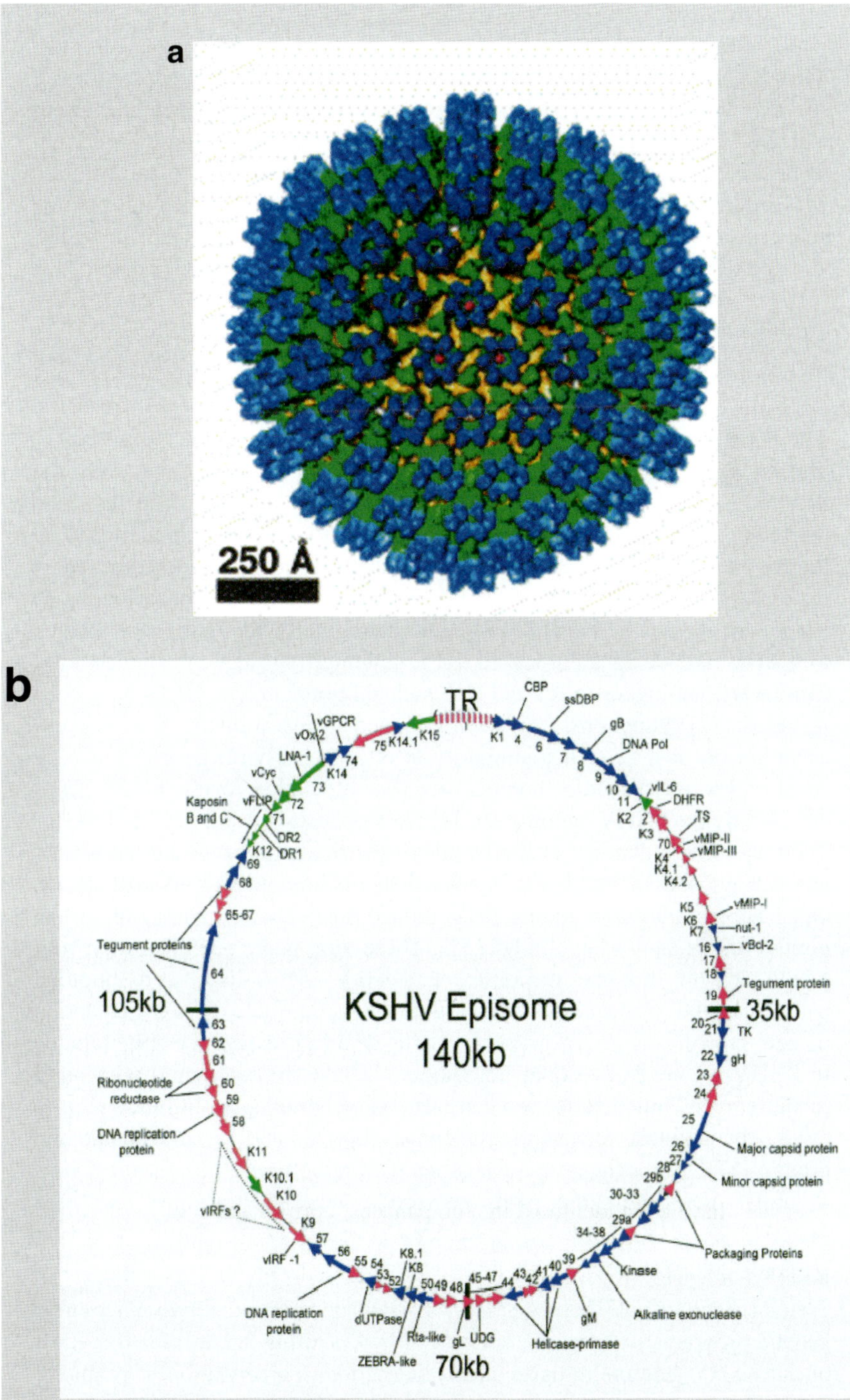
a
250 Å
b
TR
CBP
ssDBP
gB
DNA Pol
vIL-6
DHFR
TS
vMIP-II
vMIP-III
vMIP-I
nut-1
vBcl-2
Tegument protein
35kb
TK
gH
Major capsid protein
Minor capsid protein
Packaging Proteins
Kinase
Alkaline exonuclease
gM
Helicase-primase
UDG
gL
70kb
Rta-like
ZEBRA-like
dUTPase
DNA replication protein
vIRF -1
vIRFs ?
DNA replication protein
Ribonucleotide reductase
105kb
Tegument proteins
Kaposin B and C
vFLIP
DR2
DR1
vCyc
LNA-1
vOx2
vGPCR
KSHV Episome
140kb
K1
4
6
7
8
9
10
11
K2
2
K3
70
K4
K4.1
K4.2
K5
K6
K7
16
17
18
19
20
21
22
23
24
25
26
27
28
29b
30-33
29a
34-38
39
40
41
42
43
44
45-47
48
49
50
K8
K8.1
52
53
54
55
56
57
K9
K10
K10.1
K11
58
59
60
61
62
63
64
65-67
68
69
K12
71
72
73
K14
74
75
K14.1
K15

According to consensus PCR of γ2-herpesvirus-like sequences, two lineages of rhadinovirus (RV1 and RV2) have been identified in primate species from the Old World and higher primates. Only the RV1 lineage, of which KSHV is part, has been found in man. KSHV has further evolved into five main phylogenetic branches (clades A-E) that demonstrate co-evolution with different human populations [36]. Sequence variation of the viral genome between different clades is less than 3% in most regions except in hypervariable regions of the K1 gene that result in up to 40% sequence divergence between clades. This hypervariability has been attributed to selection pressure for different protein sequences by host immune responses; genetic variability may also result from recombination of viral genomes during KSHV evolution [38, 67].

The Viral Genome

The KSHV genome was mapped with cosmid and phage libraries from the BC-1 PEL cell line. It consists of about 140 kb of unique coding sequence flanked by multiple GC-rich terminal repeats (TRs) of approximately 800 bp to give a total size of 170 kb [72]. The unique region has five internal repeat regions and encodes at least 81 ORFs, of which 66 are homologous to HVS (Fig. b, adapted from [78]). ORFs originally thought to be unique to KSHV are designated K1 to K15, although subsequent studies have identified homologues to K8 and K13 and additional unique ORFs have been included [72]. Functions have been assigned to many of the KSHV genes according to their sequence similarity to other herpesviral gene products or according to their cellular homologues [For a detailed review of the genes encoded by the KSHV genome see Jenner and Boshoff 2002, reference 43].

Piracy of host genes is characteristic of the rhadinoviruses and the KSHV genome encodes a number of genes that are homologous to host genes. Other herpesviruses, including EBV, induce host cellular genes, rather than pirating them, such as cyclin D2 [57]. These viral genes are associated with modulation of immune responses, nucleotide metabolism, anti-apoptotic pathways, and cytokines and can therefore contribute to the regulation of cell growth and transformation [for a detailed review see 19]. Unique to KSHV are several cellular homologues which are not shared with other rhadinoviruses and include viral genes encoding homologues of interleukin-6, three chemokines, interferon response factors and a transmembrane protein [57]. The acquisition of host genes is hypothesised to permit KSHV to utilise host signalling pathways and avoid anti-viral host responses [42, 57].

KSHV Lifecycle

Characteristic of all herpesviruses, KSHV demonstrates two alternative genetic programs of infection: latent and lytic. During latent infection, most of the KSHV genome remains silent, likely due to methylation of promoter sequences, no infectious progeny are produced and the viral genome exists in the host cell as a circular episome [71]. During mitosis, viral DNA is tethered to the host histone H1 protein through its TR sequences via the KSHV latent nuclear antigen (LANA1), encoded by ORF73. The majority of KSHV-infected cells are latently infected and, through this mechanism, the viral genome is copied by the host cell DNA replication machinery, allowing the viral DNA to be propagated within the host following initial infection [71].

The restricted pattern of KSHV gene expression is characteristic of latent infection and is intended to minimise the number of epitopes presented by infected cells and therefore provides a means of escaping host immune responses [71]. The expression of latency-associated genes has been shown to be associated with the oncogenic properties of EBV [60] and it is likely that the latent KSHV genes are implicated in its pathogenesis. All KSHV-infected cells have been shown to express LANA, viral cyclin (v-cyclin, encoded by ORF72) and viral Fas-associated death domain (FADD) interleukin-1β-converting enzyme (FLICE) inhibitory protein (v-FLIP, encoded by ORF71). These three genes are adjacent in the viral genome and form the KSHV latency-associated transcript; they can be co-transcribed on two polycistronic mRNAs: LT1 (LANA/v-cyclin/v-FLIP) and LT2 (v-cyclin/v-FLIP) [22, 74, 83].

Latent KSHV can reactivate and enter lytic replication, during which most viral genes are expressed, viral DNA is amplified, and infectious virions are released following lysis of the infected cell [71]. The RTA gene product, encoded by ORF50, is necessary and sufficient to induce lytic replication and can activate its own promoter to generate an autocatalytic rise in RTA expression. The ORF50 promoter is heavily methylated in latently infected cells and lytic replication can be induced by TPA, sodium butyrate and ionomycin *in vitro* and humoral factors including interferon-γ and other cytokines, which stimulate demethylation [17, 69]. KSHV lytic genes are expressed in a temporal and sequential order and termed immediate-early, delayed-early or late genes. Immediate-early genes are expressed independently of *de novo* protein synthesis and encode regulators of viral gene expression; delayed-early genes are expressed slightly later, and encode products that function to replicate viral DNA. Late genes are expressed after viral DNA synthesis and encode structural proteins and those involved in virus maturation [reviewed in 25].

The original description of KS was made in 1872 by Moritz Kaposi following the observation of nodular skin lesions on five elderly men [46]. KS is now the most frequently diagnosed tumour in certain regions of Africa and was estimated to account for approximately 1% of globally diagnosed cancers in 2002 [61].

9.1 Kaposi's Sarcoma

KS was originally described as a multiple, pigmented, haemorrhagic sarcoma of the skin; it is an angioproliferative disease that gives rise to highly vascularised tumours [46, 47]. Initial presentation of KS is typically as a skin lesion (Fig. 9.1a and b) but this may be preceded by oral, visceral or nodal involvement; as the disease progresses, it can disseminate to lymphatic and visceral organs [47, 49].

Four distinct clinical and epidemiological forms of KS have been identified: Classic KS, African/endemic KS, iatrogenic (or post-transplant) KS and AIDS KS [25]. Classic KS is a rare, indolent form of the disease that occurs in older men of primarily Mediterranean or Eastern European Jewish origin; lesions usually affect the lower extremities and their genesis is not thought to be influenced by any environmental cofactor. African or endemic KS affects individuals from sub-equatorial Africa; this aggressive form of the disease involves the lymphatic and/or visceral organs of sufferers [25]. Iatrogenic KS affects patients subjected to immunosuppressive therapy, for example after an organ transplant. AIDS KS affects HIV-1-infected individuals and is the most aggressive form of the disease with lesions observed on the skin, oral cavity, gastrointestinal tract and visceral organs. The

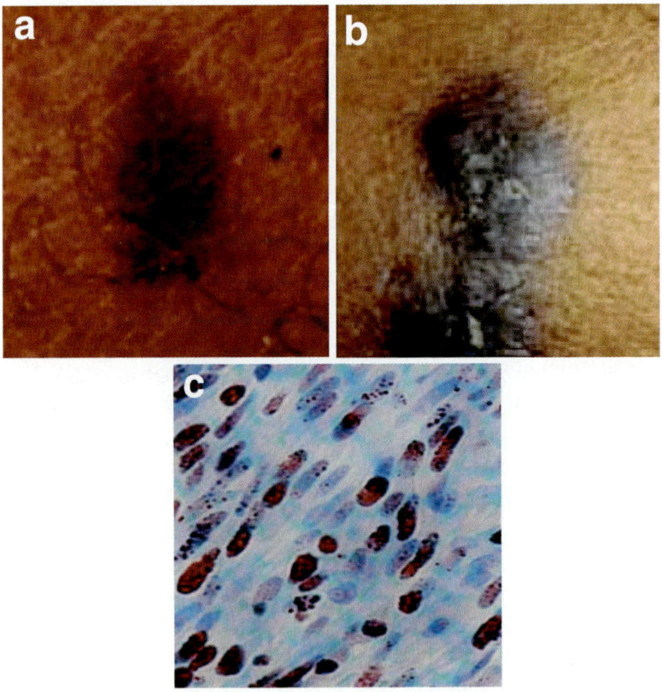

Fig. 9.1 KS lesions and histology. **a**, Patch stage KS lesion. **b**, Plaque stage KS lesion. **c**, LANA-1 staining of KS spindle cells in a nodular KS lesion. All pictures and staining are taken from Duprez et al. [29]

immunosuppression associated with HIV-1 infection is thought to result in KSHV reactivation and increased viral load, resulting in KS development. However, HIV-1 itself might also act as a contributory factor to KS progression through the actions of the HIV-Tat protein (discussed in Box 2) [4, 30, and reviewed in 34].

Box 2–Endothelial cell reprogramming by KSHV

The establishment of a spindle cell compartment and the progression of KS lesions occurs following KSHV infection of endothelial cells (Step 1 in the figure above, KSHV virions are represented as purple stars). KSHV can infect both LEC (blue) and BEC (red) *in vivo* and reprogramme the cells to form a cell with a hybrid genotype (purple) whereby mainly LEC, but also BEC, markers are expressed (2). LEC are likely to be the preferred cellular target of KSHV infection as discussed in the main text. Alternatively, endothelial precursor cells (grey) could be the *in vivo* targets of KSHV infection. These cells have the potential to differentiate into either LEC or BEC, but KSHV infection drives their differentiation towards a LEC-type cell. This hypothesis still needs investigation [40, 86].

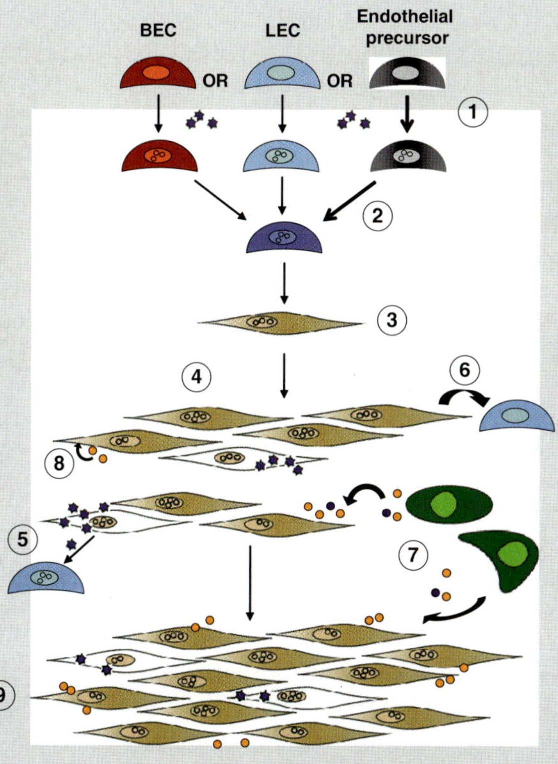

Following KSHV infection and the establishment of the initial inflammatory KS lesion, spindle cells arise from the latently infected endothelial cells (3). Early stage KS is characterised by a polyclonal proliferation of spindle cells (4) in which the majority of cells are latently infected, but a small number of lytically infected cells produce new KSHV virions that can attract and infect surrounding cells (5) [reviewed in 43]. Loss of viral infection may contribute to KS regression (6). Inflammatory cytokines, angiogenic factors and chemokines secreted from surrounding macrophages, monocytes and infiltrating lymphocytes contribute to establishing the spindle cell compartment (7, indicated by yellow circles) [reviewed in 31]. In AIDS-KS, the HIV Tat protein could also enhance KSHV infection of endothelial cells and, following its secretion from acutely infected T-lymphocytes (purple circles), influences the effects of other secreted factors on endothelial cells leading to dysregulation of adhesion, and cell growth [4, 30, and reviewed in 34]. In addition, factors secreted by spindle cells following KSHV infection may act in an autocrine or paracrine manner to contribute to the formation of KS (8) [35].

Late stage, nodular KS lesions are characterised by oligoclonal proliferation of KSHV-infected spindle cells (9), which are influenced by secreted factors in a similar manner to lesions at earlier stages of KS [35,43].

9.1.1 KS Pathology

The histology of dermal lesions associated with all four forms of KS is similar and is associated with neoangiogenesis to form slit-like vascular spaces containing erythrocytes, and inflammatory cell infiltrates leading to oedema. Activation of endothelial cells and the formation of a spindle cell component is also observed [49]. Early stage KS begins as patches of normal blood vessels surrounded by small, irregular endothelial-lined spaces accompanied by an inflammatory infiltrate. Patch lesions evolve to form plaques (plaque stage) through the expansion of spindle-celled vascular processes that form vascular channels through the dermis. Late stage lesions form nodules that can coalesce and are composed of sheets of spindle cells and slit-like vascular spaces [6, and reviewed in 31, 43]. KS spindle cells are so-called because of their characteristic morphology, observed as an elongated cytoplasm and nucleus and the presence of hyaline inclusions and hemosiderin [35]. *In situ* hybridization showed that KSHV is present in all spindle cells in KS lesions (9, 74). However, PCR against KSHV ORF 26 (encoding the minor capsid protein) have also detected KSHV in endothelial cells surrounding the vascular spaces in KS lesions [8]. The presence of KSHV in the spindle cells of nodular lesions has been confirmed by detecting the latency-associated nuclear antigen (LANA-1 protein) by immunohistochemistry (Fig. 9.1c and [29]). The spindle cells are widely considered the KS tumour cells and, although their origin is unclear, are considered to have features of endothelial cells that have been infected by KSHV (discussed below).

The underlying feature of all forms of KS is infection with KSHV (Box 1), suggesting a common aetiological agent and mechanism of development [31]. Lesions at all three histopathologic stages of the disease are positive for LANA-1 [28]. Staining is observed in the spindle cells of the lesions with the percentage of cell infected increasing from 10% in early stage lesions to 90% in late stage nodules [28]. Low levels of initial infection suggest that paracrine mechanisms influence disease progression and that the virus may provide a growth advantage to infected cells [8,28,81,82].

Unlike other soft tissue sarcomas, KS is a stationary tumour that, at least at the early stages, does not proceed to metastasis after local growth but generates multiple, independent lesions that may appear simultaneously at sites in the body according to dermatomes or symmetry [49]. In addition, particular to KS is the observation that lesions can regress, in the case of iatrogenic KS, in response to the withdrawal of immunosuppressive therapy. The risk of developing KSHV-associated disease is significantly higher in the context of host immune suppression and, concurrent with this, reduction in HIV burden through effective antiretroviral therapy has reduced KS incidents [reviewed in 31, 34]. The establishment of KS also appears to rely on inflammatory cytokines originating from the infiltrating and infected cells, as well as on immune dysregulation (Fig. 9.2) [32]. Taken together, these observations suggest that early stage KS is a reactive hyperplasia, rather than a true malignancy and is subject to a decline in the immune status of its host for disease progression.

The majority of human cancers are clonal neoplasms, indicating the origin of the tumour to be a transformed somatic cell that has acquired a selective advantage. This advantage is characterised by markers that persist in the population of cells making up the tumour [37]. Progression of KS to a malignant phenotype during nodular stages of the disease is associated with deregulation of host oncogenes and tumour suppressors, such as c-myc and p53, coincident with persistent expression of KSHV latency genes in all spindle cells [reviewed in 31]. Chromosomal abnormalities such as aneuploidy and microsatellite instability have also been detected in KS lesions [7]. Analyses of the terminal repeat sequences of KSHV DNA suggest that established lesions are monoclonal expansions of KSHV-infected cells and that advanced lesions are oligoclonal, rather than monoclonal proliferations [29,45].

9.2 KSHV-Associated Lymphomas

KSHV is closely related to the γ2-herpesvirus herpesvirus saimiri (HVS) and the γ1-herpesvirus EBV [55]. These phylogenetic relationships are indicative of a shared pathology between the gamma-herpesviruses in the development of lymphoproliferative disorders and cancer. EBV is causally linked to Burkitt's lymphoma, post-transplant lymphomas and X-linked lymphoproliferative disease and HVS can induce malignant lymphomas in animal models [reviewed in 70]. As a lymphotropic virus, KSHV is also pathogenically linked to lymphoproliferative disorders, often coincident with HIV infection. Specific manifestations of AIDS KS are termed nodal KS and are characterised by massive lymph node enlargement and/or

a

Fig. 9.2 (continued)

replacement with KS lesions [49, 58, 79]. Nodal KS has also been reported in the absence of HIV co-infection [21].

KSHV is associated with primary effusion lymphomas (PEL) [13] and multicentric Castleman's disease (MCD) [80]. In addition, KSHV has been associated with cases of angioimmunoblastic lymphadenopathy and germinal-centre hyperplasia [26, 53] as well as a subset of AIDS-related lymphomas without body cavity involvement and plasmablastic lymphoma of the oral cavity [20, 41]. In short, KSHV is present in cells belonging to the endothelial lineage in KS, in immunoblastic (or plasmablastic) cells belonging to the B-cell lineage in MCD, and in CD30-positive epithelial membrane antigen-positive lymphoma cells in PEL [28].

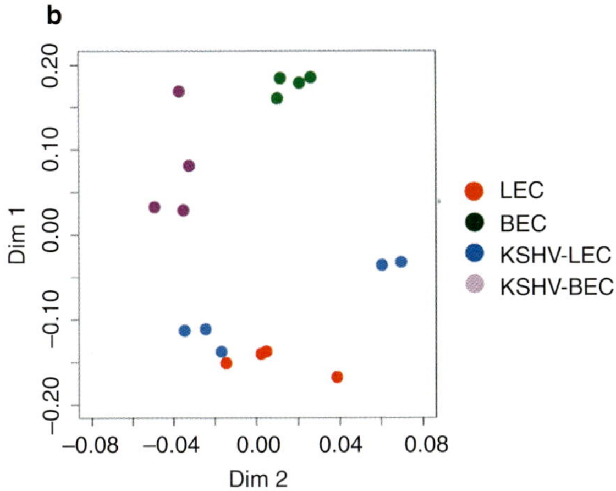

Fig. 9.2 a. The Kaposi's sarcoma expression signature. Heatmap illustrating 1,482 genes whose expression at the mRNA level significantly differentiates between biopsies of nodular KS and normal skin ($q \leq 0.05$). Genes that are down regulated in KS are shown in blue and up regulated genes are shown in red; the colour scale indicates units of standard deviation from the mean expression of each gene. Selected genes are listed and cytokines and chemokines and their receptors are underlined. The map was generated following the removal of genes from the global expression profile that were expressed at similar levels in KS and normal dermis or epidermis. This removed common tissue background present in normal skin samples to generate the KS expression signature. **b. Two-dimensional multi-dimensional scaling (MDS) plot of infected and uninfected LEC and BEC.** Within the KS expression signature, 114 genes differentiate between LEC and BEC to form a LEC-BEC discriminatory signature. Comparison of the expression levels of these genes in LEC and BEC before and after KSHV infection indicates that the gene expression profile of KSHV-infected LEC (kLEC) is closer to that of BEC than the genes expression profile of uninfected LEC; similarly, the gene expression profile of kBEC is closer to that of LEC. Each data point represents a single sample and the infected LEC and BEC samples appear to move towards each other, away from the uninfected populations. Both Fig. 9.2a and b are taken from Wang et al., and were generated by M Trotter [86]

9.2.1 Primary Effusion Lymphoma (PEL)

PEL is a rare malignant lymphoma thought to originate from post-germinal centre B-cells that typically develops in body cavities as pleural, peritoneal and pericardial effusions [13]. All cases of PEL are associated with KSHV infection although the global incidence of PEL is low, even in regions with a high prevalence of KSHV. This suggests the requirement of an additional cofactor for disease progression and, in agreement with this observation, the majority of cases of PEL occur coincident with HIV-1 and/or EBV infection [13,14,16,43,76]. PEL is discriminated from other AIDS-associated non-Hodgkin lymphomas by evidence of KSHV infection in neoplastic cells (LANA staining or DNA sequence analysis) and incidences of PEL have been described in the absence of HIV [15]. In addition, the incomplete expression of the full complement of EBV latent genes in PEL patients suggests that KSHV is the primary causative agent of this lymphoproliferative disease [reviewed in 10,

14]. Cases of PEL have been described that present with solid tissue involvement accompanied by serous effusions, either proceeding PEL development or following disease resolution [reviewed in 14]. Extracavity tumours have been reported with a tendency to localise at the lymph node [16]. Gene expression microarray analysis comparing gene expression patterns of PEL to B cell lines representative of different B cell malignancies indicates that PEL has a gene expression profile comparable to transformed plasma cells [44]. This finding suggests that KSHV may direct B cells towards a plasma cell fate through the piracy of host cellular genes [44].

9.2.2 Multicentric Castleman's Disease (MCD)

MCD is a lymphoproliferative disorder that is observed in two forms: the hyaline vascular form is a benign localised profusion of lymphoid tissue [12], the plasmablastic variant is a systemic lymphoproliferative disorder characterised by sheets of plasma cells in the lymph node interfollicular region [48]. KSHV infection is associated with nearly all incidents of MCD in HIV positive patients and approximately 50% of MCD in HIV negative patients; the specific association of KSHV infection with these incidents of MCD is a defining characteristic of the plasmablastic variant [27, 63, 80]. Plasmablastic MCD is a polyclonal tumour that originates from naïve B-cells, it is associated with lymphadenopathy and characterised by vascular proliferation of the lymph node germinal centres [14, 43]. The association between KSHV and MCD increases the susceptibility of patients to secondary tumours, such as KS, and a range of lymphoproliferative lesions including micro lymphoma and plasmablastic lymphoma [27]. It is thought that KSHV-MCD can progress from a polyclonal to a monoclonal plasmablastic lymphoma.

9.2.3 Lymph nodes as Reservoirs of KSHV Infection

KSHV has the genetic machinery of an oncogenic virus and encodes proteins required to modulate signal transduction pathways and cell cycle processes of the host cell (see [42] and Box 1). The infection of B-cells in lymph node follicles creates a reservoir for the persistence of KSHV infection that may influence the characteristics of the associated lymphomas [9, 50, 84]. Lymph node infection by KSHV has been implicated in the establishment of a case of PEL [5] and has been observed in multiple cases of KS where the lymph tumour is identical to skin lesions in appearance [58]. In addition, KSHV infection of the lymph nodes has been associated with lymphoma in HIV negative individuals in the absence of serous effusions [9].

In KS, the majority of tumour cells are latently infected with KSHV (see Box 1 and above). A small percentage of cells in KS, PEL and PEL cell lines express genes associated with the lytic life cycle of KSHV; in contrast, a greater percentage of MCD cells express lytic transcripts [54, 62]. Lytic replication can propagate viral infection by releasing new virions to further the spread of virus through host tissues [71]. These observations suggest a role for lytic infection in the pathogenesis of KSHV-related neoplasms. Early KS hyperplasia could also be promoted through paracrine mechanisms following the expression of lytic gene products that contribute to KS initiation and progression [28].

9.3 A Lymphatic Origin for KS? KSHV-Directed Reprogramming of Endothelial Cells

In isolation, the ultrastructural features of KS spindle cells, as observed by light microscopy, are not sufficient to categorize these cells as having derived from a specific lineage [35]. Cytogenetic studies, immunohistochemical profiles, DNA, RNA and protein analysis have indicated that KS spindle cells express endothelial cell markers, but the precise cellular origin is poorly defined [6, 40, 68, 86]. KSHV has been shown to infect endothelial cells in culture and in KS biopsies [11, 28].

The blood vascular phenotype is the default differentiation pathway for endothelial cells such that the lymphatic system originates from embryonic veins [reviewed in 40]. Blood vessel endothelial cells (BEC) and lymphatic endothelial cells (LEC) exhibit different functional properties and express specific genes, receptors and cytochemical markers leading to cell-specific responses to some cytokines and growth factors [39, 87]. The homeobox gene, PROX1 is necessary and sufficient to drive lymphatic differentiation and its ectopic expression down-regulates genes associated with blood vascular endothelial cell differentiation and up regulates markers associated with the lymphatic lineage [40]. cDNA microarrays have been used to examine the global changes in gene expression induced by KSHV infection of endothelial cells [11, 40, 86] and a subset of genes have been identified as describing a "KS expression signature" that characterises the gene expression pattern in KS lesions [86] (Fig. 9.2a)

The expression of LEC markers like VEGFR3, podoplanin, CAECAM1 and LYVE-1 are induced in TIME cells, a BEC-derived cell line, following KSHV infection. This indicates that KSHV can infect BEC and drive them to differentiate towards a LEC genotype. The differentiation of TIME cells from blood endothelium to lymphatic following KSHV infection is accompanied by a decrease in levels of the chemokine IL-8 [11]. Importantly, the gene expression profile induced by KSHV infection is distinct from that induced by over-expression of PROX1 in BEC, therefore suggesting that KSHV infection results in additional phenotypic (and genotypic) changes to those induced by PROX1 [11, 65]. For example, VEGFR1 is typically up-regulated in BEC compared to LEC, but is found to be greatly induced by KSHV infection [11], indicating that the virus has broad-ranging effects in terms of eliciting vasculogenic responses that are not limited to lymphatic reprogramming. VEGFR3, the receptor for VEGF-C and VEGF-D, both of which are involved in lymphangiogenesis, is up regulated in BEC after KSHV infection. The levels of VEGF-C are also increased, indicative of autocrine cellular activation mechanisms accompanying lymphatic reprogramming, and an additional KSHV-specific effect [11].

Gene expression array analysis comparing nodular KS to normal skin and various cell lines indicates that KS tumour cells are closely associated with endothelial cell lines. Furthermore, the expression profile of KS is closer to that of LEC, compared to BEC [86]. One study showed that KSHV infection induces PROX1 expression in BEC, this could explain in part the mechanism by which KSHV induces lymphatic reprogramming in BEC [40]. However, this was not confirmed in a related study [86].

Immunohistochemical staining of KS tumours for LEC and BEC markers also showed that KS tumour cells more closely resemble lymphatic endothelium than blood vessel endothelium [6, 24]. These observations are summarised in Tables 9.1 and 9.2. More recently, it was shown that spindle cells staining positive for LANA, and therefore latently infected with KSHV, showed increased expression of lymphatic markers in early and late stage KS lesions [68]. These findings included positive staining with the monoclonal antibody D2-40 against the lymphatic glycoprotein M2A, but also indicated an upregulation of CD34 in a significant percentage of KS spindle cells, a marker of BEC-derived cells [68]. While a percentage of spindle cells were CD34 negative, the identification of a spindle cell population in early lesions expressing both BEC and LEC markers suggests a hybrid phenotype in these

Table 9.1 Lymphatic-related genes up regulated by KSHV. Molecules listed in bold are lymph-specific markers up regulated by KSHV infection in tumour biopsies and cells. Data are taken from references cited in the text and [2]

Factor	Function
CC-chemokine ligand 21 (CCL21)	Chemokine that inhibits haemopoiesis and stimulates chemotaxis, may also play a role in mediating homing of lymphocytes to secondary lymphoid organs. It is a high affinity functional ligand for chemokine receptor 7 (**CCR7**) whose expression is unchanged by KSHV.
Lymphatic endothelial hyaluronan receptor-1 (LYVE1)	Transmembrane receptor expressed on all embryonic LEC but is postnatally limited to the lymphatic capillaries. Binds to the glycosaminoglycan hyaluran but its exact function is unclear. First marker of lymphatic endothelial competence and expression remains high in lymphatic capillaries in the adult.
Podoplanin (PDPN)	Transmembrane glycoprotein that promotes LEC adhesion, tube formation and migration in vitro by promoting rearrangement of the actin cytoskeleton. Recognised by the D2-40 antibody implicated in tumour progression.
Prospero- related homeobox-1 (PROX1)	Transcription factor that induces LEC-specific gene expression and is considered the most specific marker of lymphatic endothelium. Necessary and sufficient to determine progenitor-cell fate and mitosis in liver, pancreas, and lens retina.
Vascular endothelial growth factor (VEGF) receptor-3 (VEGFR3)	Receptor tyrosine kinase that specifically binds **VEGFC/VEGFD**. Expression is largely restricted to the lymphatic endothelium in the adult, and is observed in both the developing venous and presumptive lymphatic endothelia in the embryo. A specific marker of lymphatic endothelia when co expressed with LYVE1 during development. Also called FLT4.

Table 9.2 Lymphatic-related genes up regulated by KSHV. The remaining are selected lymphangiogenic factors found to be upregulated in KS compared to normal skin. Data are taken from references cited in the text and [2]

Factor	Function
Angiopoietin 2 (Ang2)	Secreted factor that destabilises blood vessels and is antagonistic in action to Ang1. Binds to the **Tie2** receptor, whose expression is unchanged by KSHV but is consistently expressed on endothelial cells.
CC-chemokine ligand 5 (CCL5)	Cytokine that functions as a chemoattractant for blood monocytes, memory T helper cells and eosinophils. Produced CD8+ cells as a major HIV-suppressive factor. A natural ligand for the **CCR5** receptor whose expression is elevated by KSHV infection.
Mannose receptor, C type 1 (MRC1)	Type I membrane protein expressed on a number of cell types including LEC. Mediates the endocytosis of glycoproteins by macrophages, thereby contributing to innate and acquired immunity. Also known as CD206.
CXCL10	Chemokine that binds to **CXCR3** whose expression is unchanged in KS. Receptor-ligand interaction results in monocyte stimulation, migration of natural killer and T-cells and modulation of adhesion molecule expression.
Insulin-like growth factor 1 (IGF-1)	Induces lymphangiogenesis in vitro and in vivo. Binds to the **IGF-1R** receptor, whose expression is unchanged by KSHV but is consistently expressed on endothelial cells.
PDGFA/PDGFB	Function as homo- or heterodimers. PDGF-BB and PDGF-AB are potent lymphangiogenic factors, PDGF-AA is weaker. PDGF-BB may act as a survival factor for newly formed lymphatics. They function through the receptor **PDGFRβ**, whose expression is up regulated in KS.
TGFβ1/TGFβ3	TGFβ is a multifunctional peptide that functions synergistically with TGFA to induce transformation. Both bind to the receptor **TGF β R2**, whose expression is up regulated in KS.

cells [68]. More advanced lesions indicate that this mixed phenotype is expressed by the majority of KS spindle cells. This concurs with our hypothesis that KSHV infects endothelial cell precursors, and drive their differentiation towards more mature lymphatic endothelial cells, or that KSHV preferentially infects LEC, and induces reprogramming to express markers associated also with BEC (described in Box 2).

Whether this hypothesis proves correct or not, it is clear that KSHV infection can drive the gene expression profiles of LEC and BEC closer to each other, generating a hybrid genotype in the spindle cell compartment, where markers from both lineages are expressed [86]. KSHV can therefore be considered to reprogramme the transcriptomes of endothelial cells [68, 86] (Fig. 9.2b).

The significant population of CD34 negative KS spindle cells observed in early lesions supports the observation that host LEC are subject to primary infection by KSHV [68]; LEC also appear to be more permissive to KSHV infection and acquire a higher KSHV copy number [86]. KS occurs most frequently at sites in which LEC are abundant (skin, lymph nodes), but not in tissues lacking lymphatic vessels, such as the brain; in addition, the blood vessels of KS lesions are not infected with

KSHV despite production of virions through lytic infection in these lesions [28]. These observations also support LEC as the cell of origin for this tumour.

KSHV infection can also influence B-cell fate, as latent or lytic infection of reactive lymph node B cells by KSHV influences pathways of B cell maturation [44,50] and may act to direct B cells towards a plasma cell fate [44]. Latent infection of naïve B cells and expression of the associated KSHV genes drives them to become resting memory B cells via the germinal center reaction [50]. Conversely, expression of KSHV lytic genes such as vIL6 (see Box 1) in naïve B cells directs their differentiation to plasmablasts without undergoing the germinal center reaction [50]. By this mechanism, the heterogeneity of KSHV-associated lymphomas can be influenced by viral infection of reactive B cells and contribute to the morphological, phenotypic, and clinical characteristics of these diseases.

Two mouse models of KSHV infection using NOD/SCID mice engrafted with human haematopoietic tissue have been reported [23, 64]. These mice are homozygous for the SCID mutation (severe combined immunodeficiency) on the NOD background (non-obese diabetic) and display impaired B- and T- lymphocyte function [reviewed in 51]. However, these models do not develop KS-like lesions. In 2007, the only mouse model of KS-like disease was reported showed that transfection of a KSHV Bacterial Artificial Chromosome (KSHVBac36) into a mouse bone marrow preparation enriched for endothelial-lineage cells generates a cell (mECK36) that forms KS-like tumours in immunodeficient (nude) mice [56]. These tumours are vascularised, spindle cell sarcomas that express a number of LEC-associated markers including VEGFR3 and podoplanin [56]. Furthermore, these mouse tumours highly express Angiopoeitin-2 (Ang2). Ang2 is expressed in human KS lesions, and in the sera of individuals with AIDS-KS [85, 86] and is a molecule linked to lymphangiogenesis [75].

9.4 Conclusions

KSHV infects endothelial cells and drives their differentiation to a lymphatic phenotype. In addition, a hypothesis currently being addressed is the possibility that KSHV interacts with circulating endothelial precursors to direct them towards a lymphatic phenotype; the mechanism by which this occurs is the subject of ongoing research. The functional implications of lymphatic differentiation in terms of the propagation of KSHV infection are poorly understood. The lymphatic phenotype may generate cells more suited to supporting viral latent infection and the upregulation of complementary factors such as VEGFR3 and VEGF-C may drive endothelial cell activation by autocrine mechanisms leading to propagation of the spindle compartment and development of the KS lesion. Lymphatically differentiated cells may also be more suited to later dissemination via the lymphatic system resulting in spread of the tumour during later stages of KS development.

Understanding the mechanisms by which KSHV influences cells following primary infection may influence treatment options available to sufferers of KS. Despite the prominence of angiogenic mechanisms in the development of KS lesions, targeting of angiogenic markers such as VEGF with inhibitors can have damaging effects on normal tissue through lack of specificity to the lesion. The identification of KS-specific markers may result in increasingly selective therapies with a reduction in cytotoxic effects. The propensity of LEC to be infected by KSHV also suggests that antilymphangiogenic therapies may be feasible treatment modalities for the management of KS and KSHV-related disease.

References

1. Proceedings of the IARC Working Group on the Evaluation of Carcinogenic Risks to Humans. Epstein-Barr Virus and Kaposi's Sarcoma Herpesvirus/Human Herpesvirus 8. Lyon, France, 17–24 June 1997. IARC Monogr Eval Carcinog Risks Hum, 1997, 70: 1–492.
2. Alitalo K, Carmeliet P. Molecular mechanisms of lymphangiogenesis in health and disease. Cancer Cell 2002, 1: 219–27.
3. Antman K, Chang Y. Kaposi's sarcoma. N Engl J Med 2000, 342: 1027–38.
4. Aoki Y, Tosato G. HIV-1 Tat enhances Kaposi sarcoma-associated herpesvirus (KSHV) infectivity. Blood 2004, 104: 810–4.
5. Ariad S, et al. Early peripheral lymph node involvement of human herpesvirus 8-associated, body cavity-based lymphoma in a human immunodeficiency virus-negative patient. Arch Pathol Lab Med 2000, 124: 753–5.
6. Beckstead JH, Wood GS, Fletcher V. Evidence for the origin of Kaposi's sarcoma from lymphatic endothelium. Am J Pathol 1985, 119: 294–300.
7. Bedi GC, et al. Microsatellite instability in primary neoplasms from HIV + patients. Nat Med 1995, 1: 65–8.
8. Boshoff C, et al. Kaposi's sarcoma-associated herpesvirus infects endothelial and spindle cells. Nat Med 1995, 1: 1274–8.
9. Carbone A. KSHV/HHV-8 associated Kaposi's sarcoma in lymph nodes concurrent with Epstein-Barr virus associated Hodgkin lymphoma. J Clin Pathol 2005, 58: 626–8.
10. Carbone A, Gloghini A. KSHV/HHV8-associated lymphomas. Br J Haematol 2008, 140: 13–24.
11. Carroll PA, Brazeau E, Lagunoff M. Kaposi's sarcoma-associated herpesvirus infection of blood endothelial cells induces lymphatic differentiation. Virology 2004, 328: 7–18.
12. Castleman B, Iverson L, Menendez VP. Localized mediastinal lymphnode hyperplasia resembling thymoma. Cancer 1956, 9: 822–30.
13. Cesarman E, et al. Kaposi's sarcoma-associated herpesvirus-like DNA sequences in AIDS-related body-cavity-based lymphomas. N Engl J Med 1995, 332: 1186–91.
14. Cesarman E, Knowles DM. The role of Kaposi's sarcoma-associated herpesvirus (KSHV/HHV-8) in lymphoproliferative diseases. Semin Cancer Biol 1999, 9: 165–74.
15. Cesarman E, et al. Kaposi's sarcoma-associated herpesvirus in non-AIDS related lymphomas occurring in body cavities. Am J Pathol 1996, 149: 53–7.
16. Chadburn A, et al. KSHV-positive solid lymphomas represent an extra-cavitary variant of primary effusion lymphoma. Am J Surg Pathol 2004, 28: 1401–16.
17. Chang J, et al. Inflammatory cytokines and the reactivation of Kaposi's sarcoma-associated herpesvirus lytic replication. Virology 2000, 266: 17–25.
18. Chang Y, et al. Identification of herpesvirus-like DNA sequences in AIDS-associated Kaposi's sarcoma. Science 1994, 266: 1865–9.
19. Choi J, et al. Molecular piracy of Kaposi's sarcoma associated herpesvirus. Cytokine Growth Factor Rev 2001, 12: 245–57.

20. Cioc AM, et al. Oral plasmablastic lymphomas in AIDS patients are associated with human herpesvirus 8. Am J Surg Pathol 2004, 28: 41–6.
21. di Gennaro G, et al. Discordant HHV8 detection in a young HIV-negative patient with Kaposi's sarcoma and sarcoidosis. Clin Infect Dis 2001, 32: 1100–2.
22. Dittmer D, et al. A cluster of latently expressed genes in Kaposi's sarcoma-associated herpesvirus. J Virol 1998, 72: 8309–15.
23. Dittmer D, et al. Experimental transmission of Kaposi's sarcoma-associated herpesvirus (KSHV/HHV-8) to SCID-hu Thy/Liv mice. J Exp Med 1999, 190: 1857–68.
24. Dorfman RF, Kaposi's sarcoma. The contribution of enzyme histochemistry to the identification of cell types. Acta Unio Int Contra Cancrum 1962, 18: 464–76.
25. Dourmishev LA, et al. Molecular genetics of Kaposi's sarcoma-associated herpesvirus (human herpesvirus-8) epidemiology and pathogenesis. Microbiol Mol Biol Rev 2003, 67: 175–212, table of contents.
26. Du MQ, et al. KSHV- and EBV-associated germinotropic lymphoproliferative disorder. Blood 2002, 100: 3415–8.
27. Dupin N, et al. HHV-8 is associated with a plasmablastic variant of Castleman disease that is linked to HHV-8-positive plasmablastic lymphoma. Blood 2000, 95: 1406–12.
28. Dupin N, et al. Distribution of human herpesvirus-8 latently infected cells in Kaposi's sarcoma, multicentric Castleman's disease, and primary effusion lymphoma. Proc Natl Acad Sci U S A 1999, 96: 4546–51.
29. Duprez R, et al. Evidence for a multiclonal origin of multicentric advanced lesions of Kaposi sarcoma. J Natl Cancer Inst 2007, 99: 1086–94.
30. Ensoli B, et al. Tat protein of HIV-1 stimulates growth of cells derived from Kaposi's sarcoma lesions of AIDS patients. Nature 1990, 345: 84–6.
31. Ensoli B, et al. Biology of Kaposi's sarcoma. Eur J Cancer 2001, 37: 1251–69.
32. Fiorelli V, et al. gamma-Interferon produced by CD8+ T cells infiltrating Kaposi's sarcoma induces spindle cells with angiogenic phenotype and synergy with human immunodeficiency virus-1 Tat protein: an immune response to human herpesvirus-8 infection? Blood 1998, 91: 956–67.
33. Flaitz CM, Hicks MJ. Molecular piracy: the viral link to carcinogenesis. Oral Oncol 1998, 34: 448–53.
34. Gallo RC, The enigmas of Kaposi's sarcoma. Science 1998, 282: 1837–9.
35. Gessain A, Duprez R. Spindle cells and their role in Kaposi's sarcoma. Int J Biochem Cell Biol 2005, 37: 2457–65.
36. Greensill J, Schulz TF. Rhadinoviruses (gamma2-herpesviruses) of Old World primates: models for KSHV/HHV8-associated disease? Aids 2000, 14: S11–9.
37. Hanahan D, Weinberg RA. The hallmarks of cancer. Cell 2000, 100: 57–70.
38. Hayward GS, Zong JC. Modern evolutionary history of the human KSHV genome. Curr Top Microbiol Immunol 2007, 312: 1–42.
39. Hirakawa S, et al. Identification of vascular lineage-specific genes by transcriptional profiling of isolated blood vascular and lymphatic endothelial cells. Am J Pathol 2003, 162: 575–86.
40. Hong YK, et al. Lymphatic reprogramming of blood vascular endothelium by Kaposi sarcoma-associated herpesvirus. Nat Genet 2004, 36: 683–5.
41. Huang Q, et al. KSHV/HHV8-associated lymphoma simulating anaplastic large cell lymphoma. Am J Surg Pathol 2004, 28: 693–7.
42. Jarviluoma A, Ojala PM. Cell signaling pathways engaged by KSHV. Biochim Biophys Acta 2006, 1766: 140–58.
43. Jenner RG, Boshoff C. The molecular pathology of Kaposi's sarcoma-associated herpesvirus. Biochim Biophys Acta 2002, 1602: 1–22.
44. Jenner RG, et al. Kaposi's sarcoma-associated herpesvirus-infected primary effusion lymphoma has a plasma cell gene expression profile. Proc Natl Acad Sci U S A 2003, 100: 10399–404.
45. Judde JG, et al. Monoclonality or oligoclonality of human herpesvirus 8 terminal repeat sequences in Kaposi's sarcoma and other diseases. J Natl Cancer Inst 2000, 92: 729–36.

46. Kaposi M. Idiopathisches multiples Pigmentsarkom der Haut. Archiv fur Dermatologie und Syphilis 1872, 4: 265–73.
47. Kaposi M. Zur Nomenclatur des iodiopathischen Pigmentsarkoms Kaposi. Archiv fur Dermatologie und Syphilis 1894, 29: 164.
48. Keller AR, Hochholzer L, Castleman B. Hyaline-vascular and plasma-cell types of giant lymph node hyperplasia of the mediastinum and other locations. Cancer 1972, 29: 670–83.
49. Krown SE. Clinical characteristics of Kaposi sarcoma. HIV InSite KNowledge Base Chapter 2006 [cited].
50. Lazzi S, et al. Kaposi's sarcoma-associated herpesvirus/human herpesvirus 8 infection in reactive lymphoid tissues: a model for KSHV/HHV-8-related lymphomas? Hum Pathol 2006, 37: 23–31.
51. Leblond V, Autran B, Cesbron JY. The SCID mouse mutant: definition and potential use as a model for immune and hematological disorders. Hematol Cell Ther 1997, 39: 213–21.
52. Liao JB. Viruses and human cancer. Yale J Biol Med 2006, 79: 115–22.
53. Luppi M, et al. Human herpesvirus-8 DNA sequences in human immunodeficiency virus-negative angioimmunoblastic lymphadenopathy and benign lymphadenopathy with giant germinal center hyperplasia and increased vascularity. Blood 1996, 87: 3903–9.
54. Luppi M, et al. Expression of cell-homologous genes of human herpesvirus-8 in human immunodeficiency virus-negative lymphoproliferative diseases. Blood 1999, 94: 2931–3.
55. Moore PS, et al. Primary characterization of a herpesvirus agent associated with Kaposi's sarcomae. J Virol 1996, 70: 549–58.
56. Mutlu AD, et al. In vivo-restricted and reversible malignancy induced by human herpesvirus-8 KSHV: a cell and animal model of virally induced Kaposi's sarcoma. Cancer Cell 2007, 11: 245–58.
57. Neipel F, Albrecht JC, Fleckenstein B. Cell-homologous genes in the Kaposi's sarcoma-associated rhadinovirus human herpesvirus 8: determinants of its pathogenicity? J Virol 1997, 71: 4187–92.
58. O'Connell KM. Kaposi's sarcoma in lymph nodes: histological study of lesions from 16 cases in Malawi. J Clin Pathol 1977, 30: 696–703.
59. Pagano JS, et al. Infectious agents and cancer: criteria for a causal relation. Semin Cancer Biol 2004, 14: 453–71.
60. Pallesen G, et al. Expression of Epstein-Barr virus latent gene products in tumour cells of Hodgkin's disease. Lancet 1991, 337: 320–2.
61. Parkin DM. The global health burden of infection-associated cancers in the year 2002. Int J Cancer 2006, 118: 3030–44.
62. Parravicini C, et al. Differential viral protein expression in Kaposi's sarcoma-associated herpesvirus-infected diseases: Kaposi's sarcoma, primary effusion lymphoma, and multicentric Castleman's disease. Am J Pathol 2000, 156: 743–9.
63. Parravicini C, et al. Expression of a virus-derived cytokine, KSHV vIL-6, in HIV-seronegative Castleman's disease. Am J Pathol 1997, 151: 1517–22.
64. Parsons CH, et al. KSHV targets multiple leukocyte lineages during long-term productive infection in NOD/SCID mice. J Clin Invest 2006, 116: 1963–73.
65. Petrova TV, et al. Lymphatic endothelial reprogramming of vascular endothelial cells by the Prox-1 homeobox transcription factor. Embo J 2002, 21: 4593–9.
66. Pisani P, et al. Cancer and infection: estimates of the attributable fraction in 1990. Cancer Epidemiol Biomarkers Prev 1997, 6: 387–400.
67. Poole LJ, et al. Comparison of genetic variability at multiple loci across the genomes of the major subtypes of Kaposi's sarcoma-associated herpesvirus reveals evidence for recombination and for two distinct types of open reading frame K15 alleles at the right-hand end. J Virol 1999, 73: 6646–60.
68. Pyakurel P, et al. Lymphatic and vascular origin of Kaposi's sarcoma spindle cells during tumor development. Int J Cancer 2006, 119: 1262–7.
69. Renne R, et al. Lytic growth of Kaposi's sarcoma-associated herpesvirus (human herpesvirus 8) in culture. Nat Med 1996, 2: 342–6.

70. Renwick N, Schulz T, Goudsmit J. Kaposi's sarcoma and Kaposi's sarcoma-associated herpesvirus/human herpesvirus 8: an overview. In Human Retroviruses and AIDS (Kuiken CL, Foley B, Hahn B, Korber B, McCutchan F, Marx PA, Mellors JW, Mullins JI, Sodroski J, Wolinksy S, Eds.) pp. 475–91. Theoretical Biology and Biophysics Group, Los Alamos National Laboratory, Los Alamos, NM, 1999
71. Roizman B. Herpesviridae. In Virology (Fields BN, et al. Eds.), pp. 2221–30, 3rd ed. Lippincott-Raven, Philadelphia, 1996.
72. Russo JJ, et al. Nucleotide sequence of the Kaposi sarcoma-associated herpesvirus (HHV8). Proc Natl Acad Sci U S A 1996, 93: 14862–7.
73. Sarid R, Olsen SJ, Moore PS. Kaposi's sarcoma-associated herpesvirus: epidemiology, virology, and molecular biology. Adv Virus Res 1999, 52: 139–232.
74. Sarid R, et al. Characterization and cell cycle regulation of the major Kaposi's sarcoma-associated herpesvirus (human herpesvirus 8) latent genes and their promoter. J Virol 1999, 73: 1438–46.
75. Scavelli C, et al. Crosstalk between angiogenesis and lymphangiogenesis in tumor progression. Leukemia 2004, 18: 1054–8.
76. Schalling M, et al. A role for a new herpes virus (KSHV) in different forms of Kaposi's sarcoma. Nat Med 1995, 1: 707–8.
77. Schulz TF. Epidemiology of Kaposi's sarcoma-associated herpesvirus/human herpesvirus 8. Adv Cancer Res 1999, 76: 121–60.
78. Sharp TV, Boshoff C. Kaposi's sarcoma-associated herpesvirus: from cell biology to pathogenesis. IUBMB Life 2000, 49: 97–104.
79. Shetty K, Giannini P, Achong R. Kaposi's sarcoma of lymph node leading to a diagnosis of HIV. Oral Oncology Extra 2005, 41: 234–7.
80. Soulier J, et al. Kaposi's sarcoma-associated herpesvirus-like DNA sequences in multicentric Castleman's disease. Blood 1995, 86: 1276–80.
81. Staskus KA, et al. Kaposi's sarcoma-associated herpesvirus gene expression in endothelial (spindle) tumor cells. J Virol 1997, 71: 715–9.
82. Sturzl M, et al. Expression of K13/v-FLIP gene of human herpesvirus 8 and apoptosis in Kaposi's sarcoma spindle cells. J Natl Cancer Inst 1999, 91: 1725–33.
83. Talbot SJ, et al. Transcriptional analysis of human herpesvirus-8 open reading frames 71, 72, 73, K14, and 74 in a primary effusion lymphoma cell line. Virology 1999, 257: 84–94.
84. Trovato R, et al. Cellular localization of human herpesvirus 8 in nonneoplastic lymphadenopathies and chronic interstitial pneumonitis by in situ polymerase chain reaction studies. J Hum Virol 1999, 2: 38–44.
85. Vart RJ, et al. Kaposi's sarcoma-associated herpesvirus-encoded interleukin-6 and G-protein-coupled receptor regulate angiopoietin-2 expression in lymphatic endothelial cells. Cancer Res 2007, 67: 4042–51.
86. Wang HW, et al. Kaposi sarcoma herpesvirus-induced cellular reprogramming contributes to the lymphatic endothelial gene expression in Kaposi sarcoma. Nat Genet 2004, 36: 687–93.
87. Weber E, et al. Cytochemical differentiation between blood and lymphatic endothelium: bovine blood and lymphatic large vessels and endothelial cells in culture. J Histochem Cytochem 1994, 42: 1109–15.
88. Wu L, et al. Three-dimensional structure of the human herpesvirus 8 capsid. J Virol 2000, 74: 9646–54.

Chapter 10

LYMPHATIC PHYSIOLOGY AND FUNCTION IN HEALTHY TISSUE AND CANCER

Jacqueline D. Shields and Melody A. Swartz
Institute of Bioengineering, École Polytechnique Fédérale de Lausanne (EPFL), CH-1015 Lausanne, Switzerland, e-mail: mailto:jacqueline.shields@epfl.ch

Abstract: The lymphatic system is the primary route of metastasis for many cancers, and it is this spread through the lymphatic vessels to lymph nodes and on to distant organs that is responsible for the majority of cancer-related deaths. Lymphatics, moreso than blood vessels, are thought to provide an overall favorable route for the survival and dissemination of tumor cells due to their anatomical features and low shear stress environment, but the mechanisms and physiological parameters governing lymphatic metastasis are only beginning to be understood. How cancer cells affect and gain access to local lymphatic vessels, travel within the vessels, and enter into the lymph nodes are all topics of recent research efforts, alongside questions of how tumor cells might mimic immune cells and escape the host adaptive immune response. In this chapter we cover the basic anatomy and physiology of the lymphatic system and how it relates to cancer metastasis through the lymphatics.

Key words: Flow · Lymph node · VEGF-C · Interstitium · Lymphangion · Capillary · Cell trafficking

10.1 Anatomy and Physiology of Lymphatic Drainage

The lymphatic system complements the venous branch of the circulation by draining excess fluid and solutes from the interstitial space and returning them to the blood (Fig. 10.1). This causes slow interstitial flow through the space between blood vessels and lymphatics that contains cells and extracellular matrix, referred to as the interstitium. Without this interstitial flow, transport of proteins and other macromolecules would occur by diffusion alone, which can be exceedingly slow (diffusion coefficients are on the order of $10^{-6}-10^{-8}$ cm^2/s for most proteins).

S.A. Stacker, M.G. Achen (eds.), *Lymphangiogenesis in Cancer Metastasis*, Cancer Metastasis – Biology and Treatment 13, DOI 10.1007/978-90-481-2247-9_10,

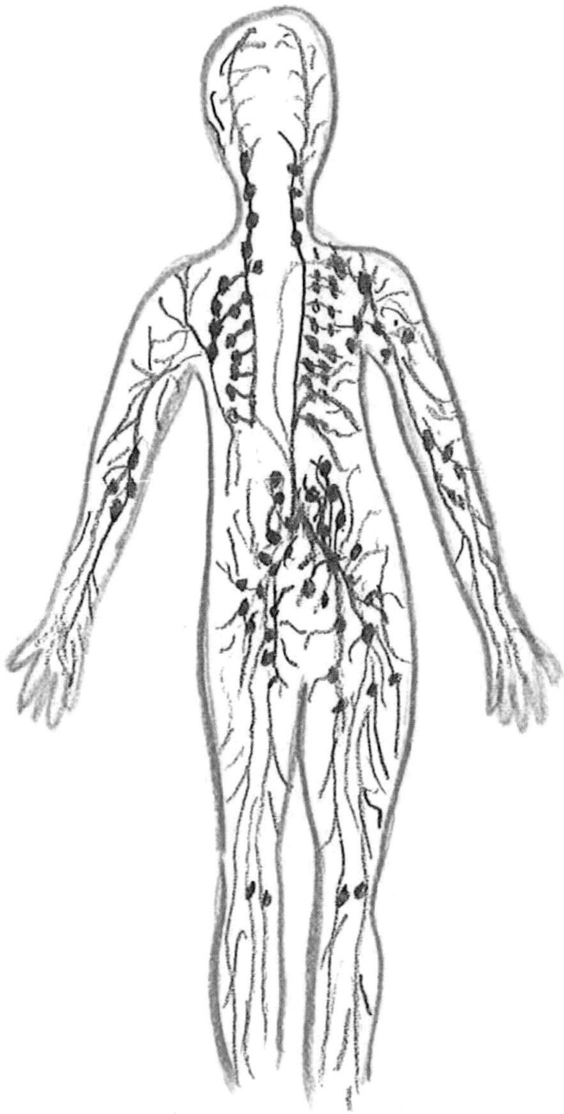

Fig. 10.1 There are 400–600 lymph nodes in the human, mostly situated in the trunk and neck, draining almost all tissues of the body. Lymph flows from periphery towards the heart, converging and draining into the thoracic duct and into the vena cava

The lymphatic system is comprised of lymphatic capillaries, collecting vessels, lymph nodes, and lymphatic ducts (Fig. 10.2). Fluid is absorbed into the capillaries or initial lymphatics from the interstitium, drains to collecting vessels that have smooth muscle to propel lymph always in the proximal direction (toward the heart) – passing through lymph nodes on the way – then draining to ducts and the thoracic duct, which eventually empties into the great veins of the neck, thus returning lymph

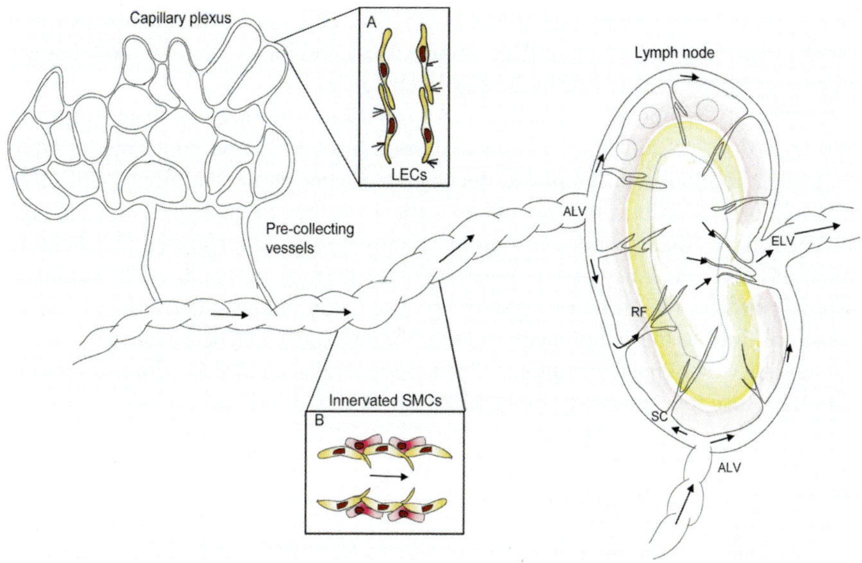

Fig. 10.2 Flow in the lymphatic system. Interconnecting lymphatic capillary networks drain via pre-collecting vessels to collecting vessels and then onto lymph nodes. **A** Lymphatic capillary morphology, where adjacent endothelial cells overlap to form a primary valve system that allows fluid drainage into the vessel when external pressure is higher but prevents backflow as the capillary fills. Endothelial cells are connected intimately to the surrounding matrix via anchoring filaments. **B** In collecting lymphatics, fluid is pumped along through individual lymphangions in sequential manner. These vessels are separated by valves and surrounded by highly contractile smooth muscle cells (SMCs), which squeeze each lymphangion sequentially. In this way, when one lymphangion contracts, the valve opens and fluid is propelled into the next lymphangion; as this second lymphangion fills, the valve closes and stretch activates local smooth muscle to contract and propel fluid into the next, etc. Lymph flows into the lymph node via afferent lymphatic vessels (ALV) and leaves via efferent lymphatic vessel (ELV), flowing around the node in the subcapsular sinus (SCS) and into the node along reticular fibers (RF)

to the blood circulation. On average, the lymphatic system returns 3 L of fluid per day to the blood, which is primarily composed of plasma proteins collected from the interstitial space. In the steady state, fluid collected in lymphatic capillaries (lymph formation) equals that produced by microvascular filtration. Disturbances to this balance result in the rapid onset of fluid accumulation in the interstitial space and the onset of edema. Under normal circumstances, it has traditionally been assumed that approximately 20 L/day of proteinaceous fluid is exuded from blood into tissue at the arterial capillary level, where pressure is high. According to the Starling principle of fluid balance where exuded fluid is constantly reabsorbed by downstream, lower pressure venous microvessels, this would leave only a small fraction of capillary filtrate ($\sim$15%) in the interstitium to enter the lymphatic system. However, the classical take on the Starling principle has recently been challenged. There is growing evidence to suggest that in contrast to the traditional hypothesis, the glycocalyx provides an additional buffer for osmotic pressure and that active protein transport by the endothelium into the subglycocalyx space helps reduce the driving forces for transendothelial flux [1, 34, 36]. This implies that much less fluid is reabsorbed into the post-capillary venules than previously thought, and that the lymphatics play a greater role in the starling balance.

Importantly, lymphatics are also a major route for cell transport, since lymphatic capillaries drain to collecting vessels and drain through lymph nodes before emptying into the blood via lymphatic ducts. Indeed, lymph nodes (along with other lymphoid organs such as the spleen) are critical sites for immune cells to exchange information, reside, expand, and initiate adaptive immune responses. By draining interstitial fluid and carrying any potential antigens to lymph nodes, the lymphatic system optimizes the immune response because it continuously delivers samples of interstitial fluid to the lymph nodes so that immune cells can sense and respond to inflammatory cytokines and antigens from peripheral sites directly drained into the node. This also has important implications for cancer, as will be explained later.

10.1.1 Lymph Formation

Lymph forms when fluid is drained from the interstitial space into lymphatic capillaries, also sometimes referred to as terminal lymphatics or initial lymphatics (although the term "terminal lymphatics" is also used sometimes to describe the lymphatic-like channels entering into lymph node from the subcapsular sinus). In the gut, these lymphatics are highly specialized for fat absorption and are called lacteals. The lymphatic capillaries are generally thought to "start" (from the perspective of interstitial fluid that becomes lymph as it enters the capillaries) as blind-ended bulb-like structures, but this may only be the case in the gut lacteals, while in tissues like skin they may simply exist as an interconnected network.

Lymphatic capillaries are optimally designed for maximal drainage and sampling of the interstitial milieu (Fig. 10.3): they are extremely thin-walled, usually encircled by a single lymphatic endothelial cell (LEC), and they have a discontinuous basement membrane and weak cell–cell junctions, making them highly permeable to fluid and high molecular weight solutes such as proteins and even nanoscale colloidal particles [53,63]. The wall of the lymphatic vessel is so thin (50–100 nm) that in traditional histology they often can be identified by only their nucleus protruding into the lumen.

Lymphatic capillaries vary in size and are not circular in cross-section, and their diameters range from 10 to 80 microns [16,33,61]. They drain fluid by virtue of their anchoring filaments, which pull the lymphatic lumen open during slight changes in tissue hydration, causing a subsequent pressure drop that drives fluid flow into the lymphatic lumen [53]. The adhesions in the overlapping cell–cell junctions of lymphatic capillaries are intermittent and comprised of focal collections of VE cadherin and tight junction-associated proteins including occludin; this is unlike cell–cell adhesions in blood endothelium, which are continuous. The periodic anchorages in lymphatics have recently been termed "buttons" [6] and are likely to function to maintain vessel integrity while allowing fluid flow through the overlapping endothelial cells into the vessel. Furthermore, the overlapping cell–cell junctions themselves

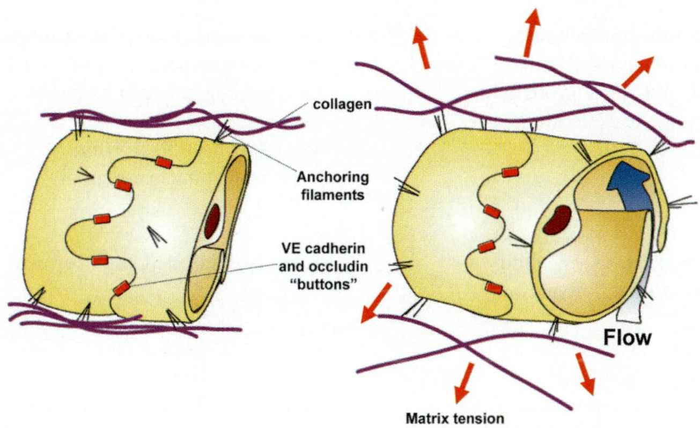

Fig. 10.3 Mechanism of lymph formation. Lymphatic capillaries consist of highly attenuated, overlapping endothelial cells (typically a single cell in diameter) with a flattened, irregular lumen. The endothelium has discontinuous basement membrane and is attached to collagen fibers by anchoring filaments, which can extend deeply into the matrix. Tissue strain caused by increased fluid pressure pulls the capillary and expands the luminal cross-sectional area, creating a pressure drop that drives fluid from the interstitium into the capillary. This is facilitated by intermittent cell–cell junctions that keep endothelial cells attached but permit fluid flow around the junctions. As the capillary fills, the valve (overlapping cell–cell junction) closes, preventing backflow; fluid then drains to collecting vessels which propel fluid by action of smooth muscle cells.

can act as valves to prevent lymph leakage from the vessel [32, 33, 65]: as the lymphatic capillary fills, they close and prevent backflow. Thus, lymphatic function causes interstitial fluid flow that is always directed towards the draining lymphatic capillary. Some differences between lymphatic and blood capillaries are summarized in Table 10.1.

10.1.2 Lymph Propulsion

Since the lymphatics are so highly permeable, lymph is essentially equal in concentration to interstitial fluid. Once lymph is formed into the initial draining lymphatic capillaries, it flows to slightly larger ducts (100–220 μm) referred to as pre-collecting vessels that drain to deeper collecting lymphatic vessels (Fig. 10.3). These collecting vessels are segmented into bulb-like structures, referred to as lymphangions, that are separated by valves to facilitate unidirectional flow of lymph. The lymphangions are surrounded by innervated smooth muscle cells that have intrinsic contractility on par with that of cardiac muscle. As with lymphatic capillaries, junctions

Table 10.1 Some differences between blood and lymphatic capillaries.

Blood	Lymphatic
Capillary structure	–
Continuous basement membrane	Discontinuous basement membrane
Small, circular lumen (5–20 μm)	Irregular, flattened lumen (40–80 μm)
Contiguous EC junctions throughout	Overlapping cell–cell junctions
Encircled by pericytes/smooth muscle	No pericytes
Wall thickness 3.5–5 μm	Wall thickness 50–100 nm (except at nuclear region)
Function	–
Highly controlled, low permeability	Loosely controlled, high permeability
Hydrostatic pressure difference: $P_{blood} \gg P_{interstitium}$	$P_{interstitium} \geq P_{lymphatic}$
Osmotic pressure difference: $\pi_{blood} \gg \pi_{interstitium}$	$\pi_{interstitium} \simeq \pi_{lymphatic}$
Attached to basement membrane filaments	Attached to ECM fibers via anchoring
Lumen diameter is controlled by pericytes	Lumen diameter is controlled by ECM strain
Abundant tight and adherins junctions (continuous across cell boundary)	Less abundant tight and adherins junctions (focal points of adhesion)
–	"Button" junctions: VE cadherin and tight junction proteins
Molecular expression	–
–	Prox1
–	Podoplanin
–	LYVE-1
VEGFR-3 on tumor associated and fenestrated vasculature*	VEGFR-3
Neuropilin 2 in some veins†	Neuropilin 2
Neuropilin 1	–
–	Macrophage mannose receptor 1
–	Desmoplakin
–	Plakoglobin
–	β-Chemokine receptor D6
–	Integrin α9
Endoglin	–
CD34	CD34 on tumor associated lymphatics‡
N cadherin	–
PAL-E	–
Collagen type XVIII	–
Collagen type IV	–
Laminin	–
CD31	CD31
VE cadherin	VE cadherin
JAM-C	JAM-C
–	CLEVER-1

* [44, 45, 52].
† [69].
‡ [15].

between collecting lymphatics are also composed of VE cadherin and the same tight junctional proteins, but unlike the "button" structures seen in lymphatic capillaries, these proteins are arranged in a continous "zipper-like" fashion between adjacent endothelial cells [6]; this is consistent with the main function of collecting vessels being lymph propulsion rather than drainage, which requires the vessel wall to be relatively impermeable. The behavior of the lymphangions depends on both pressure and flow conditions [48, 67]. When they transport lymph up a pressure gradient, they do so by contracting and pumping fluid from one lymphangion to the next by propagation of a contractile wave in a manner that is coordinated by gap junctional communication [70]. When they transport lymph down a pressure gradient (i.e., under conditions of high load), they can act as conduits with decreased contraction amplitude, modulating tone and regulating flow resistance. These prenodal collecting vessels are referred to as afferent lymphatic vessels and serve to direct fluid and activated dendritic cells to the lymph node, where molecules in the lymph may be sampled by immune cells resident within the node. In the human, there are roughly 400–600 lymph nodes ($\sim$22 in the mouse) and they are generally 1–2 cm in diameter (0.5–2 mm in mouse). After passing through several lymph nodes, the lymph is returned to the blood (from which it left via the blood capillaries) primarily through the thoracic duct.

Lymphatic vessels appear to be absent within adipose tissue – although the latter can surround lymphatic vessels, especially near lymph nodes, they do not appear to be drained by lymphatic capillaries within the tissue. Lymphatics are also absent within cartilage, cornea, epidermis, and eye lens, as well as in the central nervous system. Regarding the latter, the brain is immune privileged and the blood-brain barrier limits immune cell trafficking; that, coupled with the lack of lymphatic vessels, led to the well-accepted notion that the brain does not contain a functional lymphatic system. However, studies showed that solute injected into the brain ended up in cervical lymphatics, and a humoral immune response could be elicited to antigen injected into the brain, with antibody production in cervical lymph nodes [12]. Thus, new research suggests that fluid channels in the brain may drain directly into lymphatics, making them part of the lymphatic system.

10.2 Lymph Node Physiology

The lymph node functions in a manner to provide optimal immune cell communication and interactions, exposing antigen-presenting cells in the lymph node to any antigens or pathogenic material collected by the lymph from the peripheral tissues and additionally trafficking immune cells to the node so that antigen-specific T cells can find their cognate antigen-presenting dendritic cells (Fig. 10.2). Lymph flows through the afferent vessels into the subcapsular sinus, through capsular sinusoids and into the medullary sinusoids before leaving the lymph node via efferent vessels, as well as through fine reticular fibers consisting of collagen fibers, basement membrane components and reticular fibroblastic cells [30]. These fine reticular fibers extend into the T cell regions [4, 30] and guide lymph flow and T cell-DC interactions there [4]. Within the lymph nodes, blood flows through the high endothelial venules (HEVs), which have loose cell–cell junctions and contain adhesion receptors like ICAM-1 and L-selectin, providing a mechanism for antigen-specific lymphocytes to exit the blood into the lymph node to search for cognate antigen. Additionally, activated DCs migrate within the lymph node in the local neighborhood of the HEVs, there to present their antigens to enhance the probability of contact with their antigen-specific naïve T cell partners, which are few in number.

An increase in vascular permeability during inflammation plays an important role in enhancing the rate of antigen delivery to the lymph node from the periphery; under inflammatory conditions, tissue fluid drainage can be increased 10-fold or more [7, 14, 16]. Furthermore, cytokines that are produced in the inflamed site are carried to the draining lymph node, where they can induce changes in the lymph node that enable and enhance immune cell trafficking there, including increasing blood flow through the affected node and thus enhancing lymphocyte trafficking there via the HEVs. For example, the inflammatory cytokines IL-6 and IL-8 can increase the expression of lymphocyte receptors like ICAM-1 and the homing chemokine CCL21 by HEVs [5]. Although this behavior presumably evolved as an immunological mechanism, it also plays an important role in cancer: interstitial fluid pressure is increased in tumors, causing increased lymphatic drainage from the tumor periphery [27], and moreover many tumors express and secrete cytokines and attract immune cells. Tumor cells can also escape an initial immune response from the host, although the process by which this occurs remains unclear.

Inflammation also apparently can induce expansion of the lymphatic network or enlargement of lymphatic sinuses in the lymph node [3], which may help recruit more DCs from the periphery into the lymph node or enhance T cell trafficking out of the node. This has also been shown in cancer: lymph nodes draining a tumor demonstrate enlarged lymphatic sinuses [21,24,47,51,66], but it is unclear how this affects metastasis mechanistically.

When lymphatic vessels are blocked or compromised in their capacity to drain fluid, lymphedema can occur. This is associated with tissue fluid stagnation, enlargement of the limb, accumulation of lipids, and impaired immune function [50]. This also occurs frequently in the arms of breast cancer patients who have had lymph nodes removed during surgical resection of the tumor. There are currently very few potential treatments for lymphedema other than manual fluid drainage (massage) and compression bandages.

10.3 Local Effects of Tumor Tissue on Lymphatic Physiology

The current paradigm established in the field of lymphatic research in cancer is that tumors secrete many growth factors and cytokines that influence the behavior of blood and lymphatic vessels. Of primary interest to this field is the production and secretion of lymphatic growth factors, most notably vascular endothelial growth factor (VEGF)-C and -D, by tumor cells. In 2001 it was first shown that tumor cells engineered to overproduce VEGF-C and VEGF-D were capable of inducing growth of new lymphatic vessels (lymphangiogenesis) into xenografted tumors and that these lymphangiogenic tumors were more likely to metastasize in mice. Since these first murine experiments, there has been much interest in the role of lymphatic vessels in lymphatic metastasis in human cancers [23]; this is the subject of other chapters in this book.

Thanks to the identification of lymphatic-specific antigens and subsequent development of antibodies to these markers, investigations into the role of lymphatics in human cancer were given a much-needed helping hand. High lymphatic vessel density inside and around invasive tumors has been observed in many human cancers that metastasize via lymph nodes but not in non-invasive benign tumors [13, 28, 39, 57]. This has been attributed to the formation new lymphatic vessels following secretion of lymphangiogenic growth factors by tumor cells [29, 60, 62], stromal cells [31] and recruited macrophage subsets [40, 54], which induce peritumoral lymphangiogenesis. However, very few clinical studies of human cancers were able to successfully show proliferating, growing lymphatic vessels within the tumor, and as there are currently no known markers to distinguish immature versus mature lymphatic endothelial cells, the theories pertaining to the origin of these vessels (whether new or co-opted existing vessels) and the theories on mechanisms of lymphatic spread have begun to diverge. Indeed, the functionality, necessity, and even the presence of new lymphatic vessels is a further source of some controversy. Lymphatic structures that have been observed within tumors are frequently collapsed, most probably by virtue of their anatomy (lack of mural cells to provide structural support) and the pressures created by an expanding mass of tumor cells [43]. These "vessels" may not be functional but the mere presence of such entities may be sufficient to provide an escape route for aggressive tumor cells.

Alternative to lymphangiogenesis, VEGF-C and -D may act on peritumoral lymphatic vessels to induce lymphatic endothelial cell proliferation and hyperplasia in the absence of directed growth into a tumor mass [17]. Actively growing tumors are highly angiogenic, containing newly formed immature and structurally disorganized capillaries that lack substantial basement membrane or mural cells [27]. These vessels are "leaky", creating high interstitial fluid pressure within the mass. This fluid exudes from the tumor into surrounding tissues where it is collected by functional, draining peritumoral lymphatic vessels. There is evidence to suggest that tumor cells are shed, following loss of or changes in characteristic cell–cell adhesion molecules [11], and potentially carried from the tumor by exiting fluid and into surrounding lymphatic vessels. In fact, early studies suggested that as many as $3–4 \times 10^6$ cells per gram of tumor could be shed per day into the bloodstream [8,35], although most may be apoptotic [64]. However, a feasible mechanism for passive tumor entry into lymphatic vessels remains to be elucidated, because although the lymphatic endothelium is extremely permeable to macromolecules and have loose overlapping cell–cell junctions, these are still far too small for a cell to passively fit through and therefore a tumor cell must still actively intravasate into the lymphatic vessel unless the endothelial integrity is destroyed by the tumor cell.

Although the primary lymphatic growth factors implicated in cancer metastasis are VEGF-C and VEGF-D, there is a growing list of further potential regulators of lymphatic function, and hence changes to expression of any of these may also affect lymphatic physiology in such a way as to render vessels more susceptible to tumor cell infiltration. These include Notch signaling with VEGFR-3 [56], neural cell adhesion modecule (NCAM) [11], and many others as addressed in other chapters in this book.

10.4 Lymphatic Effects on Tumors: Is Lymphangiogenesis Necessary for Metastasis?

Although most research relating to the pathophysiological role of lymphatic vessels in cancer metastasis has focused on the phenomenon of tumor-induced lymphangiogenesis and the potential route provided by this neo-vessel formation, there is still a remaining controversy as to the necessity of lymphangiogenesis. A growing body of evidence suggests that tumors are able to metastasize to regional lymph nodes in the absence of lymphangiogenesis [9, 58, 68]. Wong et al. used techniques including RNA interference and the use of soluble VEGFR-3 to demonstrate that while VEGF-C was required for tumor-derived lymphangiogenesis, tumor lymphangiogenesis was not required for metastasis. When prostate tumor cells in which VEGF-C expression was silenced by RNAi were implanted into mice, a significantly reduced intratumoral lymphatic vessel density was observed, but the vessels around implanted tumors were unchanged; the incidence of lymph node metastasis was not significantly reduced in the absence of intratumoral vessels indicating that the surrounding, and potentially pre-existing, vessels were sufficient for metastasis to occur [68]. So if this was the case, what possible roles could tumor-derived factors be playing? How could they change lymphatic physiology to promote lymphatic metastasis without the need for new vessel formation?

First, some functional studies have suggested that lymphatic vessels found within growing tumors are not functional. Due to the anatomical features of lymphatic vessels (discussed earlier), as well as the lack of extracellular matrix integrity inside a tumor, most lymphatic vessels observed within tumors appear collapsed (not attached to the extracellular matrix) and may therefore be unable to function normally (i.e., drain fluid by virtue of subtle extracellular matrix swelling). Padera et al. demonstrated that injected ferritin failed to co-localize with such intra-tumoral vessels and instead was carried by non-endothelial-lined channels towards surrounding peritumoral vessels [42]. Furthermore, the high cell density within a tumor combined with poor extracellular matrix integrity would make it physiologically difficult for lymphatic capillaries to establish themselves within a tumor, and hence it is likely that peritumoral vessels alone, whether pre-existing, co-opted or newly formed, are sufficient for the promotion of lymph node metastasis. These newly formed peritumoral lymphatics may be functional, albeit abnormal compared with existing, mature vessels [26]. This provides evidence to further suggest, in addition to those effects described in the earlier section "Local effects of tumor tissue on lymphatic physiology", that tumor-derived factors, namely VEGF-C, are not necessarily acting to directly induce new lymphatic vessel growth into a tumor but instead may exert its effects indirectly by changing lymphatic functionality or by acting on other extra-tumoral structures or cells types.

Alternate mechanisms for tumor entry into lymphatics have been suggested. For example, there is a natural variation in lymphatic capillary density, and tumor invasion may simply occur in areas where lymphatic density is highest [57]. Also, tumors can respond to normal constitutive lymphatic signals and actively grow or migrate towards the conduits that provide them with their escape route and provide continued survival. It has been shown both *in vitro* and *in vivo* that melanoma cells are capable of actively migrating towards lymphatic, but not blood, endothelial cells, and that migration could be prevented when the lymphatic homing chemokine CCL21 was blocked [58]; CCL21 is a ligand for CCR7, a lymphocyte receptor whose signaling is required for homing to lymphatics. Such a mechanism is consistent with numerous clinical observations that patients with chemokine receptor-positive tumors have poor prognosis [37, 49]. Finally, tumor cells may exploit normal lymphatic physiology along with their biophysical microenvironment to access peritumoral lymphatics. Tumors display high interstitial fluid pressure created from newly formed capillaries that are immature and disorganized lacking any substantial basement membrane or mural cells and result in "leaky" vessels [22, 27]. Fluid from such abnormal blood capillary exudates is collected by functional, draining peritumoral lymphatic vessels. Consequently, unidirectional flow exists always *from* the tumor *towards* functional draining lymphatics. While lymphatic vessels secrete the chemokine CCL21, its broadcast distance via diffusion away from the vessel should be limited by convective forces (fluid drainage) into the vessel. To help guide them towards functional lymphatic vessels, tumor cells may also sense interstitial fluid flow by virtue of autologous chemokine signaling, as was recently shown [59]. This mechanism of "autologous chemotaxis" involves the tumor cell secreting its own directional cue (the chemokine), which becomes skewed in the direction of slow interstitial flow; thus by chemotacting up this autologously produced gradient, the tumor cell moves in the direction of flow, which leads it to the draining lymphatic vessel. These and other mechanisms may all contribute to tumor cell metastasis in addition to lymphangiogenesis.

10.5 Tumor Effects and Interactions with the Lymph Node

Recently, there has been focused interest in how the tumor affects the draining lymph node, particularly with respect to the role of newly formed or expanded lymphatics around or within the lymph node [18, 19, 24, 66]. The potential exists for tumor-secreted growth factors (especially VEGF and VEGF-C) to drain into lymphatics and be carried to the lymph node, where they may stimulate expansion of lymphatic vessels there. This has been demonstrated in mice inoculated with melanoma cells; nodes downstream from the primary tumor contained increased numbers of lymphatic sinuses throughout both the nodal cortex and medulla compared to normal nodes, in which only sparse sinuses within the cortex could be observed [21, 51].

Significantly, lymph node lymphangiogenesis has been reported to precede tumor cell dissemination and therefore may act as a preparatory signal to prepare the lymph nodes for the pending invasion of tumor cells [21, 24]. Indeed, in animal models where lymph nodes displayed both lymphangiogenesis and high levels of VEGF or VEGF-C were more prone to tumor progression and development of further metastatic lesions [19, 24, 25]. Increased lymphatic vessel densities within lymph nodes that drain tumors has been postulated to incite increases to lymph flow as compared to non-tumor draining nodes [21] and therefore would support the hypothesis that normal lymphatic function in combination with tumor-secreted growth factor enhancement of lymph nodes may facilitate tumor cell dissemination to lymph nodes and subsequent metastasis.

The structure of lymph node is such that tumor cells must enter via afferent vessel into the marginal sinuses, where cells may remain and proliferate or continue to exit the node and move to distant sites. Hence, it is probable that the marginal sinues act as an essential mechanical filter to prevent establishment of tumors within the deeper medullary regions [10, 38]. It is only after tumor cells have filled these superficial regions before deeper metastases in the pulp can be observed [38]. However, the signals and mechanisms used by the tumor to cross into the medullary regions to establish secondary tumors within a node rather than disseminate to distant nodes and organs is still not known.

10.6 Summary

In recent years, our comprehension of lymphatic physiology and pathophysiology in diseases such as cancer has dramatically improved, but basic mechanistic questions remain such as how tumor cells gain access to lymphatic vessels, how lymphatics actively regulate tumor cell entry, and how tumors survive and grow in the lymph node. In this chapter we discussed how lymphatic function is intimately involved in immune cell homing to lymph nodes, and similarly how many tumors use the lymphatics to escape and establish metastases in the lymph nodes. Recent advances in the field have indicated that further to the growth and/or remodeling of local lymphatics, the biophysical microenvironment surrounding a tumor is also critical for its progression [2, 20, 46, 55]. Such fundamental studies have hinted that tumor cells may exploit key mechanistic components of normal immune trafficking and lymphatic function to further their survival. These findings raise many new questions as to the role of the microenvironment in lymphatic metastasis such as: do inflammatory infiltrates from tumors help or hinder their ability to metastasize via lymphatics? Do lymphatic signals help recruit immune cells to the tumor? How do tumor cells survive in lymph nodes? Do tumors or tumor-associated cells modify their environment to promote migration to lymphatics? Development of such models that more accurately recapitulate the tumor-lymphatic microenvironment will undoubtedly assist the rapidly growing field of lymphatic research in the quest to unravel the complexities of lymphatic metastasis.

References

1. Adler AJ. Mechanisms of T cell tolerance and suppression in cancer mediated by tumor-associated antigens and hormones. Curr Cancer Drug Targets 2007, 7: 3–14.
2. Albini A. Tumor microenvironment, a dangerous society leading to cancer metastasis. From mechanisms to therapy and prevention. Cancer Metastasis Rev 2007, 27: 3–4.
3. Angeli V, Ginhoux F, Llodra J, Quemeneur L, Frenette PS, Skobe M, Jessberger R, Merad M, Randolph GJ. B cell-driven lymphangiogenesis in inflamed lymph nodes enhances dendritic cell mobilization. Immunity 2006, 24: 203–15.
4. Bajenoff M, Egen JG, Koo LY, Laugier JP, Brau F, Glaichenhaus N, Germain RN. Stromal cell networks regulate lymphocyte entry, migration, and territoriality in lymph nodes. Immunity 2006, 25: 989–1001.
5. Bajenoff M, Egen JG, Qi H, Huang AY, Castellino F, Germain RN. Highways, byways and breadcrumbs: directing lymphocyte traffic in the lymph node. Trends Immunol 2007, 28: 346–52.
6. Baluk P, Fuxe J, Hashizume H, Romano T, Lashnits E, Butz S, Vestweber D, Corada M, Molendini C, Dejana E, McDonald DM. Functionally specialized junctions between endothelial cells of lymphatic vessels. J Exp Med 2007, 204: 2349–62.
7. Brand CU, Hunziker T, Braathen LR. Isolation of human skin-derived lymph: flow and output of cells following sodium lauryl sulphate-induced contact dermatitis. Arch Dermatol Res 1992, 284: 123–6.
8. Butler TP, Gullino PM. Quantitation of cell shedding into efferent blood of mammary adenocarcinoma. Cancer Res 1975, 35: 512–6.
9. Clarijs R, Schalkwijk L, Ruiter DJ, de Waal RM. Lack of lymphangiogenesis despite coexpression of VEGF-C and its receptor Flt-4 in uveal melanoma. Invest Ophthalmol Vis Sci 2001, 42: 1422–8.
10. Cobb RA, Steer HW. Tumour cell trapping in rat mesenteric lymph nodes. Br J Exp Pathol 1987, 68: 461–74.
11. Crnic I, Strittmatter K, Cavallaro U, Kopfstein L, Jussila L, Alitalo K, Christofori G. Loss of neural cell adhesion molecule induces tumor metastasis by up-regulating lymphangiogenesis. Cancer Res 2004, 64: 8630–8.
12. Cserr HF, Harlingberg CJ, Knopf PM. Drainage of brain extracellular fluid into blood and deep cervical lymph and its immunological significance. Brain Pathol 1992, 2: 269–76.
13. Dadras SS, Paul T, Bertoncini J, Brown LF, Muzikansky A, Jackson DG, Ellwanger U, Garbe C, Mihm MC, Detmar M. Tumor lymphangiogenesis: a novel prognostic indicator for cutaneous melanoma metastasis and survival. Am J Pathol 2003, 162: 1951–60.
14. Drake RE, Gabel JC. Effect of outflow pressure on intestinal lymph flow in unanesthetized sheep. Am J Physiol 1991, 260: 668–71.
15. Fiedler U, Christian S, Koidl S, Kerjaschki D, Emmett MS, Bates DO, Christofori G, Augustin HG. The sialomucin CD34 is a marker of lymphatic endothelial cells in human tumors. Am J Pathol 2006, 168: 1045–53.
16. Fischer M, Franzeck UK, Herrig I, Costanzo U, Wen S, Schiesser M, Hoffmann U, Bollinger A. Flow velocity of single lymphatic capillaries in human skin. Am J Physiol Heart Circ Physiol 1996, 39: H358–63.
17. Goldman J, Le TX, Skobe M, Swartz MA. Overexpression of VEGF-C causes transient lymphatic hyperplasia but not increased lymphangiogenesis in regenerating skin. Circ Res 2005, 96: 1193–9.
18. Halin C, Detmar M. An unexpected connection: lymph node lymphangiogenesis and dendritic cell migration. Immunity 2006, 24: 129–31.
19. Halin C, Tobler NE, Vigl B, Brown LF, Detmar M. VEGF-A produced by chronically inflamed tissue induces lymphangiogenesis in draining lymph nodes. Blood 2007, 110:3158–67.
20. Hanahan D, Weinberg RA. The hallmarks of cancer. Cell 2000, 100: 57–70.

21. Harrell MI, Iritani BM, Ruddell A. Tumor-induced sentinel lymph node lymphangiogenesis and increased lymph flow precede melanoma metastasis. Am J Pathol 2007, 170: 774–86.
22. Hashizume H, Baluk P, Morikawa S, McLean JW, Thurston G, Roberge S, Jain RK, McDonald DM. Openings between defective endothelial cells explain tumor vessel leakiness. Am J Pathol 2000, 156: 1363–80.
23. He YL, Karpanen T, Alitalo K. Role of lymphangiogenic factors in tumor metastasis. Biochim Biophys Acta Rev Cancer 2004, 1654: 3–12.
24. Hirakawa S, Brown LF, Kodama S, Paavonen K, Alitalo K, Detmar M. VEGF-C-induced lymphangiogenesis in sentinel lymph nodes promotes tumor metastasis to distant sites. Blood 2007, 109: 1010–7.
25. Hirakawa S, Kodama S, Kunstfeld R, Kajiya K, Brown LF, Detmar M. VEGF-A induces tumor and sentinel lymph node lymphangiogenesis and promotes lymphatic metastasis. J Exp Med 2005, 201: 1089–99.
26. Isaka N, Padera TP, Hagendoorn J, Fukumura D, Jain RK. Peritumor lymphatics induced by vascular endothelial growth factor-C exhibit abnormal function. Cancer Res 2004, 64: 4400–4.
27. Jain RK. Barriers to drug delivery in solid tumors. Sci Am 1994, 271: 58–65.
28. Jennbacken K, Vallbo C, Wang W, Damber JE. Expression of vascular endothelial growth factor C (VEGF-C) and VEGF receptor-3 in human prostate cancer is associated with regional lymph node metastasis. Prostate 2005, 65: 110–6.
29. Jia YT, Li ZX, He YT, Liang W, Yang HC, Ma HJ. Expression of vascular endothelial growth factor-C and the relationship between lymphangiogenesis and lymphatic metastasis in colorectal cancer. World J Gastroenterol 2004, 10: 3261–3.
30. Kaldjian EP, Gretz JE, Anderson AO, Shi YH, Shaw S. Spatial and molecular organization of lymph node T cell cortex: a labyrinthine cavity bounded by an epithelium-like monolayer of fibroblastic reticular cells anchored to basement membrane-like extracellular matrix. Int Immunol 2001, 13: 1243–53.
31. Koyama H, Kobayashi N, Harada M, Takeoka M, Kawai Y, Sano K, Fujimori M, Amano J, Ohhashi T, Kannagi R, Kimata K, Taniguchi S, Itano N. Significance of tumor-associated stroma in promotion of intratumoral lymphangiogenesis. Pivotal role of a hyaluronan-rich tumor microenvironment. Am J Pathol 2007, 172: 179–93.
32. Leak LV. Permeability of lymphatic capillaries. J Cell Biol 1971, 50: 300–23.
33. Leak LV. Structure of lymphatic capillaries in lymph formation. Fed Proc 1976, 35: 1863–71.
34. Levick J. Revision of the starling principle: new views of tissue fluid balance. J Physiol 2004, 557: 704.
35. Liotta LA, Kleinerman J, Saidel GM. Quantitative relationships of intravascular tumor cells, tumor vessels, and pulmonary metastases following tumor implantation. Cancer Res 1974, 34: 997–1004.
36. Mueller L, Goumas FA, Affeldt M, Sandtner S, Gehling UM, Brilloff S, Walter J, Karnatz N, Lamszus K, Rogiers X, Broering DC. Stromal fibroblasts in colorectal liver metastases originate from resident fibroblasts and generate an inflammatory microenvironment. Am J Pathol 2007, 171: 1608–18.
37. Murakami T, Cardones AR, Hwang ST. Chemokine receptors and melanoma metastasis. J Dermatol Sci 2004, 36: 71–8.
38. Nagata H, Arai T, Soejima Y, Suzuki H, Ishii H, Hibi T. Limited capability of regional lymph nodes to eradicate metastatic cancer cells. Cancer Res 2004, 64: 8239–48.
39. Nakamura Y, Yasuoka H, Tsujimoto M, Imabun S, Nakahara M, Nakao K, Nakamura M, Mori I, Kakudo K. Lymph vessel density correlates with nodal status, VEGF-C expression, and prognosis in breast cancer. Breast Cancer Res Treat 2005, 91: 125–32.
40. Ogawa E, Takenaka K, Yanagihara K, Kurozumi M, Manabe T, Wada H, Tanaka F. Clinical significance of VEGF-C status in tumour cells and stromal macrophages in non-small cell lung cancer patients. Br J Cancer 2004, 91: 498–503.
41. Ohtani O, Ohtani Y, Carati CJ, Gannon BJ. Fluid and cellular pathways of rat lymph nodes in relation to lymphatic labyrinths and Aquaporin-1 expression. Arch Histol Cytol 2003, 66: 261–72.

42. Padera TP, Kadambi A, di Tomaso E, Carreira CM, Brown EB, Boucher Y, Choi NC, Mathisen D, Wain J, Mark EJ, Munn LL, Jain RK. Lymphatic metastasis in the absence of functional intratumor lymphatics. Science 2002, 296: 1883–6.
43. Padera TP, Stoll BR, Tooredman JB, Capen D, di Tomaso E, Jain RK. Pathology: cancer cells compress intratumour vessels. Nature 2004, 427: 695.
44. Partanen TA, Alitalo K, Miettinen M. Lack of lymphatic vascular specificity of vascular endothelial growth factor receptor 3 in 185 vascular tumors. Cancer 1999, 86: 2406–12.
45. Partanen TA, Arola J, Saaristo A, Jussila L, Ora A, Miettinen M, Stacker SA, Achen MG, Alitalo K. VEGF-C and VEGF-D expression in neuroendocrine cells and their receptor, VEGFR-3, in fenestrated blood vessels in human tissues. FASEB J 2000, 14: 2087–96.
46. Perez-Pinera P, Chang Y, Deuel TF. Pleiotrophin, a multifunctional tumor promoter through induction of tumor angiogenesis, remodeling of the tumor microenvironment, and activation of stromal fibroblasts. Cell Cycle 2007, 6: 2877–83.
47. Qian CN, Berghuis B, Tsarfaty G, Bruch M, Kort EJ, Ditlev J, Tsarfaty I, Hudson E, Jackson DG, Petillo D, Chen J, Resau JH, Teh BT. Preparing the "soil": the primary tumor induces vasculature reorganization in the sentinel lymph node before the arrival of metastatic cancer cells. Cancer Res 2006, 66: 10365–76.
48. Quick CM, Venugopal AM, Gashev AA, Zawieja DC, Stewart RH. Intrinsic pump-conduit behavior of lymphangions. Am J Physiol Regul Integr Compar Physiol 2007, 292: R1510–8.
49. Raman D, Baugher PJ, Thu YM, Richmond A. Role of chemokines in tumor growth. Cancer Lett 2007, 256: 137–65.
50. Rockson SG. Lymphedema. Am J Med 2001, 110: 288–95.
51. Ruddell A, Kelly-Spratt KS, Furuya M, Parghi SS, Kemp CJ. p19/Arf and p53 suppress sentinel lymph node lymphangiogenesis and carcinoma metastasis. Oncogene 2007, 27: 3145–55.
52. Saaristo A, Partanen TA, Arola J, Jussila L, Hytonen M, Makitie A, Vento S, Kaipainen A, Malmberg H, Alitalo K. Vascular endothelial growth factor-C and its receptor VEGFR-3 in the nasal mucosa and in nasopharyngeal tumors. Am J Pathol 2000, 157: 7–14.
53. Schmid-Schonbein GW. Microlymphatics and lymph flow. Physiol Rev 1990, 70: 987–1028.
54. Schoppmann SF, Fenzl A, Nagy K, Unger S, Bayer G, Geleff S, Gnant M, Horvat R, Jakesz R, Birner P. VEGF-C expressing tumor-associated macrophages in lymph node positive breast cancer: impact on lymphangiogenesis and survival. Surgery 2006, 139: 839–46.
55. Selleri S, Rumio C, Sabatino M, Marincola FM, Wang E. Tumor microenvironment and the immune response. Surg Oncol Clin N Am 2007, 16: 737–53, vii–viii.
56. Shawber CJ, Funahashi Y, Francisco E, Vorontchikhina M, Kitamura Y, Stowell SA, Borisenko V, Feirt N, Podgrabinska S, Shiraishi K, Chawengsaksophak K, Rossant J, Accili D, Skobe M, Kitajewski J. Notch alters VEGF responsiveness in human and murine endothelial cells by direct regulation of VEGFR-3 expression. J Clin Invest 2007, 117: 3369–82.
57. Shields JD, Borsetti M, Rigby H, Harper SJ, Mortimer PS, Levick JR, Orlando A, Bates DO. Lymphatic density and metastatic spread in human malignant melanoma. Br J Cancer 2004, 90: 693–700.
58. Shields JD, Emmett MS, Dunn DB, Joory KD, Sage LM, Rigby H, Mortimer PS, Orlando A, Levick JR, Bates DO. Chemokine-mediated migration of melanoma cells towards lymphatics – a mechanism contributing to metastasis. Oncogene 2006, 26: 2997–3005.
59. Shields JD, Fleury ME, Yong C, Tomei AA, Randolph GJ, Swartz MA. Autologous chemotaxis as a mechanism of tumor cell homing to lymphatics via interstitial flow and autocrine CCR7 signaling. Cancer Cell 2007, 11:526–38.
60. Skobe M, Hawighorst T, Jackson DG, Prevo R, Janes L, Velasco P, Riccardi L, Alitalo K, Claffey K, Detmar M. Induction of tumor lymphangiogenesis by VEGF-C promotes breast cancer metastasis. Nat Med 2001, 7: 192–8.
61. Spiegel M, Vesti B, Shore A, Franzeck UK, Becker F, Bollinger A. Pressure of Lymphatic Capillaries in Human Skin. Am J Physiol 1992, 262: H1208–10.
62. Stacker SA, Caesar C, Baldwin ME, Thornton GE, Williams RA, Prevo R, Jackson DG, Nishikawa S, Kubo H, Achen MG. VEGF-D promotes the metastatic spread of tumor cells via the lymphatics. Nat Med 2001, 7: 186–91.

63. Swartz MA. The physiology of the lymphatic system. Adv Drug Deliv Rev 2001, 50: 3–20.
64. Swartz MA, Kristensen CA, Melder RJ, Roberge S, Calautti E, Fukumura D, Jain RK. Cells shed from tumours show reduced clonogenicity, resistance to apoptosis, and in vivo tumorigenicity. Br J Cancer 1999, 81: 756–9.
65. Trzewik J, Mallipattu SK, Artmann GM, Delano FA, Schmid-Schonbein GW. Evidence for a second valve system in lymphatics: endothelial microvalves. FASEB J 2001, 15: 1711–7.
66. Van den Eynden GG, Van der Auwera I, Van Laere SJ, Huygelen V, Colpaert CG, van Dam P, Dirix LY, Vermeulen PB, Van Marck EA. Induction of lymphangiogenesis in and around axillary lymph node metastases of patients with breast cancer. Br J Cancer 2006, 95: 1362–6.
67. Von der Weid PY, Zawieja DC. Lymphatic smooth muscle: the motor unit of lymph drainage. Int J Biochem Cell Biol 2004, 36: 1147–53.
68. Wong SY, Haack H, Crowley D, Barry M, Bronson RT, Hynes RO. Tumor-secreted vascular endothelial growth factor-C is necessary for prostate cancer lymphangiogenesis, but lymphangiogenesis is unnecessary for lymph node metastasis. Cancer Res 2005, 65: 9789–98.
69. Yuan L, Moyon D, Pardanaud L, Breant C, Karkkainen MJ, Alitalo K, Eichmann A. Abnormal lymphatic vessel development in neuropilin 2 mutant mice. Development 2002, 129: 4797–806.
70. Zawieja DC, Davis KL, Schuster R, Hinds WM, Granger HJ. Distribution, propagation, and coordination of contractile activity in lymphatics. Am J Physiol 1993, 264: H1283–91.

Index